Springer Texts in Statistics

Gunnar Blom

Probability and Statistics

Theory and Applications

With 107 Illustrations

Springer-Verlag
New York Berlin Heidelberg
London Paris Tokyo

Gunnar Blom
Department of Mathematical Statistics
University of Lund and
Lund Institute of Technology
S-221 00 Lund
Sweden

Mathematics Subject Classifications (1980): 60-XX, 60-01, 62-01

Library of Congress Cataloging-in-Publication Data
Blom, Gunnar.
[Sannolikhetsteori och statistikteori med tillämpningar. English]
Probability and statistics : theory and applications / Gunnar Blom.
p. cm. — (Springer texts in statistics)
Translation of: Sannolikhetsteori och statistikteori med tillämpningar. Lund, 1980.
Bibliography: p.
Includes index.
ISBN 0-387-96852-0 (alk. paper)
1. Probabilities. 2. Mathematical statistics. I. Title. II. Series.
QA273.B592413 1989 88-29766
519.2—dc19 CIP

Original Swedish edition: *Sannolikhetsteori och statistikteori med tillämpningar*. Studentlitteratur, Lund, 1980.

Printed on acid-free paper

Typeset by Asco Trade Typesetting Ltd., Hong Kong.
Printed and bound by R.R. Donnelley & Sons, Harrisonburg, Virginia.
Printed in the United States of America.

9 8 7 6 5 4 3 2 1

ISBN 0-387-96852-0 Springer-Verlag New York Berlin Heidelberg
ISBN 3-540-96852-0 Springer-Verlag Berlin Heidelberg New York

Preface

This is a somewhat extended and modified translation of the third edition of the text, first published in 1969. The Swedish edition has been used for many years at the Royal Institute of Technology in Stockholm, and at the School of Engineering at Linköping University. It is also used in elementary courses for students of mathematics and science.

The book is not intended for students interested only in theory, nor is it suited for those seeking only statistical recipes. Indeed, it is designed to be intermediate between these extremes. I have given much thought to the question of dividing the space, in an appropriate way, between mathematical arguments and practical applications. Mathematical niceties have been left aside entirely, and many results are obtained by analogy.

The students I have in mind should have three ingredients in their course: elementary probability theory with applications, statistical theory with applications, and something about the planning of practical investigations. When pouring these three ingredients into the soup, I have tried to draw upon my experience as a university teacher and on my earlier years as an industrial statistician.

The programme may sound bold, and the reader should not expect too much from this book. Today, probability, statistics and the planning of investigations cover vast areas and, in 356 pages, only the most basic problems can be discussed. If the reader gains a good understanding of probabilistic and statistical reasoning, the main purpose of the book has been fulfilled.

Certain sections of the book may be omitted, as noted in the text. However, the reader is advised not to skip them all, for they are meant as spices in the soup.

A working knowledge of elementary calculus, in particular derivatives and Riemann integrals, is an essential prerequisite.

Many theoretical and applied problems are solved in the text. Exercises with answers are given at the end of most chapters. Most of the exercises are easy, many very easy, and are thought to be especially suitable for those using the book for self-study. For the benefit of those who like more difficult exercises there are also some rather hard problems indicated by asterisks. Furthermore, there is a collection of 80 problems of varying difficulty at the end of the book. Readers who solve the majority of these will have acquired a good knowledge of the subject.

Numerous references to the literature are given at the end of the book. They are meant to help and guide those students who wish to extend the knowledge they have acquired.

The tables at the end of the book are taken from a separate collection of tables published by AB Studentlitteratur in Lund, Sweden.

My sincere thanks are due to Herbert A. David and Theodore A. Vessey for their invaluable help with the translation. I also want to express my gratitude to Krzysztof Nowicki and Dennis Sandell for checking the answers of many exercises and to Jan-Eric Englund, Jan Lanke and Dennis Sandell for their assistance with the proof-reading.

Lund, Sweden
November, 1988

G.B.

Contents

Preface v

Chapter 1
Introduction to Probability Theory 1

1.1. On the Usefulness of Probability Theory 1
1.2. Models, Especially Random Models 3
1.3. Some Historical Notes 4

Chapter 2
Elements of Probability Theory 5

2.1. Introduction 5
2.2. Events 5
2.3. Probabilities in a General Sample Space 9
2.4. Probabilities in Discrete Sample Spaces 14
2.5. Conditional Probability 21
2.6. Independent Events 25
2.7. Some Theorems in Combinatorics 29
2.8. Some Classical Problems of Probability[1] 32
Exercises 36

Chapter 3
One-Dimensional Random Variables 40

3.1. Introduction 40
3.2. General Description of Random Variables 40
3.3. Distribution Function 42

[1] Special section, which may be omitted.

3.4. Discrete Random Variables 44
3.5. Some Discrete Distributions 47
3.6. Continuous Random Variables 49
3.7. Some Continuous Distributions 53
3.8. Relationship Between Discrete and Continuous Distributions 59
3.9. Mixtures of Random Variables 60
Exercises 61

CHAPTER 4
Multidimensional Random Variables 65

4.1. Introduction 65
4.2. General Considerations 65
4.3. Discrete Two-Dimensional Random Variables 66
4.4. Continuous Two-Dimensional Random Variables 68
4.5. Independent Random Variables 71
4.6. Some Classical Problems of Probability[1] 74
Exercises 77

CHAPTER 5
Functions of Random Variables 80

5.1. Introduction 80
5.2. A Single Function of a Random Variable 80
5.3. Sums of Random Variables 83
5.4. Largest Value and Smallest Value 89
5.5. Ratio of Random Variables 91
Exercises 92

CHAPTER 6
Expectations 96

6.1. Introduction 96
6.2. Definition and Simple Properties 96
6.3. Measures of Location and Dispersion 101
6.4. Measures of Dependence 108
Exercises 112

CHAPTER 7
More About Expectations 116

7.1. Introduction 116
7.2. Product, Sum and Linear Combination 116
7.3. Arithmetic Mean. Law of Large Numbers 119
7.4. Gauss's Approximation Formulae 123
7.5. Some Classical Problems of Probability[1] 126
Exercises 128

[1] Special section, which may be omitted.

CHAPTER 8
The Normal Distribution 131

8.1. Introduction 131
8.2. Some General Facts About the Normal Distribution 131
8.3. Standard Normal Distribution 132
8.4. General Normal Distribution 135
8.5. Sums and Linear Combinations of Normally Distributed Random Variables 138
8.6. The Central Limit Theorem 141
8.7. Lognormal Distribution 144
Exercises 144

CHAPTER 9
The Binomial and Related Distributions 147

9.1. Introduction 147
9.2. The Binomial Distribution 147
9.3. The Hypergeometric Distribution 155
9.4. The Poisson Distribution 158
9.5. The Multinomial Distribution 161
Exercises 163

CHAPTER 10
Introduction to Statistical Theory 167

10.1. Introduction 167
10.2. Statistical Investigations 167
10.3. Examples of Sampling Investigations 171
10.4. Main Problems in Statistical Theory 175
10.5. Some Historical Notes 177

CHAPTER 11
Descriptive Statistics 179

11.1. Introduction 179
11.2. Tabulation and Graphical Presentation 180
11.3. Measures of Location and Dispersion 185
11.4. Terminology 188
11.5. Numerical Computation 188
Exercises 189

CHAPTER 12
Point Estimation 191

12.1. Introduction 191
12.2. General Ideas 191
12.3. Estimation of Mean and Variance 196
12.4. The Method of Maximum Likelihood 198
12.5. The Method of Least Squares 201

12.6. Application to the Normal Distribution 203
12.7. Application to the Binomial and Related Distributions 207
12.8. Standard Error of an Estimate 209
12.9. Graphical Method for Estimating Parameters 210
12.10. Estimation of Probability Function, Density Function and Distribution Function[1] 214
12.11. Parameter with Prior Distribution[1] 217
Exercises 219

CHAPTER 13
Interval Estimation 223

13.1. Introduction 223
13.2. Some Ideas About Interval Estimates 223
13.3. General Method 226
13.4. Application to the Normal Distribution 227
13.5. Using the Normal Approximation 241
13.6. Application to the Binomial and Related Distributions 244
Exercises 249

CHAPTER 14
Testing Hypotheses 253

14.1. Introduction 253
14.2. An Example of Hypothesis Testing 253
14.3. General Method 254
14.4. Relation Between Tests of Hypotheses and Interval Estimation 261
14.5. Application to the Normal Distribution 262
14.6. Using the Normal Approximation 264
14.7. Application to the Binomial and Related Distributions 265
14.8. The Practical Value of Tests of Significance 274
14.9. Repeated Tests of Significance 275
Exercises 276

CHAPTER 15
Linear Regression 280

15.1. Introduction 280
15.2. A Model for Simple Linear Regression 280
15.3. Point Estimates 282
15.4. Interval Estimates 286
15.5. Several y-Values for Each x-Value 287
15.6. Various Remarks 289
Exercises 291

CHAPTER 16
Planning Statistical Investigations 293

16.1. Introduction 293
16.2. General Remarks About Planning 293

[1] Special section, which may be omitted.

16.3. Noncomparative Investigations 294
16.4. Comparative Investigations 301
16.5. Final Remarks 304
Appendix: How to Handle Random Numbers 305

Selected Exercises 307

References 319

Tables 323

Answers to Exercises 339

Answers to Selected Exercises 348

Index 351

CHAPTER 1

Introduction to Probability Theory

1.1. On the Usefulness of Probability Theory

We shall give three examples intended to show how probability theory may be used in practice.

Example 1. Batch with Defective Units

A car manufacturer buys certain parts (units) from a subcontractor. From each batch of 1,000 units a sample of 75 is inspected. If two or fewer defective units are found, the whole batch is accepted; otherwise, it is returned to the supplier.

Let us assume that there are 20 defective units in the whole batch. What is the probability that the batch will pass inspection? Such a question is of interest, for the answer gives the car manufacturer an idea of the efficiency of the inspection. Probability theory provides the answer. □

Example 2. Measuring a Physical Constant

At a laboratory, five measurements were made of a physical constant. The following values were obtained:

$$2.13 \qquad 2.10 \qquad 2.05 \qquad 2.11 \qquad 2.14.$$

Because of errors in measurement, there is a certain uncontrollable variation among the values, which can be termed random.

In order to find out how much the measurements deviate from the true value of the constant, the following question is natural: What is the probability that the arithmetic mean of the five measurements differs from the true value by more than a certain given quantity? To answer this question, one needs probability theory. □

Example 3. Animal Experiment

A research worker at a pharmaceutical industry gives 12 rabbits an injection of a certain dose of insulin. After a certain time has elapsed, the decrease in the percentage of blood sugar is determined. The following result is obtained (computed as a percentage decrease of the initial value):

11.2	21.2	−4.0	18.7	2.8	27.2
25.1	25.8	2.2	28.3	23.7	−2.2.

There is considerable variation among the data, as is generally the case when biological phenomena are studied. Look at the extreme values: one animal has 28% *less* blood-sugar than before the injection, and another has 4% *higher* blood-sugar than before. The variability is large, in spite of the fact that the animals have been reared in the same way and have also been treated alike in all other respects. For various reasons, the rabbits react very differently to the same dose. Nevertheless, important information can be extracted from such animal experiments. For this purpose, probability theory is again required. □

A common feature of the three examples, and of all other situations investigated by probability theory, is that *variability* is present. In Example 1 the number of defective units varies from sample to sample, if the inspection is repeated several times by taking new samples from the same batch. In Example 2, there are variations from one measurement to the next, and in Example 3, from one animal to the next.

Variability is a very general phenomenon. The variations can be small, as among cog-wheels manufactured with great precision or between identical twins "as like as two peas in a pod"; or large, as among the stones in Scandinavian moraines or among the profits in dollars of all the firms in Western Europe during a certain year. Differences among people, physical and psychological, and variations with regard to geography, mineral resources, weather and economy, affect the world in a fundamental way. Knowledge of such things is often important for the actions of individuals and society.

It can be disastrous to neglect variability. Only one, almost trivial example, will be mentioned. If you plan to wade across a river, it is not enough to know that the average depth is 3 feet; it is necessary to know something about the variations in depth!

Because of the omnipresence of variability, it is natural that probability theory—the basic science of variability—has a multitude of applications in technological research, physics, chemistry, biology, medicine and the geosciences. In recent years the interest in applied probability theory has increased considerably, in view of the investigation of models for automatic control, queueing models, models for production and epidemiological models (models for the spread of diseases). Probability theory is also of basic importance for statistical theory.

It should be added that probability theory is not only a useful, but also an enjoyable and elegant science. The enjoyment will be experienced by anyone interested in games of chance, stakes and lotteries, upon discovering that probability theory permits the solution of many associated problems. Many research problems in probability can also give one, with the right mathematical ability and interest, an opportunity for both great effort and great satisfaction.

1.2. Models, Especially Random Models

The concept of a *model* is of basic importance in science. A model is intended to describe essential properties of a phenomenon, without copying all of the details. A fundamental, but common, error is to confuse the model with the reality which it is intended to depict. It is very important to keep in mind that every model is approximate.

A good motto for anyone dealing with models is therefore: *Distinguish between the model and reality.*

Models are of different kinds. There are *physical* models (Example: a house built to the scale 1:50), *analogue* models (Examples: map, drawing, atomic model with balls on rods), and *abstract* models.

Abstract models are often used in mathematical and experimental sciences. Such models belong to the world of mathematics and can be formulated in mathematical terms. An abstract model can be *deterministic* or *stochastic*. Instead of a stochastic model we mostly say a *random* model; the term *chance* model is also used. The word stochastic derives from the Greek "stochastikos", able to aim at, to hit (the mark); "stochos" means target, conjecture.

In a deterministic model a phenomenon is approximated by mathematical functions. For example, the area of a round table is determined by conceptually replacing the table-top by a circle; we then obtain πr^2, where r is the radius of the circle. Clearly, this is a model; exactly circular tables exist only in an abstract world.

Let us consider a model with more far-reaching implications. A widely debated deterministic model of great consequence, for individuals and their views of life, presupposes that information about the positions and movements of all atoms at a given moment makes it possible to predict the whole subsequent development of the world.

Simpler, but important, examples of deterministic models are Euclidean geometry which was used for measuring areas of land in ancient Egypt and classical mechanics which made it possible for Newton to explain the movements of the planets around the sun.

In the following we consider only random models. Such models are used in probability theory for describing *random trials* (often called random experiments). By that we mean, somewhat vaguely, every experiment that can be repeated under similar conditions, but whose outcome cannot be predicted exactly in advance, even when the same experiment has been performed many times. It is just the unpredictable variation of the outcomes which is described by a random model.

Let us take a classical example of a random experiment. Throw a die repeatedly and each time note the number of points obtained. Suppose the result is 1, 5, 1, 6, 3, 2, 4, 6, 1. Clearly, this may be termed a random experiment, only if we perform the throws appropriately so that the outcome cannot be predicted. For example, if the die is held just above the table each time, the outcome is determined in advance. It follows from this simple remark that certain conditions have to be fulfilled by a random trial, and also that these conditions cannot be strictly formulated.

Another example of a random trial is afforded by radioactive decay. The result of a famous series of such experiments was published at the beginning of this century by Rutherford and Geiger. For a certain radioactive source they recorded the number of atoms that decayed in successive minutes, for example, 24, 31, 29, 16, 19, 25, ...

Before probability theory can be applied to such trials as we have described here, it is necessary to construct a random model which shows how the result may vary. In our examples, the models are rather simple, but sometimes complicated models are required; for example, when industrial processes are regulated using principles of automatic control.

In this book, we discuss some basic properties of certain important random models, and in this way lay a foundation for probability theory.

1.3. Some Historical Notes

Probability theory has its roots in sixteenth-century Italy and seventeenth-century France and was originally a theory for games of chance. At the beginning of the sixteenth century, the Italian Cardano wrote a treatise about such games, in which he solved several problems of probability. Beginning in 1654, the famous correspondence between Pascal and Fermat occurred concerning certain games of chance. James Bernoulli and de Moivre are also pioneers in the field of probability, in a somewhat later time-period. Laplace, in his *Théorie Analytique des Probabilités*, published in 1812, described probability theory as a mathematical theory of wide scope and extended its application to astronomy and mechanics. Several physicists, among them Maxwell and Boltzmann, used probability theory in the creation of statistical mechanics.

Only in our own time has probability theory become a science, in the modern sense of the word. A large number of mathematicians and probabilists have contributed to its development, among them many Russians such as Chebyshev, Markov, Liapounov and Kolmogorov. Important contributions to the theory have also been made by the Frenchman Lévy, the Austrian von Mises, the Americans Doob and Feller, and the Swede Cramér.

CHAPTER 2

Elements of Probability Theory

2.1. Introduction

This chapter is important. In order to come to grips with it rapidly, the reader is advised, on the first reading, to skip all remarks and most of the examples in the text. It is a good idea to read the chapter several times, in order to lay a solid foundation for what follows.

The most important sections are §2.2 and §2.3, which deal, respectively, with events and with the definition of probability; in the latter section, Kolmogorov's axioms are of particular importance. In §2.4 the classical definition of probability is introduced, and we give several examples of probabilities in discrete sample spaces. In §2.5 and §2.6 we discuss conditional probability and the important concept of independent events. §2.7 contains some results from combinatorics. The last section, §2.8, deals with some classical problems of probability.

2.2. Events

We begin with three definitions.

Definition. The result of a random trial is called an *outcome*.

Definition. The set of all possible outcomes is called the *sample space*.

Definition. A collection of outcomes is called an *event*.

Thus an event is a subset of the sample space (or possibly the whole sample space).

Outcomes are denoted by numbers or letters, in this book often by u_1, u_2, ... The sample space is denoted by Ω (omega). Events are denoted by capital letters A, B, C, ...

Example 1. Throwing Dice

(a) Throwing a single die

Let us throw a die. We introduce six outcomes, which we denote by 1, 2, ..., 6. The sample space Ω consists of these outcomes and can be written $\Omega = \{1, 2, \ldots, 6\}$. Examples of events are A = "number of points is odd" and B = "number of points is at most 2". Using the symbols for the outcomes, we can write these events

$$A = \{1, 3, 5\}; \qquad B = \{1, 2\}.$$

Note that we have only assumed that the die has six sides; these sides need not necessarily be alike.

(b) Throwing two dice

Let us throw two dice at the same time. It is then convenient to introduce 36 outcomes $u_1, \ldots, u_{36}$. These outcomes can also be written (1, 1), (1, 2), ..., (6, 6), where the first digit represents one die and the second digit the other. (We suppose that the dice have different colours so that they are distinguishable.) The sample space consists of these 36 outcomes. As an example of an event we can take A = "the sum of the two dice is at most 3"; this event can be written

$$A = \{(1, 1), (1, 2), (2, 1)\}.$$

Hence the event A contains three outcomes.

A figure or diagram is often useful in illustrating how Ω and A look (see Fig. 2.1, where A is represented by the enclosed points). □

Example 2. Radioactive Decay

A sample of a radioactive substance is available. A Geiger–Müller counter registers the number of particles decaying during a certain time interval. This number can be 0, 1, 2, ..., which are all possible outcomes. Thus the sample space contains a denumerably infinite number of outcomes. □

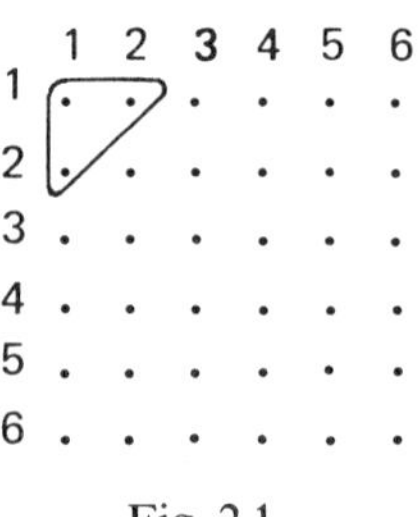

Fig. 2.1.

Example 3. Strength of Material

From a batch of reinforcing bars a certain number are taken, and their tensile strengths are determined. The result (in kp/mm^2) is, say, 62.4, 69.6, 65.0, 63.1, 76.8, ...

The possible values (= outcomes) are, at least theoretically, all numbers between 0 and ∞. We therefore let the sample space consist of all nonnegative numbers. It is impossible to write all these outcomes as one sequence. The event A might be, for example, "tensile strength of a chosen bar is between 60.0 and 65.0".

In practice, the tensile strength is stated with a certain number of decimal places, perhaps with one decimal place as we did above. Then Ω consists only of the numbers 0.0, 0.1, 0.2, ..., and the event A consists of just a finite number of outcomes 60.0, 60.1, ..., 65.0. From a mathematical point of view, it is generally easier to disregard the discreteness and to retain the continuous character of the sample space. □

We shall give a further definition.

Definition. If the number of outcomes is finite or denumerably infinite, Ω is said to be a *discrete* sample space. More particularly, if the number is finite, Ω is said to be a *finite* sample space.

If the number of outcomes is neither finite nor denumerably infinite, Ω is said to be a *continuous* sample space. □

In Examples 1 and 2 the sample space is discrete, in Example 1 it is actually finite; in Example 3 it is continuous.

Example 4. Coin Tosses

A coin is tossed repeatedly. Let a tail be denoted by 0 and a head by 1.

(a) A given number of tosses

The result of n tosses can be represented by a binary n-digit number. (For example, if $n = 3$, the possible outcomes are 000, 001, 010, 011, 100, 101, 110, 111.) We let each such number be an outcome. The number of possible outcomes is 2^n. As an illustration we use a "tree diagram" (a fallen tree); see Fig. 2.2 for the case $n = 3$.

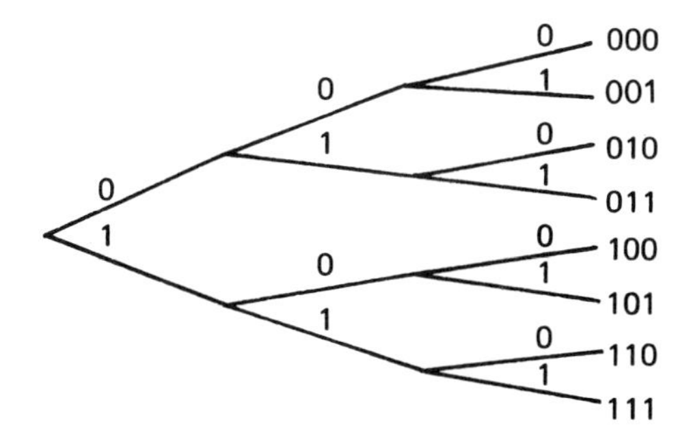

Fig. 2.2. Three tosses with a coin.

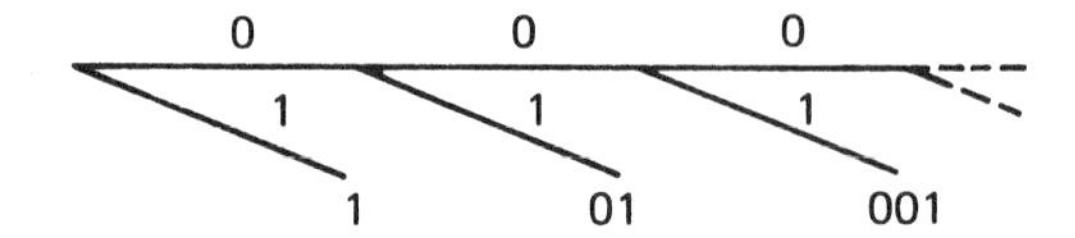

Fig. 2.3. Tosses with a coin until a head is obtained.

(b) Toss until a head is obtained

The coin is tossed until the first head is obtained. Using the representation by binary numbers, the possible outcomes can be written 1, 01, 001, 0001, ... Their number is denumerably infinite. The tree diagram can be drawn as in Fig. 2.3. □

In studying events, the notation of set theory can be used to great advantage. We assume that the reader is familiar with this notation, but nevertheless, in Fig. 2.4, we give a list of what we need in this respect, together with the corresponding Venn diagrams. On the right of Fig. 2.4 the definitions are expressed in the usual language of set theory, and on the left in the special "language of events" which we will often use in this book.

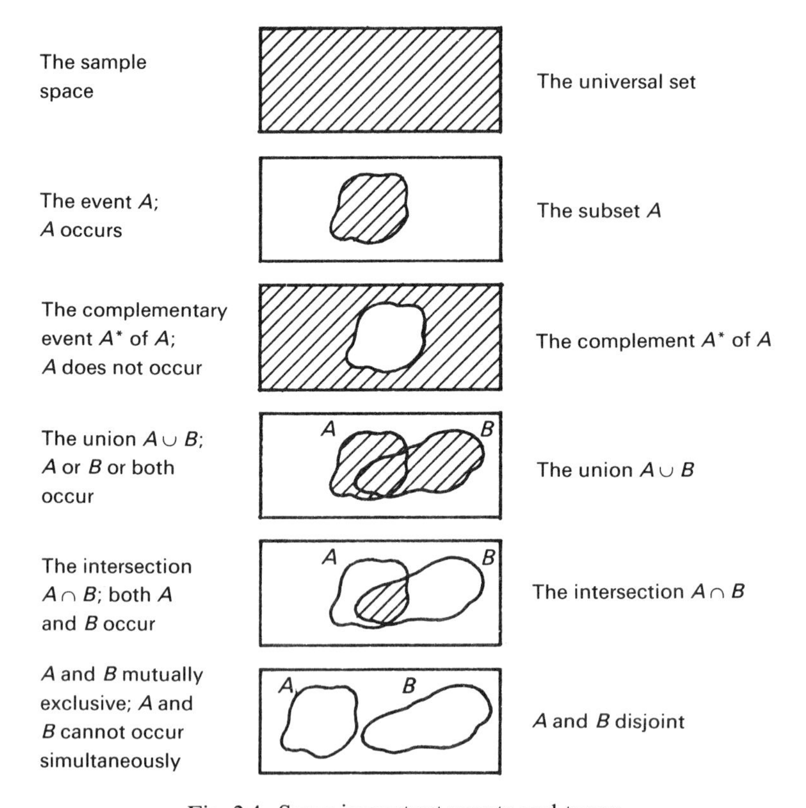

Fig. 2.4. Some important events and terms.

A set which we have not illustrated is the *null set* $\varnothing$, which is the complement of Ω and an impossible event.

The symbols $A \cup B$ and $A \cap B$ can be extended in an obvious way to more than two events. It is then sometimes convenient to use the following notation:

$$\bigcup_1^n A_i = \text{at least one of the events } A_1, \ldots, A_n \text{ occurs,}$$

$$\bigcap_1^n A_i = \text{all the events } A_1, \ldots, A_n \text{ occur.}$$

Extension to $\bigcup_1^\infty A_i$ and $\bigcap_1^\infty A_i$ is sometimes required and causes no problem.

The events $A_1, \ldots, A_n$ are said to be *mutually exclusive* (or *mutually disjoint*) if all pairs A_i, A_j are exclusive; that is, if it is impossible that two or more of these events take place at the same time.

As a symbol for an intersection we often write AB instead of the complete notation $A \cap B$.

Remark 1. Definition of Outcome

In the examples of random trials given up to now, we have introduced all the possible alternatives and called each of them an outcome. However, it is not always necessary to do so. For example, if a die is thrown, it is possible to discriminate only between the two outcomes "a one occurs" and "a one does not occur". Likewise, if a person's height is measured, one might, instead of regarding each height as an outcome, discriminate only between the two cases "the height is less than 170 cm" and "the height is greater than or equal to 170 cm". Many circumstances affect the choice of outcomes, for example, the character of the sample space and the problem posed. The most important circumstance concerns the definition of probability itself and will be discussed later (see Remark, p. 17). ☑

Remark 2. The Language of the Model and the Language of Reality

The advice on p. 3 says: "Discriminate between model and reality." To keep these concepts apart in a most efficient way a pedantic writer would perhaps introduce, in a consistent manner, different terms for something in the world of reality and something in the model world. The term "outcome" would then be used only about reality and, in the language of models, be translated to "element" in the universal set; "event" would be translated to "subset", and so on (see Fig. 2.4). In the sequel, we shall stress the model aspect, but we cannot be consistently "bilingual"; the book would then become too difficult to read—and to write. ☑

2.3. Probabilities in a General Sample Space

In this section, we consider a general sample space; that is, a sample space that can be either discrete or continuous.

After the preparations in the preceding section we now construct a random

Table 2.1.

Number of throws	Number of ones	Relative frequency
2	0	0.000
5	1	0.200
10	1	0.100
50	7	0.140
100	15	0.150
1,000	165	0.165
10,000	1,672	0.167

model for a random trial. Here, the concept *probability* is of basic importance. How can this concept be defined mathematically? Much can be said about this, and in a way it would be justified at this point to give a historical review of the definitions that have appeared up to now (see §1.3). However, we must be brief. A random model should depict reality, and when choosing our definition of probability we will utilize an important property of empirical random trials.

Suppose we throw a well-made die many times. Let us pause occasionally and compute the relative frequency of ones, say; that is, we determine the ratio of the number of ones and the total number of throws. A possible result of such calculations is reproduced in Table 2.1.

The relative frequency seems to become more and more stable and might even tend to a limit, which, by a guess, we set at $1/6 = 0.1666\ldots$ (see Fig. 2.5). If the die is badly made, the limit is perhaps equal to another value, which we cannot guess in advance.

It would be possible to define the probability of the event "one" to be the limit of the relative frequency as the number of throws increases to infinity.

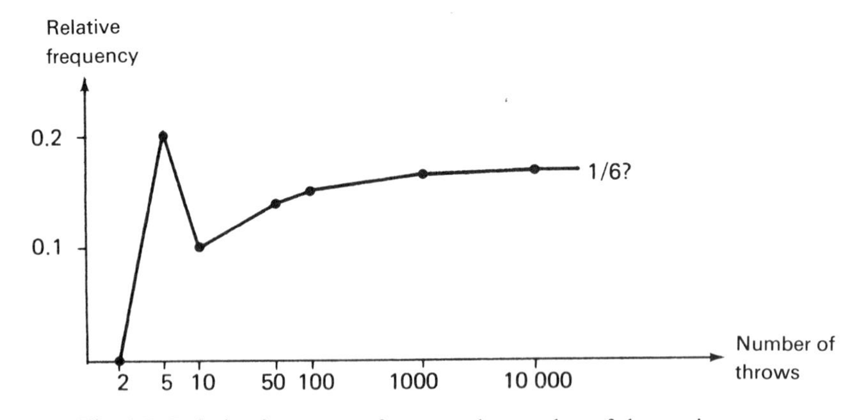

Fig. 2.5. Relative frequency of ones as the number of throws increases.

This definition would have the advantage that it is in accordance with what most people presumably mean by the sentence "the probability of a one is 1/6". However, certain mathematical difficulties arise if such a definition is chosen. It is easier to postulate that for each event in question there exists a number called the probability of that event. This means simply that *to each event a certain number is assigned*. If the event is A, the number is denoted by $P(A)$.

As to how to choose the number $P(A)$, the following can be said, speaking generally and somewhat vaguely: The numerical value of $P(A)$ should be selected so that the relative frequency of the event A falls into the neighbourhood of $P(A)$ if the number of trials is large enough. For example, if we take $P(A) = 0.2$, this statement can be given the *frequency interpretation*: If the number of trials is large, the relative frequency will presumably be near 0.2. This way of interpreting probabilities is very important, but is admittedly vague.

How to go about choosing the number $P(A)$ depends on the situation. Sometimes a sensible choice can be made beforehand. If, in the above example, the die is well made, it seems reasonable to choose $P(A) = 1/6$. In this way a fairly good approximation to reality ought to be obtained. Of course, it is only an approximation. No die is perfectly made, and sometimes one is careless when rolling it, so that one side tends to turn up too often, and so on.

The situation when the probability cannot be chosen beforehand is more common. If the event A signifies "a newborn Swedish boy lives at least 10 years", it is impossible to fix the value of $P(A)$ in advance. Instead, a certain number of newborn individuals are observed for at least 10 years and, as a probability, one takes the relative frequency of boys which are alive after this period. (For example, if we observe 5,000 boys the relative frequency might become $4{,}949/5{,}000 = 0.990$.) Such a value is always provisional and may be replaced by another value when more information has been collected. Returning to the example with the die, the procedure means that one throws the die 50 times, for example, and takes $P(A) = 0.140$, or 1,000 times and takes $P(A) = 0.165$. In such situations the accuracy of the model will depend on the number of throws, individuals or whatever the case may be, together with all other circumstances that determine the applicability of the model.

We leave now, for a while, all questions concerning the choice of numerical values of probabilities, and turn to an investigation of the general properties of probabilities in a sample space Ω.

Suppose we want to assign a probability to each event in Ω. Can we do this as we please? Evidently not, for, as seen from the frequency interpretation, certain rules must be followed if the model is to be useful.

Each probability must assume a value between 0 and 1; that is, for any event A we must have $0 \leq P(A) \leq 1$.

In particular, if A is equal to Ω, we obtain the event that one of all the possible outcomes occurs, which is, of course, a certainty; that is, we must have $P(\Omega) = 1$.

Now let A and B be two mutually exclusive events, and consider the event $A \cup B$; that is, the event that A or B takes place. The probability that $A \cup B$ occurs ought to be equal to the sum of the probability that A occurs, and the probability that B occurs; that is, we ought to have $P(A \cup B) = P(A) + P(B)$. Using the example of the die, it is easy to check that this is reasonable. If A means "one turns up" and B means "two turns up", and we put $P(A)$ and $P(B)$ equal to 1/6 each, then the event $A \cup B$, that is, "one or two turns up", ought to be given the probability $1/6 + 1/6 = 1/3$.

These considerations lead us to formulate three axioms for probabilities.

Definition ("Kolmogorov's System of Axioms"). The following axioms for the probabilities $P(\cdot)$ should be fulfilled:

Axiom 1. If A is any event, then $0 \leq P(A) \leq 1$.
Axiom 2. If Ω is the entire sample space, then $P(\Omega) = 1$.
Axiom 3 ("Addition Formula"). If A and B are mutually exclusive events, then $P(A \cup B) = P(A) + P(B)$.

The sample space Ω and the probabilities $P(\cdot)$ together constitute a *probability space*. □

Axioms 1–3 suffice if the sample space is finite. If it is denumerably infinite or continuous, Axiom 3 is usually replaced by

Axiom 3′. If $A, B, C, \ldots$ is an infinite sequence of mutually exclusive events, then

$$P(A \cup B \cup C \cup \cdots) = P(A) + P(B) + P(C) + \cdots .$$

It should be stressed that the three axioms say nothing about the choice of numerical values for probabilities—this has to be done in each particular problem. The system of axioms only describes the general structure of a random model.

A. Kolmogorov, a Russian mathematician, formulated his famous system about 1933, when he was 30 years old. His treatise on this subject, *Grundbegriffe der Wahrscheinlichkeitsrechnung* (Berlin, 1933; English translation *Foundations of the Theory of Probability*, Chelsea, New York, 1956) is nowadays regarded as a classic.

Starting from Kolmogorov's system of axioms, a multitude of interesting results can be derived. We shall prove four theorems, the first two of which are the most important.

Theorem 1 ("Complement Theorem").

$$P(A^*) = 1 - P(A).$$ □

PROOF. A and A^* are mutually exclusive and $A \cup A^* = \Omega$. If Axioms 3 and 2 are used, in this order, we obtain $P(A) + P(A^*) = P(A \cup A^*) = P(\Omega) = 1$, which gives the theorem. □

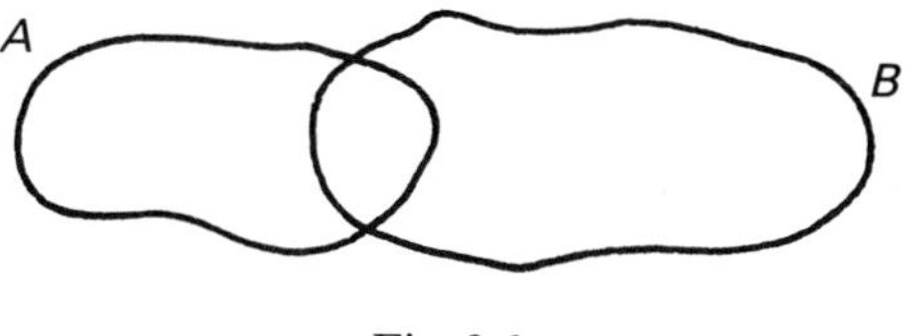

Fig. 2.6.

In particular, if $A = \Omega$, we have from Axiom 2 the

Corollary. $P(\varnothing) = 0$. □

Theorem 1 is simple and important, and many readers have, no doubt, used it without considering it as a theorem. For example, if the probability is 0.1 that A happens (for example, that you win in a lottery), the probability is $1 - 0.1 = 0.9$ that the complementary event occurs (that is, that you do not win).

Theorem 2 ("Addition Theorem for Two Events").

$$P(A \cup B) = P(A) + P(B) - P(A \cap B). \tag{1}$$

□

PROOF. We shall first give a rather vague proof (see Fig. 2.6). In order to determine the probability of the event $A \cup B$ one has to add the probabilities of all "points" within the outer contour. This can be done by first taking all points in A and adding all points in B, but then the common region $A \cap B$ has appeared twice, and the corresponding probability $P(A \cap B)$ must be subtracted.

A strict proof runs as follows. We have

$$A \cup B = A \cup A^*B,$$

$$B = AB \cup A^*B.$$

(For brevity, we have written AB instead of $A \cap B$.) Look at the right-hand members of these relations. The events A and A^*B are mutually exclusive, as are also the events AB and A^*B. The addition formula in Axiom 3 gives

$$P(A \cup B) = P(A) + P(A^*B),$$

$$P(B) = P(AB) + P(A^*B).$$

A subtraction leads to the theorem. ☑

For mutually exclusive events the third term in (1) is zero, and the expression reduces to the addition formula, as it should.

Theorem 2 can be extended by induction to more than two events. We only present the formula for three events.

Theorem 2′ ("Addition Theorem for Three Events").

$$P(A \cup B \cup C) = P(A) + P(B) + P(C) - P(A \cap B) - P(A \cap C) - P(B \cap C) + P(A \cap B \cap C). \quad (2)$$

□

A nice and simple consequence is the following:

Theorem 3 ("Boole's Inequality").

$$P(A \cup B) \leq P(A) + P(B).$$

□

PROOF. The theorem follows from Theorem 2, for $P(A \cap B)$ is always greater than or equal to zero. □

It is easy to extend the inequality to more than two events. In the case of three events one has, by applying Theorem 3 twice, that $P(A \cup B \cup C) \leq P(A) + P(B \cup C) \leq P(A) + P(B) + P(C)$.

Remark. Subjective Probability

The word probability is also used in everyday speech for phenomena that do not have the character of random trials. For example, one might say: It is quite probable that Hitler died in his bunker in Berlin. Or even: With 90% probability I claim that Hitler.... Naturally, such isolated events have no frequency interpretation (see p. 11).

There is currently a considerable interest in such subjective probabilities, and much has been written about this in the scientific literature; see, for example, L.J. Savage, *The Foundations of Statistics* (Wiley, New York, 1954) and B. de Finetti, *The Theory of Probability*, Vols. 1, 2 (Wiley, New York, 1974, 1975). These authors take a further step and maintain that *all* probabilities are subjective: If somebody asserts that "the probability is 1/5 that 'two' is obtained if a well-made die is thrown", this is his private opinion which cannot be questioned by anybody. Two persons, therefore, may well assign different numerical probabilities to the same event, since their subjective opinions about the event may differ.

A common argument against the subjective view is that science and technology seek objective knowledge and that probability theory should therefore have only objective contents in order to be useful.

In this book we will not deal further with subjective probabilities, but we are not of the opinion that such probabilities are always worthless. It might be added that results derived from Kolmogorov's axioms can be applied to any function $P(\cdot)$ that satisfies the axioms whether a frequency interpretation is possible or not. □

2.4. Probabilities in Discrete Sample Spaces

In this section, we assume that the sample space is discrete; that is, it consists of a finite or denumerably infinite number of outcomes $u_1, u_2, \ldots$. (The reader may, for example, have Example 1 or Example 2 in §2.2 in mind.) Then one can begin by assigning probabilities $P(u_1)$, $P(u_2)$, ... and has only to check that they sum to unity. In other respects the assignment is unrestricted, theoretically speaking, and is made according to the practical situation that the model is meant to represent. The probability of any event A of interest is then obtained simply by adding the probabilities of the outcomes belonging to A.

This sounds easier than it often is. To choose the probabilities, in such a way that the model is a reasonably good approximation to reality, can be rather difficult. In the following examples we avoid potential complications by studying only simple situations, where the model is easy to construct.

Example 5. Throws with a Loaded Die

A die is manufactured in such a manner that each of the outcomes "one", "two" and "three" turns up twice as often as each of the other outcomes. It is then reasonable to take

$$P(u_1) = P(u_2) = P(u_3) = 2/9; \qquad P(u_4) = P(u_5) = P(u_6) = 1/9.$$

If A is the event "odd face value", we find by the addition formula in Axiom 3, p. 12,

$$P(A) = P(u_1) + P(u_3) + P(u_5) = 2/9 + 2/9 + 1/9 = 5/9.$$ ☐

We shall now consider a special situation. Let us assume that there are m outcomes in Ω and that we assign the same probability to each of them, that is, we consider them equally likely. To make the sum of the probabilities equal to 1 we must choose each $P(u_i)$ equal to $1/m$. This special case is so important that it deserves a name of its own:

Definition. If

$$P(u_i) = 1/m \qquad (i = 1, 2, \ldots, m),$$

we have a *uniform probability distribution.* ☐

We now consider an event A with g outcomes. It is then customary to say that there are g *favourable cases.* We then also say that there are m *possible cases* in Ω. We then determine $P(A)$ from the addition formula by adding the probabilities $1/m$ of the g outcomes. Hence we obtain

$$P(A) = g/m. \tag{3}$$

See Fig. 2.7 for an illustration. This is a simple proof of an important theorem which contains the so-called classical definition of probability. This definition has played an important role in the history of probability. We retain this name, although, in this book, it is *not* a definition.

Theorem 4 ("The Classical Definition of Probability"). *If the probability distribution is uniform, the probability of an event is the ratio of the number of favourable cases to the number of possible cases.* ☐

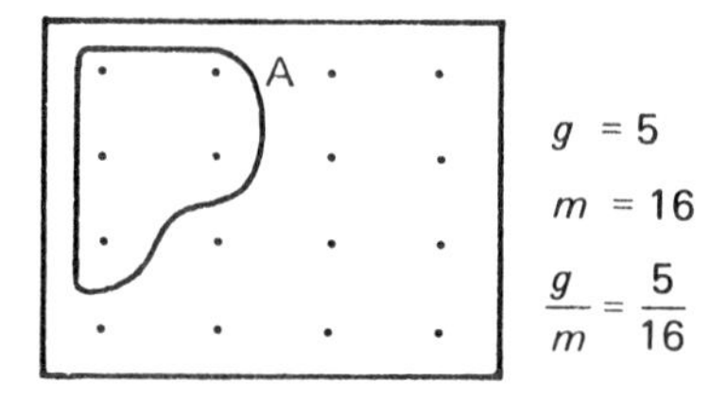

Fig. 2.7.

It should be stressed that the theorem can be used only if equal probabilities are assigned to the outcomes; for example, it cannot be applied in the case of the loaded die.

To be able to handle uniform probability distributions adroitly, a good knowledge of combinatorics is sometimes essential, especially in more complicated problems. In the older literature such problems abounded, perhaps conveying the impression that the nutcracking involved were the climax of probability theory. We will be moderate in this respect and confine ourselves to some of the most important applications. Only an elementary knowledge of combinatorial theory will be required, as contained in most mathematical textbooks. See also §2.7, where some fundamental formulae are summarized.

Example 6. Throws of Two Dice

Two well-made dice are thrown once. Let A be the event "the sum is 3 or less". There are 36 possible cases (1, 1), (1, 2), ..., (6, 6), which we regard as equally likely. The favourable cases are three: (1, 1), (1, 2), (2, 1). By Theorem 4 we obtain $P(A) = 3/36$. □

Example 7. Defective Units

In a batch of manufactured units, 2% of the units have the wrong weight (and perhaps also the wrong colour), 5% have the wrong colour (and perhaps also the wrong weight), and 1% have both the wrong weight and the wrong colour. A unit is taken at random from the batch. What is the probability that the unit is defective in at least one of the two respects?

There are as many outcomes u_i as there are units in the batch: u_i signifies that the ith unit is obtained. Such expressions as "a unit is chosen at random" or "a unit is chosen by chance" are rather vague. We mean by this, here and henceforth, that the selection is performed in such a way that it is reasonable to assign the same probability to all outcomes. Now let A be the event that "the unit has the wrong weight" and B the event that "the unit has the wrong colour". The model that we have chosen then shows that

$$P(A) = 0.02; \qquad P(B) = 0.05; \qquad P(A \cap B) = 0.01.$$

By Theorem 2 in §2.3 the desired probability is given by

$$P(A \cup B) = P(A) + P(B) - P(A \cap B) = 0.02 + 0.05 - 0.01 = 0.06.$$ □

Example 8. Random Decimal Number

Let us select one of the five-digit numbers 00000, 00001, ..., 99999 at random.

(a) What is the probability that the number only contains the digits 0 and 1?

We let each possible number correspond to one outcome. The number of possible cases is then 10^5. The number of favourable cases is seen to be 2^5, for there are two favourable digits in each of the five positions. According to Theorem 4, the desired probability is

$$2^5/10^5 = 0.00032.$$

(b) What is the probability that the digits in the random number are all different?

The number of favourable cases is now the number of permutations of 5 elements from among 10, that is $10 \cdot 9 \cdot 8 \cdot 7 \cdot 6 = 30{,}240$ (any of the 10 digits can come first, then 1 of 9 digits can come second, and so on). Hence the probability is

$$30{,}240/10^5 = 0.3024. \qquad \square$$

Example 9. A Deck of Cards

In a standard deck of 52 cards there are 13 spades. Two cards are drawn at random. Determine the probability that both are spades.

The number of favourable cases is equal to the number of combinations of 2 elements from among 13, and the number of possible cases is equal to the number of combinations of 2 elements from among 52; that is, we get

$$g = \binom{13}{2}; \qquad m = \binom{52}{2}; \qquad P(A) = \binom{13}{2} \bigg/ \binom{52}{2} = 1/17. \qquad \square$$

Remark. The Definition of Outcome

It has already been mentioned that outcomes can be defined in different ways in the same problem. We now realize that the definition of probability itself can sometimes affect the choice of definition of outcome. If it is possible to define the outcomes so that they can be given equal probabilities, we ought, of course, to use this definition to make life simple. However, this does not always lead to a unique definition. If you throw a good die and ask for the probability of the event "the number on the die is four or less", you can either take the usual six outcomes and give each the probability 1/6, or take the three outcomes "one or two", "three or four", "five or six" and give each the probability 1/3. Clearly, both definitions give the same answer, 2/3. However, the choice of definition is not always unimportant. In more complicated problems one should try to define the outcomes in such a manner that the ensuing combinatorial considerations become as simple as possible. $\square$

We shall now discuss *urn models*. Many problems in probability theory can be described in terms of drawing balls from urns. The urn is one of the favourite objects of the statistician (Fig. 2.8). However, it may be replaced by something less formal, for example, a plastic bag, or by a purely verbal description. Evidently, the urn model is meant to illustrate a certain random model, and the terminology is only an aid to thinking.

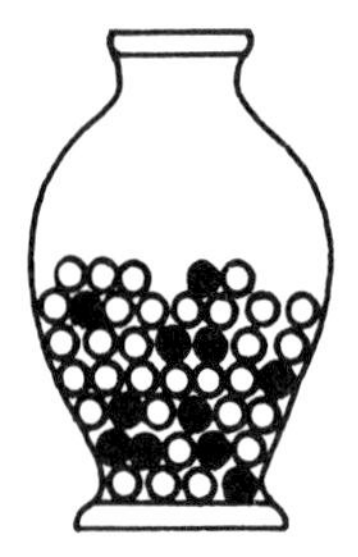

Fig. 2.8. The urn—one of the favourite objects of the statistician.

There are two sorts of objects in the urn, say a white balls and b black balls. We draw n balls from the urn. There are two ways of making the drawings which we wish to distinguish.

(a) Drawing Balls without Replacement

Definition. If one ball at a time is drawn at random without returning it to the urn, the drawings are said to be made *without replacement*. □

In this case we must have $n \leq a + b$, for otherwise the urn will be empty before the drawings are completed.

This definition may seem rather uninformative at first sight, but a moment's reflection shows that it is not. The important term is "at random". By that term we mean that each possible subset of n balls appears with the same probability.

If n balls are drawn without replacement, what is the probability that among the n balls there are exactly k white balls?

For the answer, we must first define the outcomes, which can be done in several ways. Let us, for the moment, label the balls $W_1, W_2, \ldots, W_a, B_1, B_2, \ldots, B_b$, where W denotes a white ball and B a black ball. We now define the outcomes as all possible sets of n balls drawn, *ignoring the order in which they are drawn.* (Example: if $a = 1, b = 3, n = 2$, there are six outcomes, namely W_1B_1, W_1B_2, W_1B_3, B_1B_2, B_1B_3, B_2B_3.) Because of the definition of drawing without replacement, these outcomes will have the same probability.

The number m of possible cases is the number of combinations of n objects from among $a + b$, that is $m = \binom{a+b}{n}$. In how many of these cases are there k white balls? All such favourable cases are obtained by taking k balls from the a white ones and the remaining $n - k$ balls from the b black ones. Thus the number g of favourable cases is equal to the number of combinations of k objects among a, multiplied by the number of combinations of $n - k$ objects among b, that is, we have $g = \binom{a}{k}\binom{b}{n-k}$. By the classical definition of probability the required probability is

$$\binom{a}{k}\binom{b}{n-k}\Big/\binom{a+b}{n}.$$

Example 9. A Deck of Cards (continued)

The problem in Example 9, p. 17, is a special case of the problem we have just solved; in the above expression take $a = 13$, $b = 39$, $n = 2$, $k = 2$. □

Remark. Several Kinds of Objects

The urn model just discussed can be generalized to more than two kinds of objects. Suppose that the urn contains a total of N objects of r different kinds, namely a_1 objects of the first kind, a_2 objects of the second kind, and so on. Let us draw n objects without replacement ($n \leq N$). The probability of obtaining k_1 objects of the first kind, k_2 objects of the second kind, and so on, is seen to be

$$\prod_{i=1}^{r}\binom{a_i}{k_i}\Big/\binom{N}{n} \qquad (n = k_1 + k_2 + \cdots + k_r).$$

The proof is similar to the one we have just given. ☑

(b) Drawing Balls with Replacement

Definition. If one ball at a time is drawn at random and is returned before the next ball is drawn, the drawings are said to be made *with replacement.* □

In this case, the composition of the urn is the same throughout the process. The number of drawings can be arbitrarily large.

As before, we seek the probability that, if n drawings are performed, a white ball appears exactly k times.

We label the balls as before but change the definition of the outcomes. We let the outcomes be all possible sequences of n balls drawn, *taking the order of drawing into account.* (Example: In the example on p. 18 with $a = 1$, $b = 3$, $n = 2$, there are now 16 outcomes, namely, W_1W_1, W_1B_1, W_1B_2, W_1B_3, B_1W_1, B_1B_1, B_1B_2, B_1B_3, B_2W_1, B_2B_1, B_2B_2, B_2B_3, B_3W_1, B_3B_1, B_3B_2, B_3B_3.) At each drawing of a ball there are $a + b$ possible cases, and thus the total number of possible cases for n drawings is $m = (a + b)^n$. In how many of these cases are exactly k white balls obtained? This number is determined in two stages.

(1) The white balls appear in k of the n positions, a total of $\binom{n}{k}$ possibilities.

One such possibility is

$$\underbrace{WW\ldots W}_{k}\underbrace{BB\ldots B}_{n-k},$$

that is, one obtains white balls only, followed by black balls only; the other sequences are obtained by rearranging this sequence.

(2) Consider a certain such sequence of k white and $n - k$ black balls, for example, the one above. This is obtained by placing any one of the a white balls in the first position, one in the second position, and so on, and finally one in the kth position.

There are a^k possibilities in all. Then one continues with the black balls in the last $n - k$ positions, which gives b^{n-k} possibilities. Multiplying these two numbers we have $a^k b^{n-k}$ possibilities in all. Since we found the number of sequences to be $\binom{n}{k}$, the number of favourable cases is $g = \binom{n}{k} a^k b^{n-k}$. According to the classical definition of probability, the required probability of obtaining k white balls becomes

$$\binom{n}{k}\left(\frac{a}{a+b}\right)^k\left(\frac{b}{a+b}\right)^{n-k}.$$

If we take

$$p = \frac{a}{a+b}; \qquad q = 1 - p = \frac{b}{a+b},$$

where p is the proportion of white balls and q the proportion of black balls in the urn, the expression can be written

$$\binom{n}{k} p^k q^{n-k}.$$

Example 10. Random Binary Number

Determine the probability that a six-digit random binary number contains a zero exactly twice.

We consider an urn with one white and one black ball, corresponding to the digits 0 and 1, and draw six times with replacement. The required probability is obtained by taking $a = 1$, $b = 1$, $n = 6$, $k = 2$ in the above formula, and hence the answer is

$$\binom{6}{2}\left(\frac{1}{2}\right)^2\left(\frac{1}{2}\right)^4 = \frac{15}{64}.$$

□

Remark. Several Kinds of Objects

The urn is assumed to contain a total of N objects of r kinds, of which a_1 are of the first kind, a_2 of the second kind, and so on. Let us draw n objects with replacement. The probability of obtaining k_1 objects of the first kind, k_2 objects of the second kind, and so on, is given by

$$\frac{n!}{\prod_1^r k_i!}\prod_1^r\left(\frac{a_i}{N}\right)^{k_i}.$$

The proof is left to the interested reader. If we introduce the proportion $p_i = a_i/N$ of objects of the ith kind ($i = 1, 2, \ldots, r$), the expression takes the form

$$\frac{n!}{\prod_1^r k_i!} \prod_1^r p_i^{k_i}.$$

☑

2.5. Conditional Probability

The important concept of conditional probability will first be presented in a special example. We have eight pencils

1	2	3	4	5	6	7	8
R	R	R	G	G	G	G	G

where R denotes a red pencil and G a grey pencil. One of the pencils is chosen at random with equal probability (for example, by writing the numbers $1, 2, \ldots, 8$ on slips of paper, putting them into a bag, mixing them thoroughly and drawing one of them). If A is the event that "a red pencil is obtained", the classical definition of probability shows that $P(A) = 3/8$.

Let us now also classify the pencils with respect to hardness (H = hard pencil, S = soft pencil):

1	2	3	4	5	6	7	8
R	R	R	G	G	G	G	G
H	H	S	H	H	H	S	S

If B is the event that "a hard pencil is obtained", we find $P(B) = 5/8$. It is also found that $P(A \cap B) = 2/8$ (for only no. 1 and no. 2 are both red and hard).

We now ask: If a pencil has been chosen and found to be red, what is the probability that it is hard? This probability is denoted by $P(B|A)$ and is called "the conditional probability of B given A". Since we know the pencil is red, there are now only three possible cases (nos. 1, 2, 3), two of which are hard. It is therefore reasonable to take $P(B|A) = 2/3$, which can also be written

$$P(B|A) = \frac{2/8}{3/8} = \frac{P(A \cap B)}{P(A)}.$$

One can argue similarly in other situations. We therefore choose the following definition of conditional probability:

Definition. The expression

$$P(B|A) = \frac{P(A \cap B)}{P(A)} \tag{4}$$

is called *the conditional probability of B given A.*

☑

Naturally, the definition makes sense only when $P(A)$ is positive.

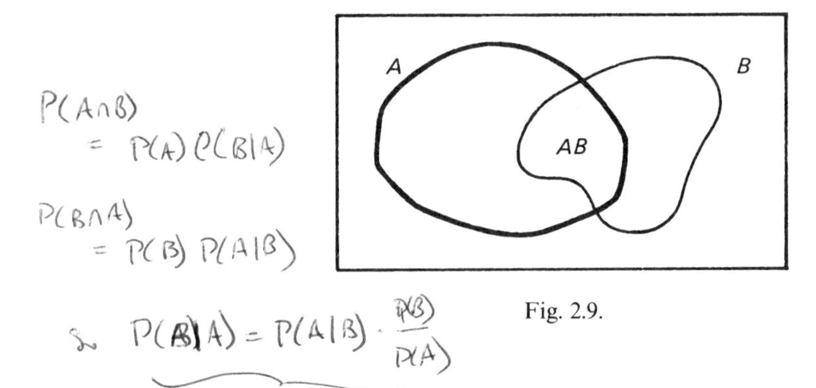

Fig. 2.9.

The idea leading to this definition can, without pencils, be described as follows (see Fig. 2.9). We know that A has taken place and we therefore move in a new, more restricted sample space A (= the heavily drawn contour in the figure). The probabilities of events within A should not change, relatively speaking. To make the probabilities of the outcomes in A sum to unity, we have therefore only to multiply each probability by a scale factor $1/P(A)$.

The formula in the definition can also be expressed in the form

$$P(A \cap B) = P(A)P(B|A) \tag{5}$$

or in words: *The probability that two events both occur is equal to the probability that one of them occurs, multiplied by the conditional probability that the other occurs, given that the first one has occurred.*

Both versions are important. It is, of course, possible to interchange A and B, and so we have the extended formula

$$P(A \cap B) = P(A)P(B|A) = P(B)P(A|B). \tag{6}$$

By employing the first formula twice, we get the following formula for three events

$$P(A \cap B \cap C) = P(A \cap B)P(C|A \cap B) = P(A)P(B|A)P(C|A \cap B) \tag{7}$$

and, analogously, for more than three events.

Example 11. Drawings Without Replacement

In a batch of 50 units there are 5 defectives. A unit is selected at random, and thereafter one more from the remaining ones. Find the probability that both are defective.

Let A be the event that "the first unit is defective" and B the event that "the second unit is defective". It is seen that $P(A) = 5/50$. If A occurs, there remain 49 units, 4 of which are defective. Hence we conclude that $P(B|A) = 4/49$, and the first formula in (6) produces the probability we seek:

$$P(A \cap B) = \frac{5}{50} \cdot \frac{4}{49} = \frac{2}{245}.$$

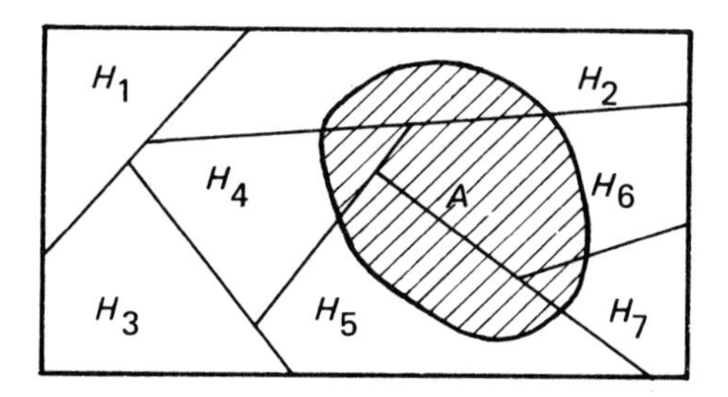

Fig. 2.10. Partition of Ω into the parts $H_1, \ldots, H_7$.

Let us now draw a third unit, and let us evaluate the probability that the first two units are defective and that the third one is good. If C is the event that "the third unit is good", formula (7) gives the answer

$$P(A \cap B \cap C) = \frac{5}{50} \cdot \frac{4}{49} \cdot \frac{45}{48} = \frac{3}{392}. \qquad \square$$

We now derive two useful theorems for conditional probabilities.

Theorem 5 ("Total Probability Theorem"). *If the events $H_1, H_2, \ldots, H_n$ are mutually exclusive, have positive probabilities, and together fill Ω completely, any event A satisfies the formula*

$$P(A) = \sum_{i=1}^{n} P(H_i)P(A|H_i). \tag{8}$$

$\square$

Proof. It is seen that

$$P(A) = P(A \cap \Omega) = P\left(A \cap \bigcup_1^n H_i\right) = P\left(\bigcup_1^n (A \cap H_i)\right).$$

Since the events H_i $(i = 1, \ldots, n)$ are mutually exclusive, this must also be true for the events $A \cap H_i$. An application of Axiom 3, p. 12, now gives

$$P\left(\bigcup_1^n (A \cap H_i)\right) = \sum_1^n P(A \cap H_i).$$

Employing the definition of conditional probability, we can write

$$P(A \cap H_i) = P(H_i)P(A|H_i).$$

Inserting this into the preceding formula, we get the theorem. $\square$

Theorem 5 is convenient for finding a probability when the event is easier to study in the subsets H_i of Ω.

Example 12. Production by Several Machines

In a factory, units are manufactured by machines H_1, H_2, H_3 in the proportions 25:35:40. The percentages 5%, 4% and 2%, respectively, of the manufactured units are defective. The units are mixed and sent to the customers. Find the probability that a randomly chosen unit is defective.

Taking, in Theorem 5, H_i = "unit produced by machine H_i" and A = "unit is defective", we find

$$P(A) = 0.25 \cdot 0.05 + 0.35 \cdot 0.04 + 0.40 \cdot 0.02 = 0.0345. \qquad \square$$

Theorem 6 ("Bayes' Theorem"). *Under the same conditions as in Theorem 5*

$$P(H_i|A) = \frac{P(H_i)P(A|H_i)}{\sum_{j=1}^{n} P(H_j)P(A|H_j)}. \tag{9}$$

$\square$

Proof. By using the definition of conditional probability we obtain

$$P(H_i|A) = \frac{P(H_i \cap A)}{P(A)} = \frac{P(H_i)P(A|H_i)}{P(A)}.$$

If the expression for $P(A)$ in (8) is substituted in the denominator, we obtain formula (9). $\square$

Theorem 6 was already established in the eighteenth century by the English clergyman Bayes. Seldom has a formula been misused so often by so many. The reason for this is, among others, that the probabilities $P(H_i)$ are often difficult to determine in practice. An easy way to get rid of them is simply to assume that they are equal, for if we make this assumption, they just disappear! That this is not generally justified is apparent from the following two examples.

Example 12. Production by Several Machines (continued)

Suppose that a customer discovers that a certain unit is defective. What is the probability that it has been manufactured by machine H_1? Bayes' theorem gives the answer (A = defective unit)

$$P(H_1|A) = \frac{0.25 \cdot 0.05}{0.25 \cdot 0.05 + 0.35 \cdot 0.04 + 0.40 \cdot 0.02} = 0.36.$$

(If we had taken each $P(H_i)$ equal to 1/3, we would have obtained the erroneous result $P(H_1|A) = 0.05/(0.05 + 0.04 + 0.02) = 0.45$.) $\square$

Example 13. Mixed Population

In a certain country there are two nationalities living together, the Bigs and the Smalls. Among the Bigs 80% are tall, and among the Smalls 1%. A visiting tourist encounters a person at random who turns out to be tall. Determine the probability that this person belongs to the Bigs.

Let H_1 = "the person belongs to the Bigs", H_2 = "the person belongs to the Smalls", A = "the person is tall". Bayes' theorem shows that

$$P(H_1|A) = \frac{P(H_1)\cdot 0.80}{P(H_1)\cdot 0.80 + P(H_2)\cdot 0.01}.$$

The formulation of the problem is inadequate: It is necessary to know the probabilities $P(H_1)$ and $P(H_2)$, the proportions of Bigs and Smalls in the country.

If the proportions are the same, so that $P(H_1) = P(H_2) = 1/2$, the probability becomes $80/81 = 0.99$. But if the Bigs are so few that $P(H_1) = 0.001$ and $P(H_2) = 0.999$, the probability is instead

$$0.001\cdot 0.80/(0.001\cdot 0.80 + 0.999\cdot 0.01) = 80/1079 = 0.08. \qquad \square$$

2.6. Independent Events

If two events A and B are such that $P(B|A) = P(B)$, that is, if the probability that B occurs is the same whether or not it is known that A has occurred, then it is reasonable to say that A and B are independent. In this case, formula (4) gives

$$P(B) = \frac{P(A \cap B)}{P(A)}.$$

Multiplication of both sides by $P(A)$ results in the following basic definition.

Definition. If

$$P(A \cap B) = P(A)P(B), \tag{10}$$

then A and B are said to be *independent* events. □

Example 14. Throwing Two Dice

Find the probability that a throw with two well-made dice results in "one" on both dice.

Let A and B denote the events "one die shows one" and "the other die shows one". We have $P(A) = P(B) = 1/6$. If the throw is performed in such a manner that A and B may be assumed independent, formula (10) gives

$$P(A \cap B) = \frac{1}{6}\cdot\frac{1}{6} = \frac{1}{36}.$$

Clearly, the definition gives a sensible result. In fact, we might have argued in a more direct manner. There are 36 possibilities in all, 1 of which is favourable; that is, we have $P(A \cap B) = 1/36$. Note the difference between the arguments. In the first solution, we introduce probabilities for each die separately, and hence there are 6 outcomes for each; in the second solution, we consider outcomes for both dice simultaneously, which leads to 36 outcomes. □

Example 15. Tall and Wealthy Statisticians

Compute the probability that a randomly chosen statistician has both the properties A = "over 6 feet tall" and B = "assets exceeding 100,000 dollars".

We assume that it is known somehow that 10% of all statisticians have property A and 5% have property B. We therefore take

$$P(A) = 0.10; \qquad P(B) = 0.05.$$

If the two properties are taken to be independent, which seems reasonable, we get

$$P(A \cap B) = 0.10 \cdot 0.05 = 0.005.$$

It is easy to check that the statistical definition of independence used here agrees with independence as used in everyday speech. Independence in the latter sense ought to imply that the proportion of wealthy people among the tall ones is the same as among those who are not tall; that is, 5% in both cases. Now then, the tall people constitute 10% of all statisticians, and hence the fraction of these that are also wealthy ought to be 5% of these 10%, that is, 5 in 1,000. Consequently, if a statistician is chosen at random, the probability of finding a person who is both tall and wealthy ought to be 0.005, which is in accord with the definition. □

Quite generally, experience shows that if two events are independent in the everyday meaning of the word, reasonable results are obtained if they are also regarded as independent in the sense used in probability theory.

The definition of independence can be extended to more than two events. Here is the definition in the case of three events:

Definition. If

$$\begin{aligned} P(A \cap B) &= P(A)P(B), \\ P(A \cap C) &= P(A)P(C), \\ P(B \cap C) &= P(B)P(C), \\ P(A \cap B \cap C) &= P(A)P(B)P(C), \end{aligned} \tag{11}$$

then A, B and C are said to be *independent* events. ☑

(In the general case, all possible subgroups of two, three, ... events must satisfy analogous formulae.)

If $A, B, C, \ldots$ are independent, then the complements $A^*, B^*, C^*, \ldots$ are also independent; further, $A, B^*, C^*, \ldots$ are independent, and so on. This is not difficult to prove, and is recommended as an exercise for the reader.

The definition of independence has many important consequences; here is one:

Theorem 7. *If the events $A_1, A_2, \ldots, A_n$ are independent and $P(A_i) = p_i$, then the probability that at least one of them occurs is equal to*

$$1 - (1 - p_1)(1 - p_2) \cdot \ldots \cdot (1 - p_n).$$ □

Proof. The event we want to study can be written $A_1 \cup A_2 \cup \cdots \cup A_n$. Its complement is that none of the events A_i occurs; that is, the event $A_1^* A_2^* \ldots A_n^*$.

But since $P(A_i^*) = 1 - p_i$ and the events A_i^* are independent, it follows that

$$P(A_1^* A_2^* \ldots A_n^*) = (1 - p_1)(1 - p_2) \cdot \ldots \cdot (1 - p_n).$$

By the complement theorem, p. 12, it now follows that

$$P(A_1 \cup A_2 \cup \cdots \cup A_n) = 1 - (1 - p_1)(1 - p_2) \cdot \ldots \cdot (1 - p_n)$$

and the theorem is proved. □

Corollary. *If the events A_i are independent and each one of them occurs with probability p, then the probability that at least one of them occurs is equal to $1 - (1 - p)^n$.* □

This corollary has important applications, as shown by

Example 16. Risks

A person is exposed, on 1,000 independent occasions, to a risk of accident amounting to 1/1,000 each time. The probability that an accident occurs on one or more of these occasions is, by the corollary,

$$1 - (1 - 0.001)^{1000} \approx 1 - e^{-1} = 0.63.$$

In spite of the small risk each time, the overall risk is greater than 1/2. ☑

A particularly important application of independent events is to *independent trials.* A strict mathematical definition of this concept cannot be given, for trials belong to the empirical world. We must therefore give a rather loose description. It seems reasonable to regard two trials as independent, if the result of one does not affect, or is not affected by, the result of the other.

An important special case is provided by *repeated trials.* By this term we mean trials which are performed under the same conditions, over and over again, and independently of one another. Repeated measurements of a physical constant (for example, the velocity of light, the melting-point of a given substance) illustrate the meaning of the term. In experimental work, the researcher strives to attain the situation described here.

Example 17. Target-Shooting

A person repeatedly shoots at a target and stops as soon as he hits it. The probability of hitting the target is 2/3 each time. The shots are fired independently, and hence may be regarded as a sequence of repeated trials. Let us determine the probability that the target is hit on the kth attempt.

The event in which we are interested happens if the target is missed $k - 1$ times and then is hit the kth time. The probability of missing the target with the first shot is $1 - 2/3 = 1/3$, and likewise in each of the following shots. Because of the independence, the probability we seek is $(1/3)^{k-1}(2/3)$. ☑

We shall now take a more difficult example of a type which is rather common in the practical applications of probability theory.

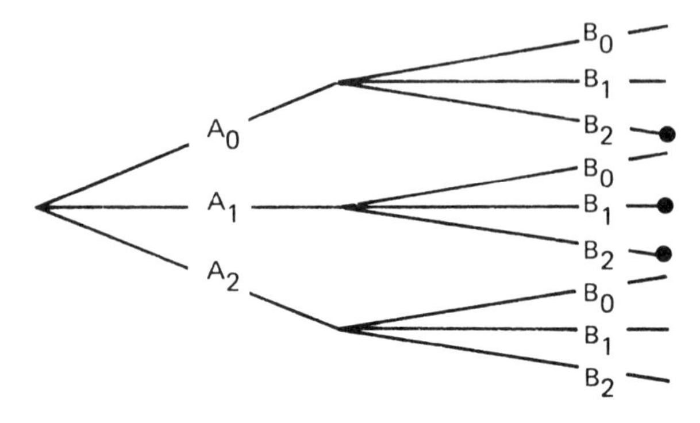

Fig. 2.11.

Example 18. Inventory Problem

A shop has two refrigerators in stock when it opens on a Monday morning. More refrigerators (units) cannot be obtained until Wednesday. The probability that 0, 1, 2 units are wanted by customers on Monday is 0.5, 0.3 and 0.2, respectively, and the corresponding probabilities for Tuesday are 0.7, 0.2 and 0.1. The demands on the two days are assumed to be independent, and hence may be regarded as two independent, but not repeated, trials. Find the probability that the stock of units becomes depleted some time on Tuesday. We call this event H.

It is instructive to draw a tree diagram, where A_0, A_1, A_2 and B_0, B_1, B_2 denote the events that 0, 1, 2 units are wanted by customers on Monday and Tuesday, respectively; see Fig. 2.11.

The stock becomes empty on Tuesday if one of the following three events occur: (1) Monday 0 unit is sold, Tuesday 2 units; (2) Monday 1 unit is sold, Tuesday 1 unit; (3) Monday 1 unit is sold, Tuesday 2 units are wanted, one of which is sold and one cannot be provided since the stock is depleted. (See Fig. 2.11, where the three events have been marked with dots.) Thus the event H can be written

$$H = A_0B_2 \cup A_1B_1 \cup A_1B_2.$$

Since the three events in this union are mutually exclusive, we have

$$P(H) = P(A_0B_2) + P(A_1B_1) + P(A_1B_2).$$

Furthermore, since A_0 and B_2 are independent, as also are the two other pairs, we obtain

$$P(H) = P(A_0)P(B_2) + P(A_1)P(B_1) + P(A_1)P(B_2)$$
$$= 0.5 \cdot 0.1 + 0.3 \cdot 0.2 + 0.3 \cdot 0.1 = 0.14. \qquad \square$$

Remark 1. Improbable Events

The newspapers often publish notices about improbable events; for example, that a person has been dealt 13 spades; or that a second person has encountered his twin brother, hitherto unknown, at the northernmost point of Lake Victoria; or that a third person has happened to get the same bank account number as his wife's birth date. Viewed as isolated events, each of these and other peculiarities is perhaps improbable.

The probability that, in a bridge hand, someone will hold 13 spades is of the order 10^{-11}. It is therefore very improbable that the reader, on the day he reads this remark, will get 13 spades. On the other hand, it is not improbable (because of computations similar to those in Example 16 above) that this event will occur for one of all of the bridge-players in the world during a period of, say, 20 years; after all, bridge is played a great deal.

A remark with similar consequences can be made about a given individual. He may be disposed to observe coincidences of the type we have just described, and does this day after day. By Theorem 7, it is quite natural that one of these events, although improbable individually, will occur one day. Furthermore, the set of all such peculiarities is usually not given beforehand, but is unspecified. One may suddenly notice something strange that *has just occurred*, and notice that its probability is very small. This may inspire the imaginative person to quite remarkable discoveries. For example, he may report as something extraordinary that 17 tosses of a coin, which he has just performed, resulted in seven heads, one tail, two heads, three tails, and four heads. The chance that this will occur again in 17 tosses is not large; in fact, it is equal to $(1/2)^{17} = 1/131{,}072$. □

Remark 2. Statistical Superstition

While on the subject of repeated trials, we shall try to dispel a false notion. If one tosses a well-made coin repeatedly and gets, let us say, heads 10 times in succession, many people think that the probability of obtaining a tail at the eleventh throw is greater than 1/2. Similarly, if 10 boys in succession have been born in a parish, some people are of the opinion that the probability of a girl next time is greater than if such a sequence of births of the opposite sex had not occurred. This misconception presumably arises because it is argued that the two events, head–tail or boy–girl, ought to take place equally often (which is true, apart from the fact that there are slightly more male than female births), and that a sequence of like outcomes causes a "deficiency" which Nature strives to compensate. The concept of independent trials reveals the falsity of all such ideas. The chance that a head appears is the same in each toss and is not affected by the outcomes of earlier trials.

As a curiosity, it might be added that there is a reverse form of superstition in the case of events which are advantageous to the person involved. If a lottery ticket, sold by some agency, turns out to be a winning ticket, the agency is thought to be favoured by Lady Luck; hence the chance that it will sell a winning ticket next time is believed to be greater than for other agencies! □

2.7. Some Theorems in Combinatorics

As has already been stated, combinatorial methods are sometimes useful when solving problems in probability theory. We shall give a brief survey of some basic combinatorial results.

Consider a set of N elements $A_1, A_2, \ldots, A_N$. From this set we take n elements. We wish to know in how many ways this can be done if various conditions are fulfilled. We utilize the terminology known from probability theory: "drawing elements with replacement" and "drawing elements without replacement".

The following notation is used:

$$n! = n(n-1)(n-2)\cdot\ldots\cdot 2\cdot 1,$$

$$\binom{a}{b} = \frac{a!}{b!(a-b)!} = \frac{a(a-1)\cdot\ldots\cdot(a-b+1)}{b!}.$$

We begin with drawings with replacement. In how many ways can n elements be drawn with replacement? If $N = 3$, $n = 2$, the number of ways is nine, viz.

$$A_1A_1, A_1A_2, A_1A_3, A_2A_1, A_2A_2, A_2A_3, A_3A_1, A_3A_2, A_3A_3.$$

Note that the elements are given in the order drawn.

Generally, the number of ways is equal to N^n. Indeed, there are N possible elements to be placed in the first position, N in the second, and so on, and they can be combined with one another without any restriction. This gives

$$\underbrace{N\cdot N\cdot\ldots\cdot N}_{n} = N^n$$

ways. This proves

Theorem 8. *Drawing n elements from N with replacement, with regard to order, can be done in N^n different ways.* □

For completeness we shall give a further theorem which, however, will not be applied in this book. Sometimes one is not interested in the order of the elements drawn. In the above example, A_1A_2 and A_2A_1 are then not distinguished, but counted as one single result. The number of ways is then reduced as follows:

Theorem 9. *Drawing n elements from N with replacement, without regard to order, can be done in* $\binom{N+n-1}{n}$ *different ways.* □

This theorem will not be proved.

We now consider drawing elements without replacement. In how many ways can a sequence of n elements be drawn without replacement from the given set of N elements? If $N = 3$ and $n = 2$, the number of ways is equal to 6, viz.

$$A_1A_2, A_1A_3, A_2A_1, A_2A_3, A_3A_1, A_3A_2.$$

Generally, the number of ways can be found in the following way. First, one element is drawn and placed in the first position, which can be done in N ways; then, one of the $N-1$ remaining elements is drawn and placed in the second position, and so on; finally, there remain $N - n + 1$ elements to choose from in the nth position. The total number of ways is the product of these numbers, that is, $N(N-1)\cdot\ldots\cdot(N-n+1)$. This proves

Theorem 10. *Drawing n elements from N without replacement, with regard to order, can be done in* $N(N-1)\cdot\ldots\cdot(N-n+1)$ *different ways.* ☐

In this situation, the number of ways is usually called the number of *permutations of n elements among N*.

Corollary. *The number of permutations of N elements among N is equal to* $N(N-1)\cdot\ldots\cdot 2\cdot 1 = N!$, *or more briefly, N elements can be ordered in N! different ways.* ☐

In Theorem 10 we took into account the order of the elements. We shall now disregard this order. How is the number of ways then reduced?

Let us again consider the special case $N = 3$, $n = 2$. Since A_1A_2 and A_2A_1 are now no longer distinguished, there are three different subsets, viz.

$$A_1A_2,\ A_1A_3,\ A_2A_3.$$

Hence the number of cases is reduced from 6 to 3, that is, divided by a factor 2.

Generally, n drawn elements can be ordered in $n!$ ways (according to the above Corollary). We do not distinguish between these any longer, and hence obtain the number of different subsets by dividing the number shown in Theorem 10 by $n!$. Hence we have proved

Theorem 11. *Drawing n elements among N without replacement, without regard to order, can be done in*

$$\frac{N(N-1)\cdot\ldots\cdot(N-n+1)}{n!} = \binom{N}{n} \quad \textit{different ways.}$$ ☐

The number of ways in this situation is called the number of *combinations of n elements among N*.

Example 19. Three-Letter Code

Let us consider a three-letter code with 26 different letters in each position. How many code words are possible? Theorem 8 gives the answer, $26^3 = 17{,}576$. How many are there if the letters in the word are required to be different? Theorem 10 gives the answer $26\cdot 25\cdot 24 = 15{,}600$. ☐

Example 20. Handshakes

In a party of 10 persons, all shake hands. The number of handshakes is equal to the number of combinations of two elements among 10; by Theorem 11 the number is

$$\binom{10}{2} = \frac{10 \cdot 9}{2 \cdot 1} = 45. \qquad \square$$

Theorems 8–11 are summarized in Table 2.2.

Table 2.2. If n elements are taken from a set of N elements the number of ways is given in the table.

	With replacement	Without replacement
With regard to order	N^n	$N(N-1)\cdot\ldots\cdot(N-n+1)$
Without regard to order	$\binom{N+n-1}{n}$	$\binom{N}{n}$

Finally, we shall mention the simple but important *multiplication principle.*

Example 21. Menu

At a restaurant, a guest may choose between four hot dishes and three sweets. A dinner with a hot dish and a sweet can be composed in $4 \cdot 3 = 12$ ways. (A condition for this is, of course, that the guest begins with the hot dish and has a free choice.) $\square$

Generally, the multiplication principle states: If choice 1 can be performed in a_1 ways and choice 2 in a_2 ways, then there are $a_1 a_2$ ways to perform both choices. In the case of three choices, the number of ways is $a_1 a_2 a_3$, and so on.

2.8. Some Classical Problems of Probability[1]

In this section we shall discuss some problems of probability which are well known in the literature, and well deserve the title of classical problems. They are reproduced here as illustrations of the theory, but are not necessary for the understanding of the book. Some of the problems have historical interest, and it may perhaps be part of an all-round statistical education to be familiar with them.

Example 22. Russian Roulette

Russian roulette is played with a revolver equipped with a rotatable magazine of six shots. The revolver is loaded with one shot. The first duellist, A, rotates the magazine at random, points the revolver at his head and presses the trigger. If, afterwards, he is still alive, he hands the revolver to the other duellist, B, who acts in the same way as A. The players shoot alternately in this manner, until a shot goes off. Determine the probability that A is killed.

[1] Special section, which may be omitted.

Let H_i be the event that "a shot goes off at the ith trial". The events H_i are mutually exclusive. The event H_i occurs if there are $i - 1$ "failures" and then one "success". Hence we get (see Example 17)

$$P(H_i) = \left(\frac{5}{6}\right)^{i-1} \frac{1}{6}.$$

The probability we want is given by

$$P = P(H_1 \cup H_3 \cup H_5 \cup \cdots) = \frac{1}{6}\left[1 + \left(\frac{5}{6}\right)^2 + \left(\frac{5}{6}\right)^4 + \cdots\right]$$

$$= \frac{1}{6} \cdot \frac{1}{1 - (\frac{5}{6})^2} = \frac{6}{11}.$$

Hence the probability that B loses his life is $1 - 6/11 = 5/11$; that is, the second player has a somewhat greater chance of surviving, as might be expected.

The problem can also be solved in another surprisingly simple way. Let T be the event that "A is killed on the first trial". We have

$$P(A \text{ is killed}) = P(T) + P(T^*)P(A \text{ is killed} | T^*).$$

But if A survives the first trial, the roles of A and B are interchanged and so $P(A \text{ is killed} | T^*) = P(B \text{ is killed}) = 1 - P(A \text{ is killed})$. Inserting this above, we find

$$P(A \text{ is killed}) = \frac{1}{6} + \frac{5}{6}[1 - P(A \text{ is killed})].$$

Solving this equation we find $P(A \text{ is killed}) = 6/11$. □

Example 23. Poker

In a deck of cards there are 52 cards consisting of 4 suits with 13 denominations in each. A poker deal contains 5 randomly selected cards. A "full house" means that the player receives 3 cards of one denomination and 2 cards of another denomination; "three-of-a-kind" means that he gets 3 cards of one denomination, 1 of another denomination and 1 of a third denomination. Determine the probability of a full house and the probability of three-of-a-kind.

(a) Full house

The number of possible poker deals is $\binom{52}{5}$. The favourable cases are found as follows. The denominations for the 3-cards and the 2-cards can be selected in $13 \cdot 12$ ways. There are $\binom{4}{3}$ ways of selecting 3 cards from 4 cards with the same denomination; analogously, there are $\binom{4}{2}$ ways of taking out 2 cards. Hence the number of favourable cases is $13 \cdot 12 \cdot \binom{4}{3} \cdot \binom{4}{2}$, and the probability we want becomes

$$13 \cdot 12 \binom{4}{3}\binom{4}{2} \bigg/ \binom{52}{5} \approx \frac{1}{694}.$$

(b) Three-of-a-kind

The denominations for the 3-cards, the 1-card and the 1-card can be chosen in $13 \cdot \binom{12}{2}$ ways (not $13 \cdot 12 \cdot 11$ ways!). Hence we find, in about the same way as in (a), the probability

$$13 \cdot \binom{12}{2}\binom{4}{3}\binom{4}{1}\binom{4}{1} \Big/ \binom{52}{5} \approx \frac{1}{47}.$$

□

Example 24. The Birthday Problem

Find the probability that at least two of n randomly chosen persons have the same birthday.

We use the following model: Assume that there are 365 possible birthdays, all equally probable. (Clearly, the model is approximate, for not all years have 365 days, and the birth rate varies somewhat during the year.)

Denote by A the event we are interested in and by A^* its complement, that is, the event that the n persons have different birthdays. An extension of (7) in §2.5 shows that $P(A^*) = p_n$, where

$$p_n = \left(1 - \frac{1}{365}\right)\left(1 - \frac{2}{365}\right) \cdot \ldots \cdot \left(1 - \frac{n-1}{365}\right).$$

To realize this, let us "draw" the persons one at a time. The probability that the second person does not have the same birthday as the first one is $1 - 1/365$; the probability that the third person does not have the same birthday as either the first person, or the second one (given that these two have different birthdays) is $1 - 2/365$, and so on. Hence the required probability is given by $P(A) = 1 - p_n$.

For n not too large it turns out that

$$\ln p_n \approx -n(n-1)/730.$$

This follows from

$$\ln p_n = \sum_{j=1}^{n-1} \ln(1 - j/365) \approx -\sum_{j=1}^{n-1} j/365 = -n(n-1)/(2 \cdot 365).$$

For $n = 23$ it is found that

$$P(A) \approx 1 - e^{-23 \cdot 22/730} = 0.50.$$

Thus the probability is approximately 1/2 that, among 23 persons chosen at random, at least two have the same birthday. It is amazing that the probability is so large! □

Example 25. The Ruin Problem

A and B have a and b dollars, respectively. They repeatedly toss a fair coin. If a head appears, A gets 1 dollar from B; if a tail appears, B gets 1 dollar from A. The game goes on until one of the players is ruined. Determine the probability that A is ruined.

The solution is easier to find if we attack a more general problem. Suppose that, at a given moment, A has n dollars, and hence B has $a + b - n$ dollars. Let P_n be the probability that, given this condition, A is eventually ruined. Some reflection reveals that

$$P_n = \tfrac{1}{2}P_{n+1} + \tfrac{1}{2}P_{n-1}.$$

(This is a consequence of Theorem 5 in §2.5 with H_1 = "head appears in the next throw", H_2 = "tail appears in the next throw".) This recursive equation is solved by making use of the fact that the points (n, P_n) lie on a straight line. From this observation we conclude that

$$P_n = C_1 + C_2 n.$$

The constants C_1 and C_2 are determined by noting that $P_{a+b} = 0$ (for if A possesses all the money, he cannot be ruined) and $P_0 = 1$ (for if A has no money, he is ruined). Using these conditions a simple calculation shows that

$$P_n = 1 - \frac{n}{a+b}.$$

By taking $n = a$, we find that the probability is $b/(a + b)$ that A is ruined. In the same way, it is found that the probability is $a/(a + b)$ that B is ruined.

Suppose that A starts the game with $a = 10$ dollars and B with $b = 100$ dollars. The probability of ruin for A is then $100/110 = 0.91$. Conclusion: it is inadvisable to gamble with rich people! □

Example 26. The Problem of Rencontre

In an urn there are N slips of paper marked $1, 2, \ldots, N$. Slips are drawn at random, one at a time, until the urn is empty. If slip no. i is obtained at the ith drawing, we say that a "rencontre" has occurred (French rencontrer: to meet, to encounter). Find the probability of at least one rencontre. This problem is called the problem of rencontre or the matching problem.

If A_i = "rencontre at the ith drawing", we can write the required probability P as

$$P = P\left(\bigcup_1^N A_i\right).$$

By Theorem 2′ in §2.3, generalized to N events, we have the general addition formula

$$P\left(\bigcup_1^N A_i\right) = \sum_i P(A_i) - \sum_{i<j} P(A_iA_j) + \sum_{i<j<k} P(A_iA_jA_k) + \cdots + (-1)^{N-1}P\left(\bigcap_1^N A_i\right).$$

In the first sum there are N terms, in the second sum $\binom{N}{2}$ terms, and so on. Consider a general term $P(A_{i_1}A_{i_2}\ldots A_{i_r})$, which expresses the probability of rencontres in drawings $i_1, i_2, \ldots, i_r$. Let us compute this probability. The total number of possible cases for the N drawings is $N!$. Favourable cases are where the slips $i_1, \ldots, i_r$ appear in the drawings with these numbers, while the remaining slips can appear in any order in the other drawings; this gives $(N - r)!$ possibilities. Hence we have

$$P(A_{i_1}A_{i_2}\ldots A_{i_r}) = (N - r)!/N!.$$

If this is inserted into the expression given before, we find

$$P\left(\bigcup_1^N A_i\right) = N\cdot\frac{(N-1)!}{N!} - \binom{N}{2}\cdot\frac{(N-2)!}{N!} + \cdots + (-1)^{N-1}\binom{N}{N}\frac{1}{N!}$$

or, after a reduction,

$$P = 1 - \frac{1}{2!} + \frac{1}{3!} - \cdots + (-1)^{N-1}\frac{1}{N!}.$$

For large N this is approximately equal to $1 - e^{-1} = 0.63$.

The problem of rencontre was first discussed by Montmort at the beginning of the eighteenth century. □

We shall now present a problem which can be solved by a related method, although the situation is quite different.

Example 27. The Prize in the Food Package

In a food package there is a small prize which can be of N different types. All types are distributed into the packages at random with the same probability. A person buys n packages in a shop. What is the probability that he obtains a complete collection of presents?

Denote the event in question by H and consider its complement H^*, which is the event that at least one type is missing. We then have

$$P(H^*) = P\left(\bigcup_1^N A_i\right),$$

where A_i = "type no. i is missing". We now utilize the same addition formula as in the previous problem. It is evident that

$$P(A_{i_1}A_{i_2}\ldots A_{i_r}) = \left(1 - \frac{r}{N}\right)^n$$

for the probability that a given package does not contain types nos. $i_1, \ldots, i_r$ is $1 - r/N$. Hence we find

$$1 - P(H) = P(H^*) = N\left(1 - \frac{1}{N}\right)^n - \binom{N}{2}\left(1 - \frac{2}{N}\right)^n + \cdots + (-1)^{N-1}\binom{N}{N}\left(1 - \frac{N}{N}\right)^n$$

which furnishes the probability $P(H)$. □

Exercises

201. In a hen-house there are 10 hens. Assign outcomes and sample space to the trial "enter the poultry-yard and count the hens that sit on their roosts". (§2.2)

202. Give suitable sample spaces for the following random trials:
   (a) Throw a die and count the tosses until one of the six outcomes has appeared twice.
   (b) Throw a die and count the tosses until one of the six outcomes has appeared twice in succession. (§2.2)

203. At a sawmill boards are cut into 3-feet lengths, and the remaining pieces smaller than 3 feet are thrown on a heap. Give a suitable sample space for the trial: Choose one of the small parts and note its length (unit: feet). (§2.2)

204. Let A and B be two events. Express in words the events AB, AB^*, and A^*B^*. Represent them by Venn diagrams. (§2.2)

205. Let A, B, C and D be four events. Let E be the event that at least three, and F the event that exactly three, of the four events occur. Express E and F in terms of A, B, C and D (use suitable symbols). (§2.2)

206. The events A and B are mutually exclusive events such that $P(A) = 0.25$, $P(A \cup B) = 0.75$. Determine $P(B)$. (§2.3)

207. Let A and B be two events such that $P(A) = 0.6$, $P(B) = 0.7$ and $P(A \cup B) = 0.8$. Find $P(AB)$. (§2.3)

208. From a box containing the letters in the word PROBABILITY two letters are taken at random. Using the classical definition of probability, compute the probability of obtaining the letters P and R. (§2.4)

209. Three cards are taken at random without replacement from a deck of cards. Using the classical definition of probability, find the probability that:
(a) all three are hearts;
(b) none is a heart;
(c) all three are aces. (§2.4)

210. From an urn containing three white balls and four black balls, two balls are drawn at random without replacement. Determine the probability that the balls have different colours. (§2.4)

211. From an urn containing three white balls and four black balls, two balls are drawn at random with replacement. Determine the probability that the balls have different colours. (§2.4)

212. The triplets A, B and C play together with seven other children.
(a) They arrange themselves randomly in a row one behind the other. Find the probability that the triplets are in the front.
(b) They again line up randomly in a row. Find the probability that the triplets are together.
(c) They arrange themselves randomly in a circle. Find the probability that the triplets are together. (§2.4)

213. In a lottery there are five tickets left, exactly one of which is a winning ticket. A and B decide to buy one ticket each. A draws first, then B. Compute the conditional probability that B obtains the winning ticket, given that A has not obtained it. (§2.5)

214. From a deck of cards four cards are drawn one at a time without replacement.
(a) If the three first cards are hearts, what is the conditional probability that the fourth is not a heart?
(b) What is the probability that the three first cards are hearts and the fourth is a spade? (Use the answer to Exercise 209(a).) (§2.5)

215. Three measuring instruments, numbered 1, 2 and 3, function with probabilities 0.9, 0.8 and 0.4, respectively. One instrument is selected at random.
(a) What is the probability that the chosen instrument functions?
(b) Suppose that the chosen instrument functions. Determine, for $k = 1, 2, 3$, the conditional probability that instrument k has been chosen. (§2.5)

216. A and B have eleven fruits, three of which are poisonous. A eats four fruits and B six, all chosen at random; their pig gets the remaining fruit. Determine:
(a) the probability that the pig is not poisoned;
(b) the conditional probability that A and B are both poisoned given that the pig is not poisoned;
(c) the probability that A and B are both poisoned and the pig is not poisoned. (§2.5)

217. From a signpost marked MIAMI two randomly chosen letters fall down. A friendly illiterate puts the letters back. By means of the formula for total probability, determine the probability that the signpost again reads MIAMI. (§2.5)

218. Suppose that the probability of a male birth is p and that the sexes of the children in a family are independent. (The latter assumption is somewhat dubious.) A couple has four children. Determine:
(a) the conditional probability that they have two children of each sex given that their eldest child is a boy;
(b) the conditional probability that they have two children of each sex given that they have at least one boy. (§2.5)

219. In a lottery there are n tickets, two of which give prizes. The tickets are sold one by one, and it is immediately announced if a buyer wins. A and B buy tickets as follows. A buys the first ticket sold. B buys the first ticket sold after the first winning ticket. For each person, determine his probability of winning a prize. (§2.5)

220. The events A and B are independent. Prove that A and B^* are independent and that A^* and B^* are also independent. (§2.6)

221. The events A and B are independent, and $P(A) = 0.1$, $P(B) = 0.05$. Compute $P(A^*B^*)$. (§2.6)

222. The families A, B and C are invited to dinner. The probabilities that they will come are 0.8, 0.6 and 0.9, respectively, and these events are independent. Find the probability that:
(a) all the families come;
(b) no family comes;
(c) at least one family comes. (§2.6)

223. A well-made die is thrown twice. Consider the events: A = "the first throw results in a two or a five"; B = "the sum for the two throws is at least 7". Are the events A and B independent or dependent? (§2.6)

*224. A and B shoot at a target. Each shot has hit probability p, and the shots are independent. The persons shoot in the order $A, B, A, B, A, \ldots$, until the target has been hit twice. Determine the probability that these two shots have been fired by the same person.

*225. A well-made die is thrown three times in a first sequence of trials, and then again three times in a second sequence. Determine the probability that both sequences give the same result, apart from the order.
Hint: Divide into three cases according to whether the first sequence results in three equal numbers, two equal numbers or different numbers.

*226. A deck of cards is thoroughly shuffled and divided at random into four parts of equal size. Determine the probability that each part contains one ace.

*227. A bridge player announces that, among the 13 cards which he has received:
(a) there is at least one ace;
(b) there is the ace of hearts.
In each of these two cases, find the conditional probability that he has received more than one ace. (It is tempting to believe that the probability is the same in both cases, but this is false.)

*228. A signal that can be either 0 or 1, and that assumes these values with probabilities 2/3 and 1/3, respectively, passes two stations. Independently of the other, each station correctly receives an incoming signal with probability 6/7, and incorrectly with probability 1/7. The received signal is 0. Find the conditional probability that it was really the signal 0 that was sent.

CHAPTER 3

One-Dimensional Random Variables

3.1. Introduction

In this chapter, we become acquainted with one-dimensional random variables. In §3.2, a general presentation is given. In §3.3, the fundamental concept of a distribution function is introduced. In §3.4 and §3.5, there is a discussion of discrete random variables; and in §3.6 and §3.7 there is a discussion of continuous random variables. The relationship between discrete and continuous random variables is illustrated in §3.8. In §3.9, mixtures of random variables are introduced. Upon the first reading, the reader is advised to study §3.2–3.4 and §3.6.

3.2. General Description of Random Variables

Random variables are very important in probability theory. The concept is uncomplicated, but it takes some time and energy to become familiar with it. Examples aid understanding, and the first two examples below should be read at a leisurely pace.

A random trial often produces a number associated with the outcome of the trial. This number is not known beforehand; it is determined by the outcome of the trial, that is, by chance. It is therefore called a *random variable* (abbreviated *rv*).

Sometimes several numbers at a time are associated with the outcome of each trial. The rv then becomes multidimensional. We discuss this concept in the next chapter. When one wants to stress that there is one single number for each outcome, the term one-dimensional rv is used.

Typical examples of a one-dimensional rv are the number of heads in a sequence of tosses of a coin, the gain of a player in a roulette game in Monte Carlo, the number of children in a randomly chosen family in New York and the length of life of a randomly chosen English citizen.

We have been somewhat careless in these examples. Strictly speaking, the random variable has nothing to do with the random trials themselves, for these trials are phenomena in the empirical world, while the variables belong to the model we build by means of probability theory.

In this book, random variables are denoted by capital letters such as X, Y and Z.

Example 1. Throwing a Die

A well-made die is thrown once. A person receives one dollar for a "one", two dollars for a "two" or a "three", and four dollars for a "four", a "five" or a "six". The amount he gets is a rv X which can take the values 1, 2 and 4.

Let u_i denote the outcome "die shows i". Formally, we have assigned numbers to the outcomes as follows:

$$u_1 \to 1, \qquad u_2 \to 2, \qquad u_3 \to 2, \qquad u_4 \to 4, \qquad u_5 \to 4, \qquad u_6 \to 4.$$

This is also illustrated in Fig. 3.1. The rv X defines a function from the sample space to the real numbers. It has the domain $\Omega = \{u_1, \ldots, u_6\}$ and the range $\{1, 2, 4\}$. □

Example 2. Measuring a Physical Constant

When a physical constant is measured, we may let X be the value obtained. Because of errors of measurement, X may deviate from the true value of the constant.

In this case, the sample space is continuous and the outcomes are u_x = the value x is obtained. The function is simply $u_x \to x$.

The rv is not always the measurement itself: it can be a function of this value. For instance, if the radius of a circle is measured and the quantity of interest is the area, we take $Y = \pi X^2$, where X is the measured radius. Hence the function defined by the rv Y is $u_x \to \pi x^2$. □

After this preliminary discussion we are ready for the following basic definition.

Definition. A *random variable* (rv) is a function defined on a sample space. □

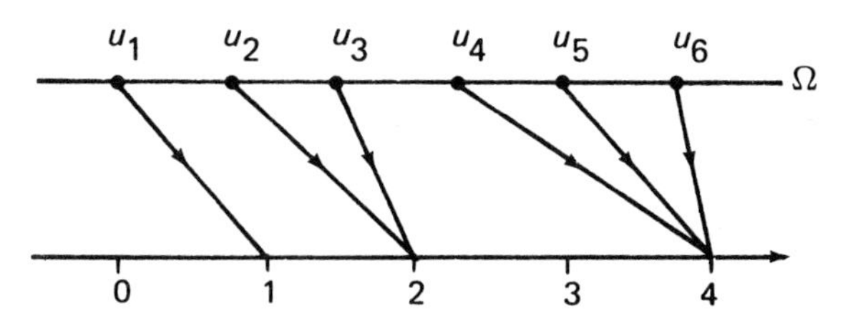

Fig. 3.1.

Fig. 3.2.

Since we consider real-valued rv's, the rv X is a function from Ω to R^1 (see Fig. 3.2). In order to emphasize that X is a function, we may write $X(u)$, where u is an outcome in Ω. However, we use this notation only rarely.

The term "random variable" is not very good; random function or chance function would be better, but the terminology is sanctified by usage.

If a rv can take on only a finite or denumerably infinite number of different values, the variable is said to be *discrete* (see Example 1 above). In Example 2 a *continuous* rv appears.

3.3. Distribution Function

The variation of a rv can be expressed in different ways. One very general way is to use the distribution function of the variable. It is obtained as follows: For a given real number x we compute the probability $P(X \leq x)$ that X is smaller than or equal to x. This procedure is performed for any x. We then obtain a function $F_X(x) = P(X \leq x)$ defined for all x in the interval $-\infty < x < \infty$.

Definition. $F_X(x) = P(X \leq x)$ is called the *distribution function* of the rv X. □

We sometimes say "cumulative distribution function". Please note carefully the difference between X, which denotes the rv, and x, which is used as an argument in the function $F_X(x)$ and hence is a mathematical variable. When no confusion is possible, we may simply write $F(x)$.

Example 1. Throwing a Die (continued)

The rv X was defined as the payment following a throw of a symmetric die, according to the rule shown in Fig. 3.1.

Let us determine the distribution function of X. Since the amount paid cannot be less than 1, the probability $P(X \leq x)$ is equal to zero for any $x < 1$, that is, $F_X(x) = 0$ for any such value of x. Further, we have $P(X \leq 1) = 1/6$, for this event occurs either if $X < 1$ (which has probability zero) or if $X = 1$ (which has the probability $1/6$); hence for $x = 1$ we have $F_X(x) = 1/6$, that is, the function increases by a step at this point.

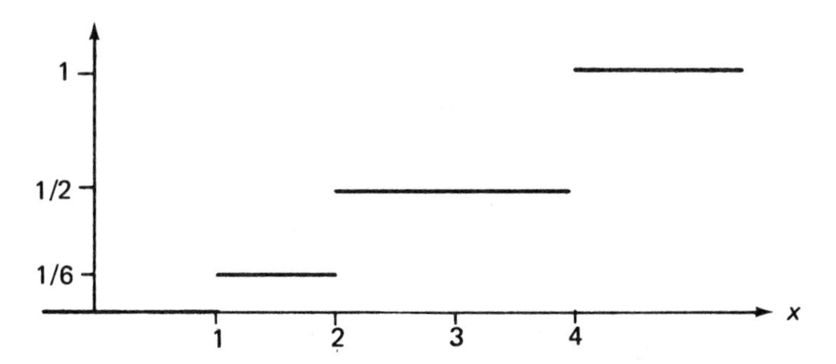

Fig. 3.3. Distribution function of the discrete rv in Example 1.

The next step is taken for $x = 2$, and the function then increases to 1/2 (for $X = 2$ occurs when either 2 or 3 appears on the die, which happens with probability $P(u_2) + P(u_3) = 1/6 + 1/6 = 1/3$, and $1/3 + 1/6 = 1/2$). For $x = 4$ the function increases to 1 (for $X = 4$ occurs with probability $P(u_4) + P(u_5) + P(u_6) = 1/2$, which causes an increase to $1/2 + 1/2 = 1$). The function then remains at this value. A graph of the function is given in Fig. 3.3. □

Example 2. Measuring a Physical Constant (continued)

If X is the result of measuring a physical constant it is, in general, impossible to derive theoretically how the distribution function looks in detail. However, at least a little can be said about its graph. Suppose that the measurement always lies in the interval (c, d) and that all values in this interval are possible. We must then have $F_X(x) = 0$ for $x < c$ and $F_X(x) = 1$ for $x > d$. Furthermore, the function must be strictly increasing in the interval, for the probability that $X \leq x$ must increase with x. Hence the graph of $F_X(x)$ looks something like the curve in Fig. 3.4. □

These two examples suggest that a distribution function is zero if x is small enough and 1 is x is large enough; also, the function seems to be increasing. In, fact, we have the following general theorem which we present without proof:

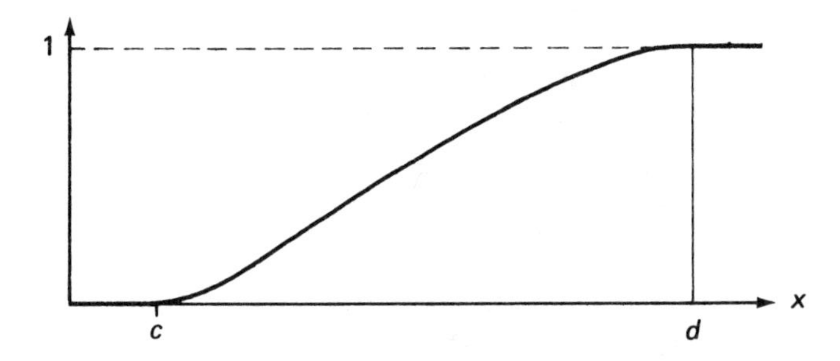

Fig. 3.4. Distribution function of the continuous rv in Example 2.

Theorem 1. *The distribution function $F_X(x)$ of a* rv *X has the properties*:

$$F_X(x) \to \begin{cases} 0 \\ 1 \end{cases} \quad \text{when} \quad x \to \begin{cases} -\infty \\ \infty; \end{cases} \tag{1}$$

$$F_X(x) \text{ is a nondecreasing function of } x; \tag{2}$$

$$F_X(x) \text{ is continuous from the right for any } x. \tag{3}$$

□

It may be added that any function satisfying the conditions (1)–(3) is a distribution function of some rv.

A simple consequence of the definition of a distribution function is the following:

Theorem 2. *If $a \leq b$, then*

$$F_X(b) - F_X(a) = P(a < X \leq b). \tag{4}$$

□

Proof. We know that $F_X(b)$ is the probability that $X \leq b$. If $X \leq b$, we must have either $X \leq a$ or $a < X \leq b$, and these events are mutually exclusive. Hence, it follows from the addition formula in Axiom 3, p. 12, that $F_X(b) = P(X \leq a) + P(a < X \leq b)$. This implies that $F_X(b) = F_X(a) + P(a < X \leq b)$, which is equivalent to (4). □

Theorem 2 is important, for it tells us how to use the distribution function for calculating the probability that the values of the rv lie in a given interval.

Example. Suppose that the probability is 0.93 that a measured value X is ≤ 56.0, and that the probability is 0.13 that it is ≤ 51.5. Relation (4) shows that the probability of $51.5 < X \leq 56.0$ is $0.93 - 0.13 = 0.80$. □

3.4. Discrete Random Variables

We mentioned at the end of §3.2 how a discrete rv is defined. This section is devoted to such variables, and we begin by repeating the definition.

Definition. A random variable is *discrete* if it can assume a finite or a denumerably infinite number of different values. □

We shall now introduce the reader to discrete rv's in some detail, but will discuss mainly the case when they take on nonnegative integer values 0, 1, 2,

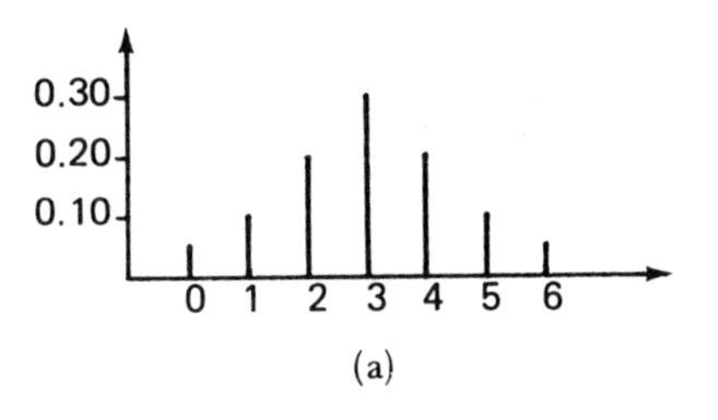

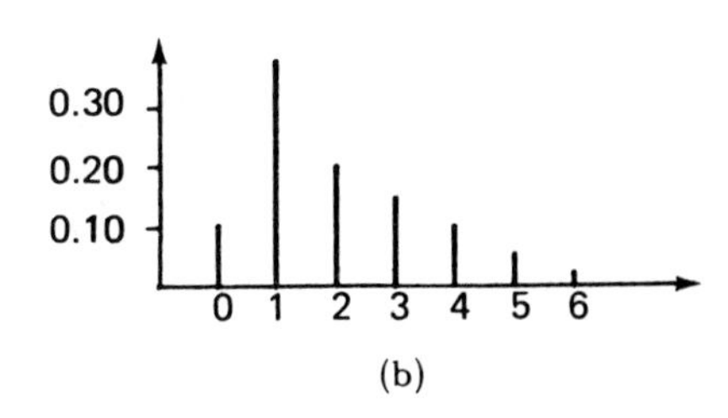

Fig. 3.5. (a) Symmetric probability function. (b) Skew probability function.

... .This is a common situation: all sorts of random trials, where counting is involved, give rise to such variables.

We shall introduce a new symbol. The probability $P(X = k)$ that X is equal to a given value k is usually denoted by $p_X(k)$.

Definition. The quantities $p_X(k)$ $(k = 0, 1, \ldots)$ are jointly called the *probability function* of the rv X. □

The probability function can be illustrated by a bar diagram (see the examples in Fig. 3.5).

In Fig. 3.5(a) the probability function is *symmetric*, in Fig. 3.5(b) it is *skew*. The meaning of these terms is obvious.

The distribution function can be determined from the probability function using the formula

$$F_X(x) = \sum_{j \leq x} p_X(j), \tag{5}$$

that is, the probability function is summed over all j such that j is at most equal to x. Needless to say, these computations are made only for integer values $k = 0, 1, \ldots$:

$$F_X(k) = \sum_{j=0}^{k} p_X(j). \tag{5'}$$

The step function $F_X(x)$ is then immediately obtained for all x.

Conversely, it follows from (5′) that the probability function is computed from the distribution function according to the expression

$$p_X(k) = \begin{cases} F_X(0) & \text{if} \quad k = 0, \\ F_X(k) - F_X(k-1) & \text{otherwise.} \end{cases}$$

Example. Determining a Distribution Function

In the following two examples we show how a distribution function is obtained according to (5′) by adding the probabilities $p_X(k)$ successively.

(a)

k	$p_X(k)$	$F_X(k)$
0	0.05	0.05
1	0.10	0.15
2	0.20	0.35
3	0.30	0.65
4	0.20	0.85
5	0.10	0.95
6	0.05	1.00
Sum	1.00	

(b)

k	$p_X(k)$	$F_X(k)$
0	0.10	0.10
1	0.38	0.48
2	0.20	0.68
3	0.15	0.83
4	0.10	0.93
5	0.05	0.98
6	0.02	1.00
Sum	1.00	

The probability functions of these rv's are given in graphical form in Fig. 3.5. □

We shall mention two other consequences of (5). If in this expression we let x tend to infinity, we get

$$\sum_{j=0}^{\infty} p_X(j) = 1. \tag{6}$$

Hence the bars in a bar diagram sum to 1; that is, the sum of all the probabilities is 1.

If (4) and (5′) are combined we find

$$P(a < X \leq b) = \sum_{k=a+1}^{b} p_X(k). \tag{7}$$

(Here a and b are integers.) Hence the probability that X lies between two given values is computed by summing the probability function for appropriate values of k. One must handle the equality and inequality signs on the left-hand side with great care. For instance, we have

$$P(a \leq X \leq b) = \sum_{k=a}^{b} p_X(k). \tag{8}$$

Sometimes we talk about the *probability distribution* or, shorter, the *distribution* of a rv. By such expressions we mean a function showing, in one way or another, how the variable is distributed over its range of values. The distribution of a discrete rv can be described by means of the probability function or by means of the distribution function.

As an alternative to the bar diagram, a probability function may be illustrated by using a well-known concept from mechanics. The total mass of 1 is distributed with masses $p_X(0)$, $p_X(1)$, ... over the points 0, 1, ..., respectively. This picture of a distribution of a rv is of great value for grasping the idea of a rv. In the sequel, we will therefore occasionally speak of the "mass" or "probability mass" of a discrete rv.

3.5. Some Discrete Distributions

We shall present some discrete distributions in a list which is useful for reference, but is perhaps hard to absorb at first reading. The most important of these distributions are the binomial distribution, the hypergeometric distribution and the Poisson distribution.

(a) One-Point Distribution

The total mass of the rv X is concentrated in one point a, that is,

$$p_X(a) = 1.$$

This distribution, which may seem uninteresting, occurs, for example, if a measurement X is made without error.

(b) Two-Point Distribution

Definition. If the rv X assumes only two values a and b with probability p and $q = 1 - p$, respectively, then X is said to have a *two-point distribution.* □

The probability function is (see Fig. 3.6)

$$p_X(a) = p; \qquad p_X(b) = q.$$

This distribution is encountered if a person tosses a balanced coin and receives a dollars for a head and b dollars for a tail. The payment X is then distributed according to a two-point distribution: X assumes the values a and b with the same probability, 1/2. If the coin is unbalanced, X also has a two-point distribution, but with $p \neq 1/2$.

(c) Uniform Distribution

Definition. If the rv X assumes the values $1, 2, \ldots, m$ with the same probability, $1/m$, then X is said to have a *uniform distribution.* □

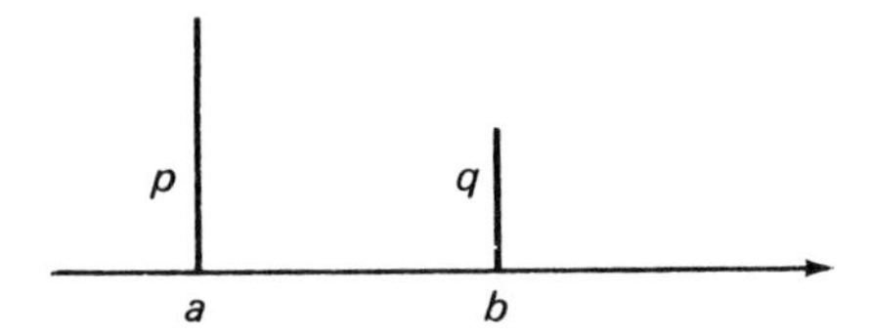

Fig. 3.6. Two-point distribution.

(The term was mentioned in §2.4.) Hence we have, in this case,

$$p_X(k) = 1/m \qquad (k = 1, 2, \ldots, m).$$

The same term will later be used in the continuous case; for that reason it is preferable to write *uniform distribution over* $\{1, 2, \ldots, m\}$.

(d) Geometric Distribution

Definition. If the rv X has probability function

$$p_X(k) = q^k p \qquad (k = 0, 1, \ldots),$$

where $q = 1 - p$, then X is said to have a *geometric distribution*. □

Consider independent trials such that a certain event may happen at any given trial with probability p. The trials continue until the event occurs for the first time. The number, X, of trials performed *before* the event occurs has a geometric distribution.

(e) fft-Distribution

Definition. If a rv X has probability function

$$p_X(k) = q^{k-1} p \qquad (k = 1, 2, \ldots),$$

where $q = 1 - p$, then X is said to have an *fft-distribution*. □

See the comment to (d). If the number of trials counted includes the trial in which the event occurs *f*or the *f*irst *t*ime, this distribution is obtained, and this explains its name.

(f) Binomial Distribution

Definition. If the rv X has probability function

$$p_X(k) = \binom{n}{k} p^k q^{n-k} \qquad (k = 0, 1, \ldots, n),$$

where $q = 1 - p$, then X is said to have a *binomial distribution*. □

Code name. $X \sim \text{Bin}(n, p)$.

We encounter here the first example of a "code name" of a common distribution; it saves space to use such a name. We know already that the binomial distribution appears when drawings are made with replacement from an urn with two kinds of objects (see §2.4).

(g) Hypergeometric Distribution

Definition. If the rv X has probability function

$$p_X(k) = \binom{a}{k}\binom{b}{n-k}\bigg/\binom{a+b}{n},$$

where k assumes all integer values such that $0 \leq k \leq a$, $0 \leq n - k \leq b$, then X is said to have a *hypergeometric distribution.* □

This distribution has also appeared before; it occurs when drawings are made without replacement from an urn with two kinds of objects (see §2.4).

(h) Poisson Distribution

Definition. If the rv X has probability function

$$p_X(k) = e^{-m}m^k/k! \qquad (k = 0, 1, \ldots),$$

then X is said to have a *Poisson distribution.* □

Code name. $X \sim \text{Po}(m)$.

The binomial distribution, the hypergeometric distribution and the Poisson distribution will be examined in more detail in Chapter 9.

3.6. Continuous Random Variables

We have already suggested what a continuous rv is (see Example 2, p. 41), and shall now give a more comprehensive exposition.

A continuous rv X can assume all values in an interval (or in several distinct intervals). The interval may be unbounded, comprising, say, the positive part of the x-axis, or the whole x-axis. The possible outcomes are "infinitely close" to each other, and no single outcome has a positive probability. Hence no probability function exists, nor is formula (5) valid.

Instead, we resort to the following representation of the distribution function:

$$F_X(x) = \int_{-\infty}^{x} f_X(t)\,dt. \tag{9}$$

Here the summation of the probability function in (5) has been replaced by an integration. When such a representation of $F_X(x)$ is possible, we say that the rv X is continuous.

This leads us to introduce the following:

Definition. If a function $f_X(x)$ exists such that (9) is valid, we say that X is a *continuous* rv. The function $f_X(x)$ is called the *density function* of X. □

Another common name is frequency function, which has exactly the same meaning as density function.

We shall mention some consequences of the definition. Since $F_X(x)$ is increasing, it follows from (9) that $f_X(x)$ is nonnegative. Furthermore, by a well-known property of the Riemann integral it is seen that $F_X(x)$ is everywhere continuous. One more consequence will be given in a theorem:

Let $F'_X(x)$ be the derivative of $F_X(x)$ with respect to x.

Theorem 3. *At any point of continuity x of $f_X(x)$:*

$$F'_X(x) = f_X(x).$$ □

This is an important result from integral calculus. Hence $F_X(x)$ and $f_X(x)$ are closely connected: $F_X(x)$ is the integral of $f_X(x)$, as seen from (9); furthermore, $f_X(x)$ is the derivative of $F_X(x)$, as seen from Theorem 3.

Examples of density functions are given in Fig. 3.7; in (a) the density function is *symmetric*, in (b) it is *skew*.

Using the mechanical analogy which we introduced earlier, it can be said that the density function shows how the total probability mass of 1 is distributed over the infinitely many x-values. (Remember that no value has a positive mass.)

It follows immediately from the definition of a continuous rv that

$$F_X(b) - F_X(a) = \int_a^b f_X(t)\,dt. \tag{10}$$

Hence we have from Theorem 2 that

$$P(a < X \leq b) = \int_a^b f_X(t)\,dt \qquad (a < b). \tag{11}$$

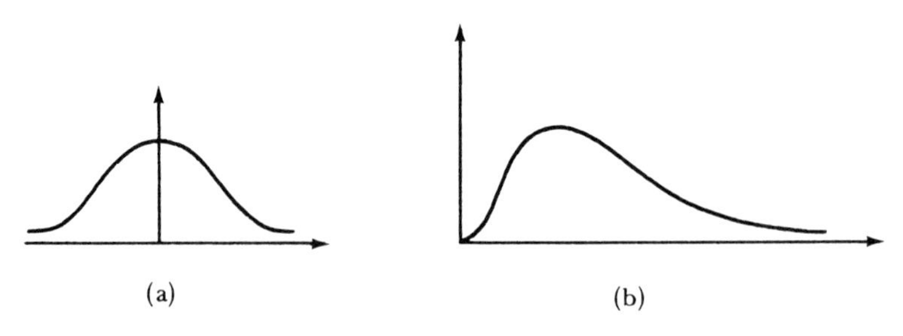

Fig. 3.7. Examples of density functions. (a) Symmetric density function. (b) Skew density function.

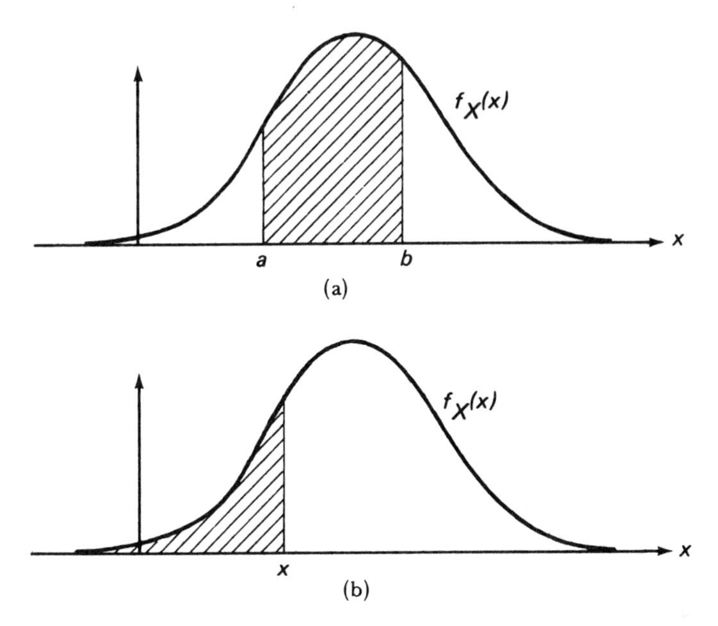

Fig. 3.8. (a) Shaded area shows $P(a < X \leq b)$. (b) Shaded area shows $F_X(x)$.

This expression is illustrated in Fig. 3.8(a). The probability that $a < X \leq b$ is equal to the area under the density function between a and b. Further, Fig. 3.8(b) illustrates that the distribution function at the point x is equal to the area under the density function to the left of x.

If x tends to ∞ in (9) we find, noting that $F_X(x)$ then tends to 1,

$$\int_{-\infty}^{\infty} f_X(t)\, dt = 1. \tag{12}$$

Hence the total area under the density function is 1.

As has already been mentioned, the probability that a continuous rv assumes any particular given value x is zero. Hence we are allowed to remove the equality sign in (11) and write $P(a < X < b)$, or add an equality sign and write $P(a \leq X \leq b)$, without affecting the value of the probability. (This was not possible in the discrete case.)

If, in particular, we have $b = a + \Delta a$ and $f_X(x)$ is taken to be continuous, we obtain by the mean value theorem

$$P(a < X < a + \Delta a) = \int_{a}^{a+\Delta a} f_X(t)\, dt = \Delta a f_X(a + \theta \Delta a) \approx \Delta a f_X(a), \tag{13}$$

where $0 < \theta < 1$. This is illustrated in Fig. 3.9. The probability that X lies within a small interval at a is approximately equal to the area of a rectangle, with the interval as a base and the value of the density function as a height.

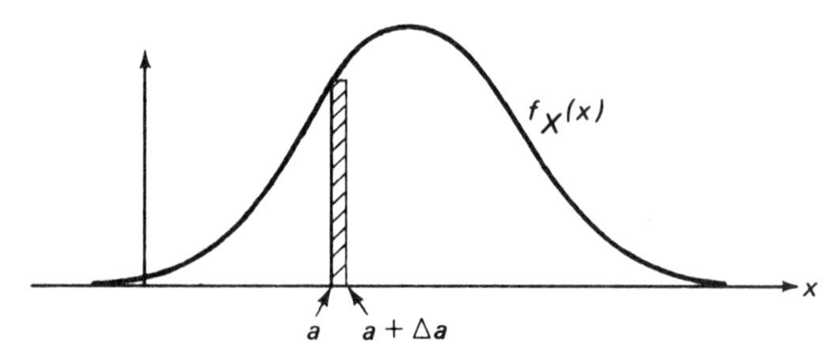

Fig. 3.9. The shaded area is approximately equal to $P(a < X < a + \Delta a)$.

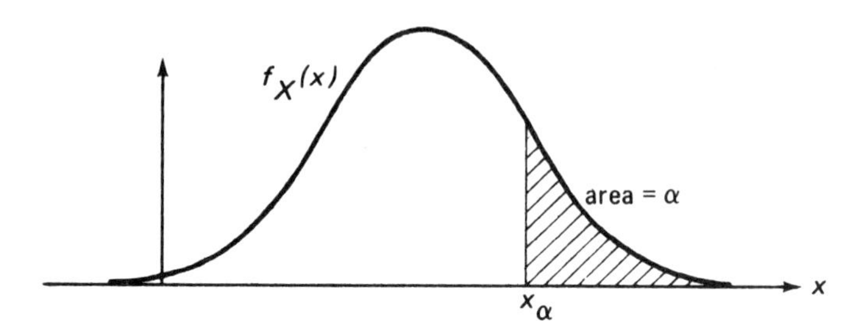

Fig. 3.10. Quantile.

If x is determined such that the area under the density function *to the right* of x is equal to a given quantity α, $0 < \alpha < 1$, one obtains the so-called α-quantile of the distribution (see Fig. 3.10). The α-quantile can also be defined using the distribution function:

Definition. The solution $x = x_\alpha$ of the equation

$$F_X(x) = 1 - \alpha$$

is called the *α-quantile* of the rv X. □

The reader ought to check that this definition is in accordance with Fig. 3.10. The value α is often expressed as a percentage (5%-quantile, 0.1%-quantile, and so on).

Quantiles are widely used in the theory of statistics. Other symbols than x_α are then often introduced.

The quantiles $x_{0.25}, x_{0.50}, x_{0.75}$ are called (in this order) the *upper quartile* (note the r!), the *median* and the *lower quartile*. The median divides the area under the density function into two equal parts, and the three quartiles divide it into four equal parts (see Fig. 3.11).

In the literature one encounters different terms for quantiles. Common terms are fractiles or, if α is expressed as a percentage, percentiles. Sometimes a quantile is defined

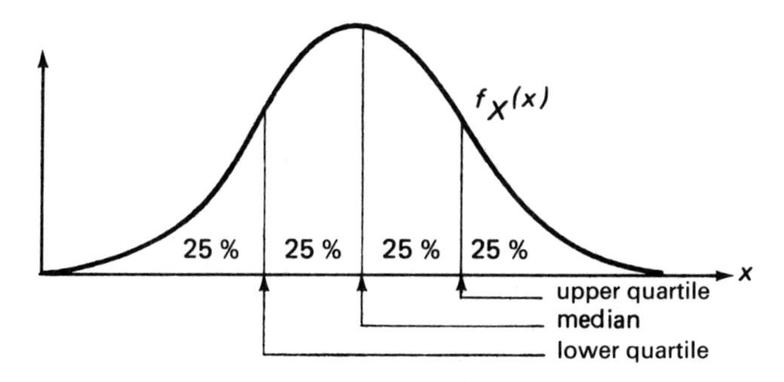

Fig. 3.11. Quartiles and median.

so that α is the area to the left; this adds to the confusion. It is therefore necessary to find out, in each book, what the author means.

3.7. Some Continuous Distributions

We shall present five continuous distributions. The most important of these are the uniform distribution, the exponential distribution and the normal distribution.

(a) Uniform Distribution

Definition. If the rv X has density function

$$f_X(x) = \begin{cases} 1/(b-a) & \text{if} \quad a < x < b, \\ 0 & \text{otherwise,} \end{cases}$$

then X is said to have a *uniform distribution*. □

Since the same term has already been used in the discrete case, it is better to write *uniform distribution on* (a, b) (and use the shorter name when no confusion can arise).

An integration according to (9) shows that:

$$F_X(x) = \begin{cases} 0 & \text{if} \quad x < a, \\ \dfrac{x-a}{b-a} & \text{if} \quad a \le x \le b, \\ 1 & \text{if} \quad x > b. \end{cases}$$

See also Fig. 3.12.

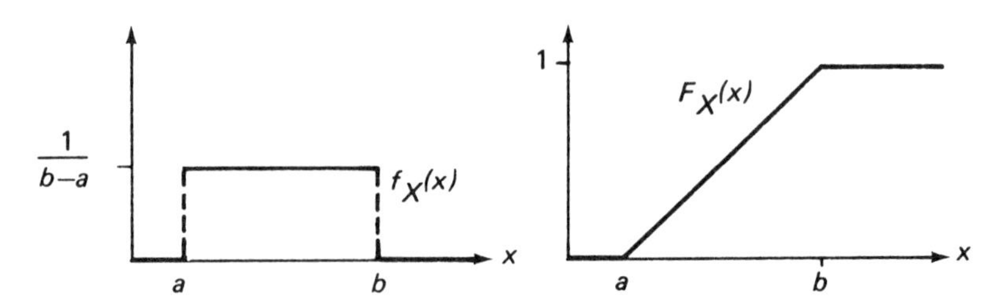

Fig. 3.12. Density function and distribution function of a uniformly distributed rv.

Example 3. Round-Off Errors

A value obtained by a measurement is rounded in the usual way to whole millimeters. The round-off error (that is, the difference between the rounded and the original value) is called X; this error can assume any value between -0.5 and 0.5. It is often reasonable to assume that X is uniformly distributed on the interval $(-0.5, 0.5)$ so that

$$f_X(x) = \begin{cases} 1 & \text{if } \; -0.5 \le x \le 0.5, \\ 0 & \text{otherwise.} \end{cases}$$ □

Example 4. Waiting Time

Buses pass a bus stop punctually every ten minutes. A person arrives at a randomly chosen time. His waiting time X is a random variable, uniformly distributed on the interval $(0, 10)$. □

(b) Exponential Distribution

Definition. If the rv X has density function

$$f_X(x) = \begin{cases} \dfrac{1}{m} e^{-x/m} & \text{if } \; x \ge 0, \\ 0 & \text{if } \; x < 0, \end{cases}$$

where $m > 0$, then X is said to have an *exponential distribution.* □

Code name. $X \sim \text{Exp}(m)$.

The shape of the density function is shown in Fig. 3.13. The greater m, the more spread out is the probability mass over the interval $(0, \infty)$. By (9) the distribution function becomes

$$F_X(x) = \begin{cases} 0 & \text{if } \; x < 0, \\ 1 - e^{-x/m} & \text{if } \; x \ge 0. \end{cases}$$

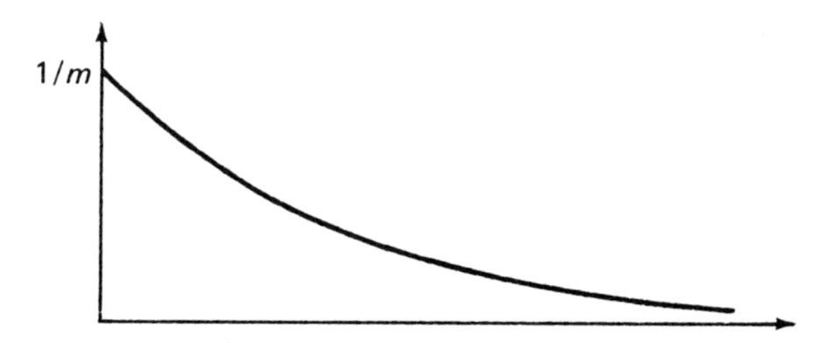

Fig. 3.13. Density function of an exponentially distributed rv.

Example 5. Radioactive Decay

For a radioactive substance, it can be shown that the time X between two consecutive times when a particle disintegrates has an exponential distribution. If the decay intensity is λ particles per unit of time, we have $f_X(x) = \lambda e^{-\lambda x}$ $(x \geq 0)$; that is, the proper code name is Exp$(1/\lambda)$. A derivation of this result is given in the following remark. (This remark is instructive but not very easy to read, and may be skipped.) □

***Remark. Theoretical Derivation**

Example 5 is a special case of the following general situation. The exponential distribution will arise whenever we examine the time intervals between events A that occur at random in time (radioactive decay, calls to a telephone station, arrivals at a shop, and so on). We shall explain this mathematically though not entirely rigorously.

The following model is employed. The probability that an event A occurs in the interval $(t, t + h)$ is supposed to be $\lambda h + o(h)$, where λ is called the intensity. (We have here assumed that the probability is, approximately, proportional to the length of the interval, which is a reasonable assumption. By $o(h)$ we mean a quantity that goes to zero when $h \to 0$, but faster than h.) Now let $P(t)$ denote the probability that, counted from a given time point T, no event occurs before t. Then the probability that no event occurs before $t + h$ equals the product of the probability that no event occurs before t and the probability that no event occurs between t and $t + h$, that is,

$$P(t + h) = P(t)\{1 - \lambda h + o(h)\}.$$

If we subtract $P(t)$ from both sides, divide by h and let h approach 0, we find, noting that $o(h)/h \to 0$, that

$$\frac{dP(t)}{dt} = -\lambda P(t),$$

whence

$$P(t) = e^{-\lambda t}.$$

Here we have used the self-evident condition $P(0) = 1$. Now let X be the time it takes,

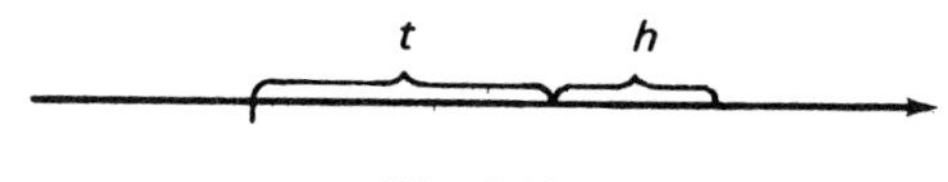

Fig. 3.14.

counted from T, until an A occurs. Since the event $X > x$ is equivalent to the event that "no A occurs before x (counted from T)" we get

$$P(X > x) = e^{-\lambda x},$$

and hence

$$F_X(x) = P(X \leq x) = 1 - e^{-\lambda x}.$$

Thus we have proved that the time X is Exp$(1/\lambda)$. The result is true for any choice of T. If we now take T to be the time of disintegration of a particle, we realize that the time from T to the next disintegration has the given exponential distribution. □

(c) Normal Distribution

Definition. If the rv X has density function

$$f_X(x) = \frac{1}{\sigma\sqrt{2\pi}} e^{-(x-m)^2/2\sigma^2} \qquad (-\infty < x < \infty),$$

where m and σ are given quantities ($\sigma > 0$), then X is said to have a *normal distribution.* □

Code name. $X \sim N(m, \sigma^2)$.

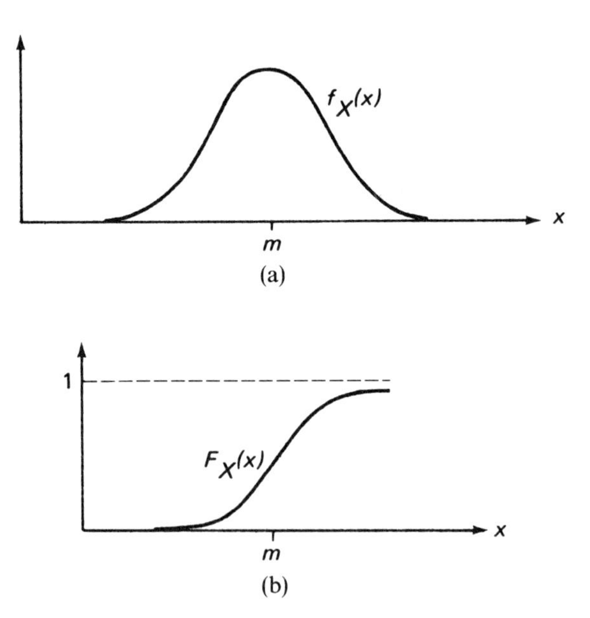

Fig. 3.15. Normal distribution. (a) Density function of a normally distributed rv. (b) Distribution function of a normally distributed rv.

The name Gaussian distribution is also often used. The graphs of the density function and the distribution function are shown in Fig. 3.15.

We shall prove that the area under the density function is 1, a problem known to generations of students in mathematics. Hence we want to prove that

$$\int_{-\infty}^{\infty} f_X(x)\,dx = 1,$$

where $f_X(x)$ is shown in the above definition. Substituting $t = (x - m)/\sigma$ we see that the problem is solved if we can prove that

$$\int_{-\infty}^{\infty} e^{-t^2/2}\,dt = \sqrt{2\pi}.$$

For that purpose we employ the following device. Call the integral I, write

$$I^2 = \int_{-\infty}^{\infty}\int_{-\infty}^{\infty} e^{-(t^2+u^2)/2}\,dt\,du$$

and introduce polar coordinates $t = r\cos v$, $u = r\sin v$. The double integral then becomes

$$I^2 = \int_0^{2\pi}\left[\int_0^{\infty} re^{-r^2/2}\,dr\right]dv = 2\pi.$$

Since $I > 0$ the result follows.

(d) Weibull Distribution

Definition. If the rv X has density function

$$f_X(x) = \begin{cases} \dfrac{c}{a}(x/a)^{c-1}e^{-(x/a)^c} & \text{if } x \geq 0, \\ 0 & \text{if } x < 0, \end{cases}$$

where a and c are positive, then X is said to have a *Weibull distribution*. □

The distribution is named in honour of the Swedish physicist W. Weibull, who was a specialist in applied mechanics. It has proved useful in describing various characteristics of materials (for example, yield strengths, ultimate strengths and fatigue limits). By giving a and c different values a whole family of distributions is generated (see Fig. 3.16). For $c = 1$ the exponential distribution is obtained as a special case, for $c = 2$ the Rayleigh distribution. (Rayleigh was an English physicist.)

By integration we obtain

$$F_X(x) = \begin{cases} 0 & \text{if } x < 0, \\ 1 - e^{-(x/a)^c} & \text{if } x \geq 0. \end{cases}$$

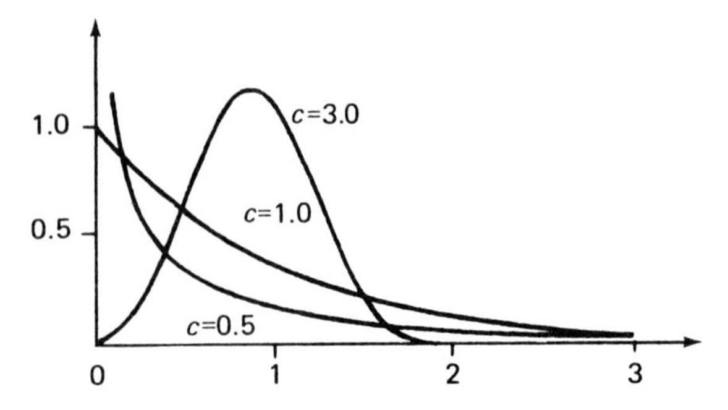

Fig. 3.16. Density functions of Weibull distributed rv's ($a = 1$; $c = 0.5, 1.0, 3.0$).

(e) Gamma Distribution

Definition. If the rv X has density function

$$f_X(x) = \begin{cases} \dfrac{1}{a^p \Gamma(p)} \cdot x^{p-1} e^{-x/a} & \text{if} \quad x \geq 0, \\ 0 & \text{if} \quad x < 0, \end{cases}$$

where $p > 0$ and $a > 0$, then X is said to have a *gamma distribution*. □

The name "gamma distribution" arises from the gamma function:

$$\Gamma(p) = \int_0^\infty x^{p-1} e^{-x} \, dx \qquad (p > 0).$$

If p is an integer, a sequence of partial integrations shows that $\Gamma(p) = (p-1)!$. The density function is shown for some values of p in Fig. 3.17. In the particular case $p = 1$ the gamma distribution reduces to the exponential distribution. The gamma distribution occurs frequently in probability theory.

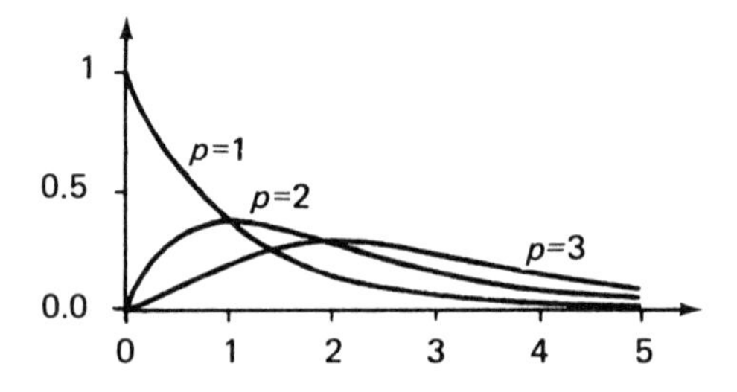

Fig. 3.17. Density functions of gamma distributed rv's ($a = 1$; $p = 1, 2, 3$).

3.8. Relationship Between Discrete and Continuous Distributions

As we have already seen, there are certain analogies between discrete and continuous distributions. In this section we shall further illustrate this relationship.

The height of a randomly chosen 20-year-old Swedish male can be regarded as a continuous rv with a density function of the form sketched in Fig. 3.18(a).

If the height is rounded to the nearest 5 cm value, we obtain instead a discrete rv with the probability function of Fig. 3.18(b). It has some resemblance to the density function, in spite of its discreteness. (Note how this probability function is obtained. All heights between 167.5 and 172.5, say, are rounded off to 170. Hence the area under the density function between 167.5 and 172.5 is the value assigned to the probability function for 170.)

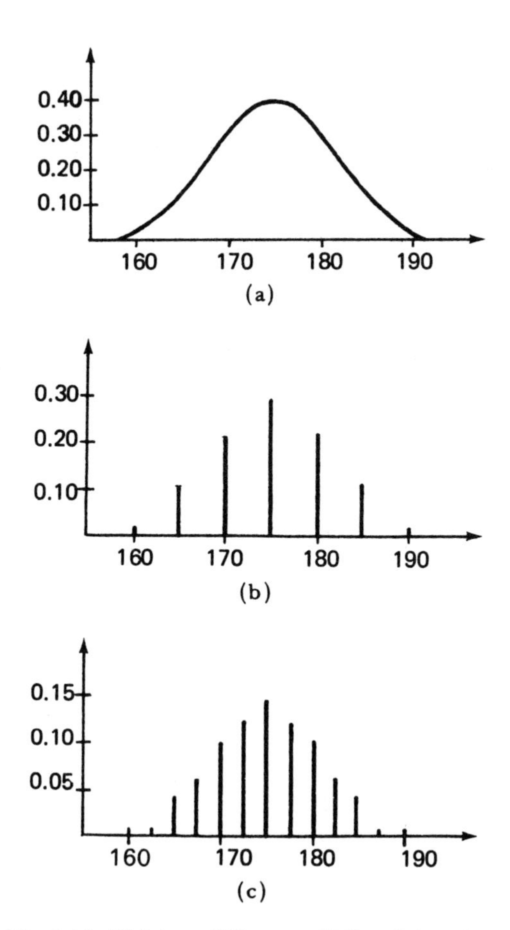

Fig. 3.18. Heights of 20-year-old Swedish males.

If the rounding is less coarse, and is made, for instance, to the nearest 2.5 cm value, the resemblance to the density function is still closer (see Fig. 3.18(c)). If the rounding is made finer and finer, we get a sequence of discrete distributions which look more and more like the continuous distribution of Fig. 3.18(a).

The relationship here described can be used to approximate a distribution. Any continuous distribution can be approximated by a discrete one that assumes values at equally spaced points with probabilities that can be computed from the density function. Conversely, a discrete distribution may sometimes be approximated by a continuous one; how that is done will not be discussed here (for an example see Fig. 9.3).

3.9. Mixtures of Random Variables

Besides discrete and continuous rv's there exist "hybrids" which we call mixtures of discrete and continuous rv's. Two examples will be given which are intended to demonstrate the practical background.

Example 6. Waiting Time at Crossing

At a crossing there is a traffic light showing alternately green and red light for a seconds. A car driver who arrives at random has to wait for a time period Z. Find the distribution of Z.

(1) If the driver arrives during a green period, his waiting time is zero. Since the green and the red lights have the same durations, we have $P(Z = 0) = 1/2$. (2) If the driver arrives during the first $a - z$ seconds of a red period, his waiting time is greater than z (see Fig. 3.19).

Hence for $0 < z < a$ we have

$$P(Z > z) = \frac{a - z}{2a} \qquad \text{and} \qquad P(Z \leq z) = 1 - \frac{a - z}{2a} = \frac{1}{2} + \frac{z}{2a}.$$

The distribution function is (see Fig. 3.20)

$$F_Z(z) = \begin{cases} 0 & \text{if} \quad z < 0, \\ 1/2 & \text{if} \quad z = 0, \\ \dfrac{1}{2} + \dfrac{z}{2a} & \text{if} \quad 0 < z < a, \\ 1 & \text{if} \quad z \geq a. \end{cases}$$

□

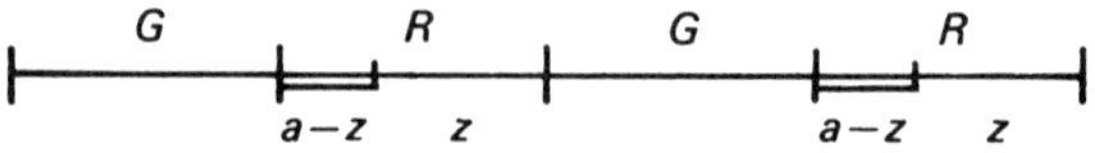

G R G R

$a-z$ z $a-z$ z

Fig. 3.19.

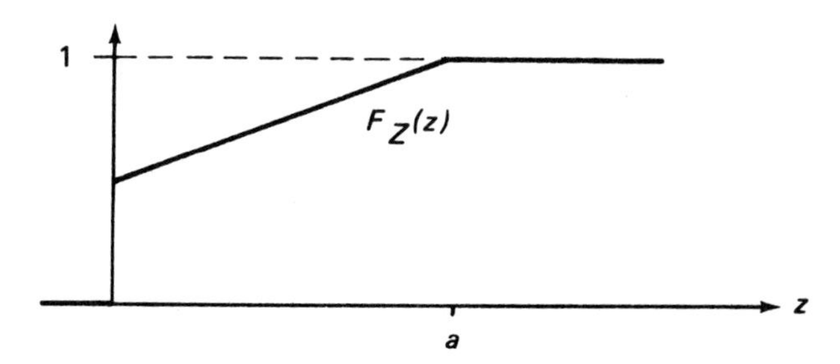

Fig. 3.20. Distribution of waiting time.

Example 7. Duration of Local Anaesthesia

A dentist gives a local anaesthetic to a patient. Let Z be the duration of the anaesthesia. There is a positive probability that the patient does not react at all, that is, that the duration is zero. If the patient reacts, the duration is a continuous rv. The distribution is of the same type as in the preceding example, but its form for positive z is of course different. □

Examples 6 and 7 contain special cases of a general situation, which we mention in passing without giving any details. Let X and Y be rv's with distribution functions $F_X(x)$ and $F_Y(y)$. If X and Y are mixed in the proportions p and $1 - p$, the distribution function of the mixture Z is given by

$$F_Z(z) = pF_X(z) + (1 - p)F_Y(z).$$

In the examples, one discrete and one continuous rv were mixed, but mixtures of two continuous or two discrete distributions are also possible.

Exercises

301. A rv X assumes the values a and b $(a < b)$ with the same probability. Find the distribution function of X and draw its graph. (§3.3)

302. Draw the graph of the distribution function

$$F_X(x) = \begin{cases} 0 & \text{if} \quad x < 1, \\ \frac{1}{3} + \frac{2}{3}(x - 1) & \text{if} \quad 1 \leq x \leq 2, \\ 1 & \text{if} \quad x > 2. \end{cases}$$

Compute $P(X \leq 5/3)$, $P(X > 3/2)$ and $P(4/3 < X \leq 5/3)$. (§3.3)

303. A rv can assume only the values 3, 4, 7, 8 and 9. It is known that

$$p_X(3) = 1/3, \qquad p_X(4) = 1/4, \qquad p_X(7) = 1/6, \qquad p_X(8) = 1/6.$$

Compute:
(a) $p_X(9)$;
(b) $F_X(5)$;
(c) $P(4 \leq X \leq 8)$ and $P(X \geq 8)$. (§3.4)

304. In a box there are two 10-cent coins and one 50-cent coin. Two of the three coins are chosen at random. Let X be the total value of the two coins.
 (a) Which values can X assume?
 (b) Find the probability function of X. (§3.4)

In Exercises 305–308 there occur distributions with special names. Give, in each case, the name of the relevant distribution.

305. From an urn with four white balls and six black balls, one ball is drawn repeatedly with replacement until a white ball is obtained. Let X be the number of drawings resulting in a black ball.
 (a) Determine the probability function of X.
 (b) Compute the probability $P(X \geq 3)$. (§3.5)

306. In the previous exercise let Y be the total number of drawings.
 (a) Determine the probability function of Y.
 (b) Compute the probability $P(Y \geq 2)$. (§3.5)

307. An urn contains four white balls and six black balls. Three balls are drawn with replacement. Let X be the number of white balls among the three balls.
 (a) Determine the probability function of X.
 (b) Find the probability that the three drawings result in more white than black balls. (§3.5)

308. In the previous exercise let the drawings be made without replacement. The questions are the same as before. (§3.5)

309. The rv X has probability function $p_X(k) = e^{-m}m^k/k!$ for $k = 0, 1, \ldots$ and is such that $P(X = 0) = 1/2$. Compute $P(X \geq 2)$. (§3.5)

310. Determine the constant c such that the function

$$f(x) = cx^2 \qquad (0 \leq x \leq 6)$$

becomes a density function. (§3.6)

311. Determine the constant c such that

$$f(x) = c/\sqrt{x+1} \qquad (-1 < x \leq 1)$$

becomes a density function. Find the probability that a rv with this density function assumes a positive value. (§3.6)

312. The velocity of a randomly chosen gas molecule in a certain amount of gas is a rv with density function

$$f_X(x) = cx^2e^{-\beta x^2} \qquad \text{if} \quad x \geq 0.$$

Here $\beta = M/2RT$, where $R =$ the gas constant, $M =$ the molecular weight of the gas and $T =$ the absolute temperature. Determine the constant c expressed in β. It may be assumed known that

$$\int_0^\infty e^{-x^2}\,dx = \sqrt{\pi}/2. \quad (\S3.6)$$

313. A rv has distribution function

$$F_X(x) = \begin{cases} 0 & \text{if } x < 1, \\ 1 - 1/x^2 & \text{if } x \geq 1. \end{cases}$$

Determine the median of the distribution. (§3.6)

314. Determine the three quartiles for a rv with density function

$$f_X(x) = \begin{cases} \frac{3}{2}e^{3x} & \text{if } x < 0, \\ \frac{1}{2}e^{-x} & \text{if } x \geq 0. \end{cases} \quad (\S 3.6)$$

315. A rv X with density function

$$f_X(x) = \frac{1}{a}e^{-x/a}\exp(-e^{-x/a}), \qquad -\infty < x < \infty,$$

where $a > 0$ is said to have an extreme value distribution. Determine the median of X. (§3.6)

In Exercises 316–319 there occur distributions with special names. In each case give the name of the relevant distribution.

316. A train is scheduled to arrive at a station at 13:03, but is generally somewhat late. The delay may be considered as a rv with density function $f_X(x) = 1/5$ $(0 \leq x \leq 5)$.
(a) Determine the probability that the train arrives later than 13:06.
(b) Determine the probability that the train arrives between 13:04 and 13:05. (§3.7)

317. Let X be the waiting time in minutes from the opening time of a shop until the first customer arrives. The distribution function of X is given by

$$F_X(x) = \begin{cases} 0 & \text{if } x < 0, \\ 1 - e^{-0.4x} & \text{if } x \geq 0. \end{cases}$$

Determine the probability that the waiting time is:
(a) at most 3 minutes;
(b) at least 4 minutes;
(c) between 3 and 4 minutes;
(d) at most 3 minutes or at least 4 minutes;
(e) exactly 2.5 minutes. (§3.7)

318. Let X denote the maximum snow depth during a winter at a certain place. We assume that X has density function

$$f_X(x) = 2xe^{-x^2} \qquad \text{if } x \geq 0.$$

Find the quantile x_α for $\alpha = 0.5$, 0.1 and 0.01. (§3.7)

319. A rv X has density function $f_X(x) = cx^2e^{-x}$ $(x \geq 0)$.
(a) Find the constant c.
(b) Determine the value of x for which the density function is as large as possible.
(c) Draw a graph of the density function. (§3.7)

320. A rv with density function $f_X(x) = (1/a)e^{-x/a}(1 + e^{-x/a})^{-2}$ $(-\infty < x < \infty)$ is said to have a logistic distribution. Show that this distribution is symmetric and determine its α-quantile. (§3.7)

*321. The rv X has an exponential distribution. Prove that the conditional probability $P(X > t + a|X > a)$ does not depend on a. This result implies that if a new unit, say an electric bulb, has a lifetime with an exponential distribution, then the *remaining* lifetime of a unit functioning after time a has the same distribution as the total lifetime of a new unit. Informally, we may say that the exponential distribution has no memory.

*322. A person tosses a symmetric coin until both head and tail have appeared at least twice. Determine the probability function of the number of tosses.

*323. An urn contains one white ball and one black ball. One ball is drawn at random and is returned to the urn together with one ball of the same colour. This operation is performed $n - 2$ times ($n \geq 3$). The urn then contains n balls, X of which are white. Find the probability function of X.

CHAPTER 4

Multidimensional Random Variables

4.1. Introduction

This chapter is a direct continuation of the preceding one. In §4.2 its main theme is given. In §4.3 we discuss discrete two-dimensional rv's and in §4.4 continuous two-dimensional rv's. §4.5 treats independent rv's. In §4.6 we solve some classical problems in probability.

This chapter is important but not quite as central as the previous one. To understand the later chapters it is, however, necessary to read §4.5.

4.2. General Considerations

The rv's that we studied in the previous chapter were one-dimensional, and were introduced by associating one single number with each outcome of a random trial. When two or more numbers are associated with each outcome we obtain a *multidimensional* rv.

As an illustration we take a batch of units produced by a factory. Let us select a number of units at random from the batch and count the units which are unsatisfactory in length and also the units which are unsatisfactory in breadth. The result may be given as a pair (X, Y), where X and Y are the numbers obtained. The pair (X, Y) is a two-dimensional rv.

We will now generalize the earlier results to multidimensional rv's and will mostly restrict the exposition to the two-dimensional case.

Definition. A *two-dimensional random variable* is a function (X, Y) defined on a sample space Ω. □

As before, the distribution function is of basic importance:

Definition. The function

$$F_{X,Y}(x, y) = P(X \leq x, Y \leq y)$$

is called the *distribution function* of the rv (X, Y). □

The distribution function is a function of two variables, x and y. Sometimes we call this function the *joint distribution function* of X and Y. We then stress the fact that the two-dimensional rv consists of two one-dimensional rv's which are studied jointly.

4.3. Discrete Two-Dimensional Random Variables

If both X and Y assume a finite or denumerably infinite number of different values, the rv (X, Y) is said to be discrete. For simplicity, we might suppose that these values are nonnegative integers, which is the most common situation. The probability function is then defined in the following way:

Definition. The quantities

$$p_{X,Y}(j, k) = P(X = j, Y = k) \qquad (j = 0, 1, \ldots; k = 0, 1, \ldots),$$

together constitute the *probability function* of the discrete two-dimensional rv (X, Y). □

We sometimes speak of the *joint probability function*. The sum of the probabilities is 1, that is,

$$\sum_{j=0}^{\infty} \sum_{k=0}^{\infty} p_{X,Y}(j, k) = 1.$$

The probability function can be illustrated by means of a bar diagram (see Fig. 4.1). The distribution function can be determined from the probability function by a summation:

$$F_{X,Y}(x, y) = \sum_{j \leq x} \sum_{k \leq y} p_{X,Y}(j, k). \tag{1}$$

Using some imagination the reader will be able to form a picture of the distribution function—it is not very easy to draw the graph. The distribution function consists of rectangular horizontal plane sections which, in the second, third and fourth quadrants, have height zero above the (x, y)-plane, and which increase in height stepwise with x and y in the first quadrant, and tend to 1 when x and y both go to infinity.

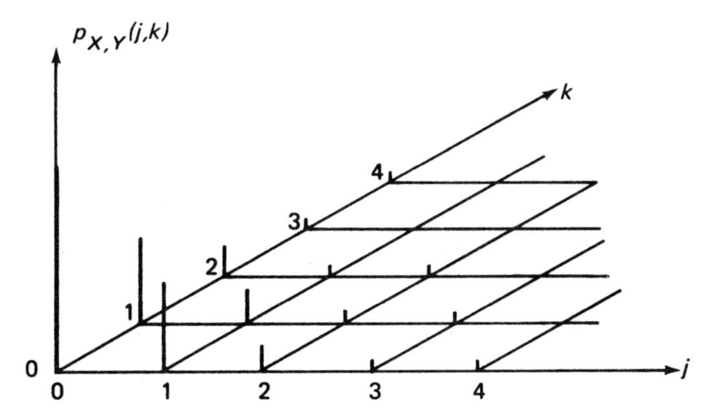

Fig. 4.1. Probability function of a discrete two-dimensional rv.

When X and Y are viewed separately, they are one-dimensional rv's, as has earlier been indicated. Their distributions, which are called *marginal distributions*, are obtained by handling the joint distribution of X and Y in an appropriate way. For example, the *marginal probability function* of X is given by

$$p_X(j) = \sum_{k=0}^{\infty} p_{X,Y}(j, k). \tag{2}$$

The *marginal distribution function* of X is obtained by adding the probabilities $p_X(j)$ in the usual manner or, directly, by setting $y = \infty$ in $F_{X,Y}(x, y)$.

Example 1. Children in a Family

Let X and Y, respectively, denote the number of boys and girls in a randomly chosen Swedish family. Let us assume that the probability function $p_{X,Y}(j, k)$ takes the values given in the following table. (Actually, this is the probability function depicted in Fig. 4.1.)

j \ k	0	1	2	3	4	Sum
0	0.38	0.16	0.04	0.01	0.01	0.60
1	0.17	0.08	0.02			0.27
2	0.05	0.02	0.01			0.08
3	0.02	0.01				0.03
4	0.02					0.02
Sum	0.64	0.27	0.07	0.01	0.01	1.00

For example, the table shows that the probability $p_{X,Y}(1, 0)$ of 1 boy and 0 girls is 0.17. As is easily seen, the vertical column to the right contains the marginal probability function $p_X(j)$ of X, and the last horizontal row contains the marginal probability function $p_Y(k)$ of Y.

The distribution function is obtained by a summation. For instance, it is found that

$$F_{X,Y}(1, 1) = P(X \leq 1, Y \leq 1) = 0.38 + 0.16 + 0.17 + 0.08 = 0.79. \qquad \square$$

Multinomial Distribution

A good example of a multidimensional discrete distribution is the *multinomial distribution.*

We present it first in two dimensions: If a rv (X, Y) has probability function

$$p_{X,Y}(j, k) = \frac{n!}{j!k!(n-j-k)!} p_1^j p_2^k (1 - p_1 - p_2)^{n-j-k}$$

$$(j \geq 0, k \geq 0, j + k \leq n), \quad (3)$$

where n is a positive integer and $p_1 > 0$, $p_2 > 0$, $1 - p_1 - p_2 > 0$, then this rv is said to have a trinomial distribution (cf. the binomial distribution).

We now pass to the general case and present the probability function of a multinomial rv $(X_1, \ldots, X_r)$:

$$p_{X_1,\ldots,X_r}(k_1, \ldots, k_r) = \frac{n!}{k_1! \cdot \ldots \cdot k_r!} p_1^{k_1} \cdot \ldots \cdot p_r^{k_r},$$

where $\sum p_i = 1$. In this expression, $k_1, \ldots, k_r$ take on all nonnegative integer values with sum n.

This looks like an r-dimensional distribution, but is due to our predilection for symmetry. In fact, the sum $\sum_1^r X_j$ is always n. Hence any one of the r variables can be expressed as a function of the remaining $r - 1$ variables, and the correct number of dimensions is $r - 1$. In the trinomial case, we have used the nonsymmetrical representation and omitted the third "unnecessary" variable.

4.4. Continuous Two-Dimensional Random Variables

Definition. If there exists a function $f_{X,Y}(x, y)$ such that

$$F_{X,Y}(x, y) = \int_{-\infty}^{x} \int_{-\infty}^{y} f_{X,Y}(t, u) \, dt \, du, \quad (4)$$

then the rv (X, Y) is said to be *continuous*. The function $f_{X,Y}(x, y)$ is called the *density function* of (X, Y). $\square$

For completeness, the density function is sometimes called the *joint density function*. It is obtained by differentiating the distribution function with respect

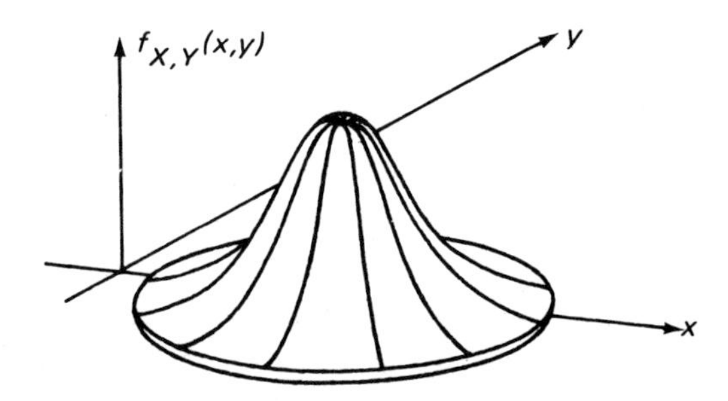

Fig. 4.2. Density function of a continuous two-dimensional rv.

to both x and y, that is, we have

$$\frac{\partial^2 F(x, y)}{\partial x\, \partial y} = f_{X,Y}(x, y). \tag{5}$$

We suppose here that the density function is continuous at the point (x, y). The density function is always greater than or equal to zero. The graphical picture often looks like a heap of sand (see Fig. 4.2). The volume between the surface and the (x, y)-plane is 1, that is,

$$\int_{-\infty}^{\infty} \int_{-\infty}^{\infty} f_{X,Y}(x, y)\, dx\, dy = 1.$$

Figures like Fig. 4.2 are difficult to draw. Often it is just as informative to draw contour curves $f_{X,Y}(x, y) = c$, where c takes on different values (see Fig. 4.3 for some examples).

By integrating one variable in the density function from $-\infty$ to ∞ we obtain the marginal distribution of the other. For example, we get the *marginal density function* of X from

$$f_X(x) = \int_{-\infty}^{\infty} f_{X,Y}(x, u)\, du \tag{6}$$

and the *marginal distribution function* of X from

$$F_X(x) = F_{X,Y}(x, \infty). \tag{7}$$

It is sometimes required to find the probability that (X, Y) satisfies some condition of the type $(X, Y) \in A$, where A is a two-dimensional region. We then have (see Fig. 4.4)

$$P[(X, Y) \in A] = \iint_A f_{X,Y}(x, y)\, dx\, dy. \tag{8}$$

We shall present two continuous distributions.

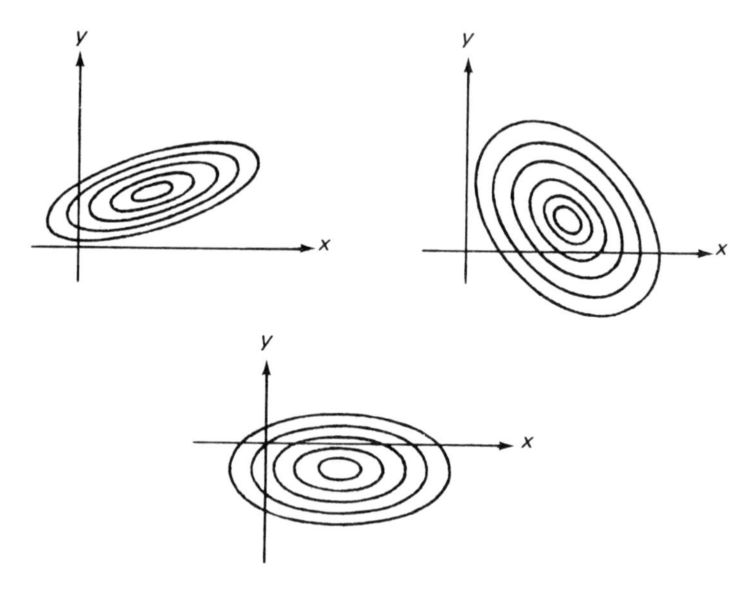

Fig. 4.3. Contour curves of a two-dimensional density function.

(a) Uniform Distribution

A two-dimensional rv (X, Y) is *uniformly distributed over the* region Ω, if for each region A of Ω it is true that

$$P[(X, Y) \in A] = \frac{\text{area of } A}{\text{area of } \Omega}. \tag{9}$$

Example 2. Target Shooting

A person shoots at a circular target of radius r. Let us make the assumption (admittedly rather unrealistic) that he is so clever that he always hits the target, but at the same time so unskilful that he hits all points on the target with equal probability. We want to determine the probability that the shot hits the target at a point at most distance

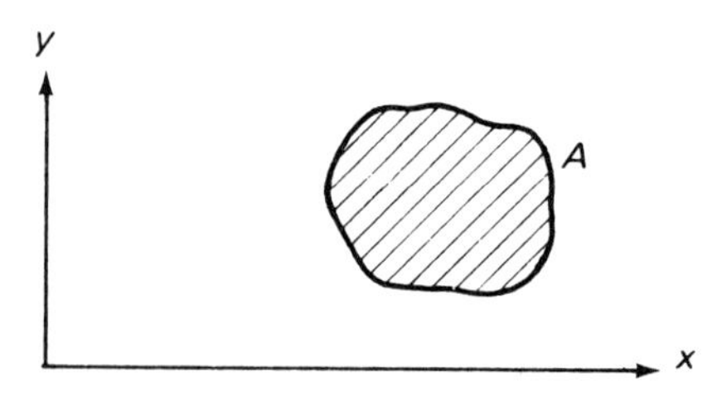

Fig. 4.4.

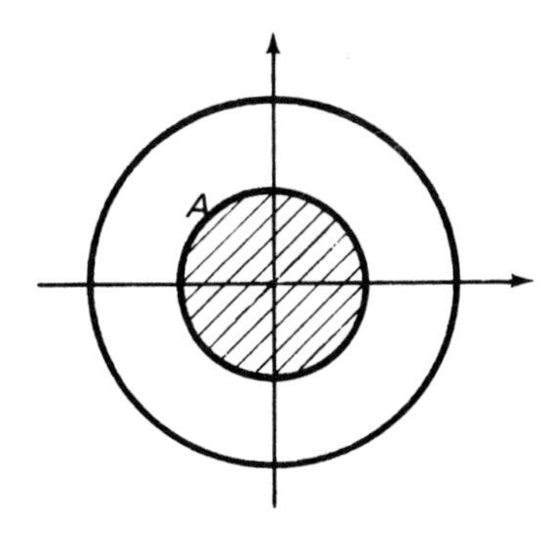

Fig. 4.5.

$r/2$ from the centre. Expression (9) immediately shows that the answer is (see Fig. 4.5)

$$\frac{\pi(r/2)^2}{\pi r^2} = \frac{1}{4}. \qquad \square$$

(b) Bivariate Normal Distribution

We now mention briefly an important two-dimensional distribution, namely, the *bivariate normal distribution*. The density function has the form

$$f_{X,Y}(x, y) = ke^{-Q(x,y)/2} \qquad (-\infty < x < \infty,\ -\infty < y < \infty).$$

Here Q is a polynomial of the second degree which can be written

$$Q = a(x - m_1)^2 + 2b(x - m_1)(y - m_2) + c(y - m_2)^2,$$

where a, b, c, m_1, m_2 are constants ("parameters"), while k is chosen so that

$$\iint_{R^2} f_{X,Y}(x, y)\, dx\, dy = 1.$$

(In fact, it can be shown that $k = \sqrt{(ac - b^2)}/2\pi$.) By giving the parameters different values, a whole family of distributions is obtained. Figure 4.2 shows how a bivariate normal distribution looks; however, it has been cut off at the edges.

4.5. Independent Random Variables

It is of basic importance to be familiar with independent rv's. Without exaggeration it may be stated that no other concept has contributed more to the development of probability theory and its applications.

When defining independent rv's, one uses the definition of independent events (see §2.6).

Let (X, Y) be the result of two throws of a symmetrical die, where X corresponds to the first throw and Y to the second throw. It seems reasonable to consider X and Y independent if the events $X = j$ and $Y = k$ are independent for all j and k; that is, if $P(X = j, Y = k) = P(X = j)P(Y = k)$, or in other notations, if $p_{X,Y}(j, k) = p_X(j)p_Y(k)$. Since, in this relation, each factor on the right is $1/6$, this implies that each combination of results (j, k) is given the probability $1/36$, which seems very natural.

If the rv (X, Y) is continuous, it is not so obvious how to proceed. In this case, X and Y are said to be independent if the joint density function can be written as a product of the marginal density functions, that is, if $f_{X,Y}(x, y) = f_X(x)f_Y(y)$.

We sum up in the following important definition.

Definition. The rv's X and Y are called *independent* if, for all j and k or x and y, we have

$$\begin{aligned} p_{X,Y}(j, k) &= p_X(j)p_Y(k) \qquad \text{(discrete rv's)},\\ f_{X,Y}(x, y) &= f_X(x)f_Y(y) \qquad \text{(continuous rv's)}. \end{aligned} \tag{10}$$

□

The following definition is equivalent: The rv's X and Y are called independent if, for all x and y, we have

$$F_{X,Y}(x, y) = F_X(x)F_Y(y). \tag{11}$$

To show the equivalence is not difficult, but we shall not do this.

The definitions can easily be extended to more than two rv's. For example, the discrete rv's X, Y and Z are independent if $p_{X,Y,Z}(i, j, k) = p_X(i)p_Y(j)p_Z(k)$. Note that only one condition is needed, in contrast to the definition of three independent events in §2.6, where several conditions appear.

When a random model is to be constructed for some phenomenon in the empirical world, it is important to decide whether measurements, say, can be represented by independent rv's. If so, the model is simplified. If a sequence of independent trials is performed (see §2.6), there is no doubt about the validity of assuming independence. Indeed, independence of trials means precisely that the result of one trial does not affect the result of the following ones.

Example 3. Lifetimes of Electric Bulbs

Two new bulbs are switched on. Their lifetimes X and Y are assumed to be independent and Exp(a) and Exp(b), respectively; hence the density functions are:

$$f_X(x) = \begin{cases} \dfrac{1}{a}e^{-x/a} & \text{if} \quad x \geq 0,\\ 0 & \text{if} \quad x < 0, \end{cases}$$

$$f_Y(y) = \begin{cases} \dfrac{1}{b} e^{-y/b} & \text{if} \quad y \geq 0, \\ 0 & \text{if} \quad y < 0. \end{cases}$$

According to the definition of independent rv's, the joint density function of X and Y is given by

$$f_{X,Y}(x, y) = f_X(x) f_Y(y) = \begin{cases} \dfrac{1}{a} e^{-x/a} \cdot \dfrac{1}{b} e^{-y/b} & \text{if} \quad x \geq 0, y \geq 0, \\ 0 & \text{otherwise.} \end{cases}$$

Let us evaluate the probability that both bulbs fail before the time t. This probability is seen to be

$$\iint\limits_{\substack{0 \leq x \leq t \\ 0 \leq y \leq t}} \frac{1}{a} e^{-x/a} \frac{1}{b} e^{-y/b} \, dx \, dy = \int_0^t \frac{1}{a} e^{-x/a} \, dx \int_0^t \frac{1}{b} e^{-y/b} \, dy$$
$$= (1 - e^{-t/a})(1 - e^{-t/b}).$$

We shall use this opportunity to show that the definition of independent rv's is sensible, by proving that exactly the same result is obtained if we compute the probability directly without using the joint density function. The event in question can be written $A \cap B$, where A is the event that the first bulb functions at most for the time t, and analogously for B. Now we see that

$$P(A) = P(X \leq t) = \int_0^t \frac{1}{a} e^{-x/a} \, dx = 1 - e^{-t/a},$$

$$P(B) = P(Y \leq t) = \int_0^t \frac{1}{b} e^{-y/b} \, dy = 1 - e^{-t/b},$$

and hence, since A and B are independent,

$$P(A \cap B) = P(A)P(B) = (1 - e^{-t/a})(1 - e^{-t/b}).$$

There is another lesson to be learned from this example. When studying independent rv's, the joint density function should not be used unless it is really needed, since usually the second approach is more convenient. □

***Example 4. Lifetimes of Electronic Components**

The lifetimes of two electronic components of an instrument are X and Y. Their joint density function is taken to have the form

$$f_{X,Y}(x, y) = \begin{cases} k(xy + c)e^{-x-y} & \text{if} \quad x > 0, y > 0, \\ 0 & \text{otherwise,} \end{cases}$$

where $c \geq 0$. Double integration shows that $k = 1/(1 + c)$. The rv's X and Y are independent if and only if $c = 0$, for the joint density function can then be written as a product of the marginal density functions xe^{-x} and ye^{-y}. □

4.6. Some Classical Problems of Probability[1]

Example 5. Fermi–Dirac and Bose–Einstein Statistics

(a) Fermi–Dirac statistics

Consider n identical and indistinguishable particles. The particles are distributed with the same probability among m cells ($m \geq n$), with the restriction that at most one particle is placed in one cell. Find the m-dimensional probability function of the numbers $X_1, \ldots, X_m$ of particles in the cells.

The particles can be placed in $\binom{m}{n}$ ways in the cells, all equally probable. Hence the probability function that we seek has the simple form

$$p_{X_1,\ldots,X_m}(k_1, \ldots, k_m) = 1\bigg/\binom{m}{n},$$

where all k_i are 0 or 1. This uniform distribution is due to Fermi and Dirac and appears in statistical mechanics.

(b) Bose–Einstein statistics

Consider the same situation as before but remove the restriction that only one particle is placed in each cell. The condition $m \geq n$ is then no longer necessary.

The number of possible cases is now evaluated by representing each such case by a sequence of bars and circles, where a bar signifies a borderline between two neighbouring cells and a circle signifies a particle:

○ ○ | | ○ | ○ ○ ○ | | ○

Evidently, there are $m - 1$ bars and n circles, that is, $m - 1 + n$ possible positions, n of which must be filled by circles. Hence the number of possible cases is $\binom{m-1+n}{n}$, and the probability function is given by

$$p_{X_1,\ldots,X_m}(k_1, \ldots, k_m) = 1\bigg/\binom{m-1+n}{n},$$

where $0 \leq k_i \leq n$ for all $i = 1, 2, \ldots, m$. This distribution is also uniform. It was first given by Bose and Einstein. □

Example 6. Bertrand's Paradox

Draw a circle with centre at the origin and with radius r. Also draw an equilateral triangle inscribed in the circle. The triangle has sides of length $r\sqrt{3}$, and the distance from the sides to the origin is $r/2$ (see Fig. 4.6). Draw a chord at random in the circle and determine the probability P that the length of the chord exceeds that of a side of the triangle. Before the solution can be found, we must set up a suitable model.

(a) First model

Let Z be the distance from the mid-point of the chord to the origin (see Fig. 4.7). Suppose that Z has a uniform distribution over the interval $(0, r)$. The length of the

[1] Special section which may be omitted.

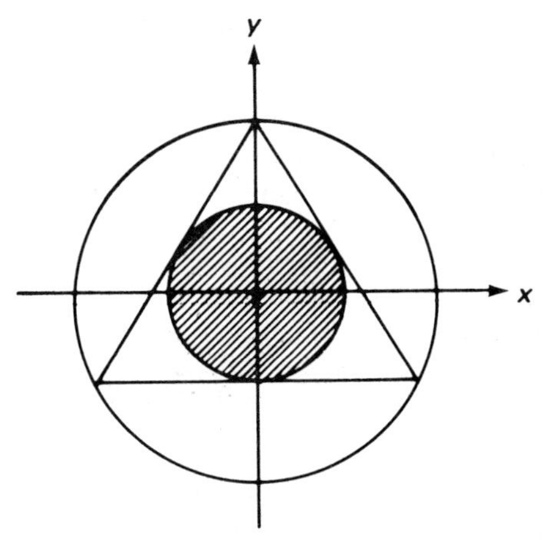

Fig. 4.6.

chord is greater than the side of the triangle if $Z < r/2$, that is,

$$P = P(Z < r/2) = 1/2.$$

(b) Second model

Let θ be the angle subtended by the chord. Suppose that θ is uniformly distributed over the interval $(0, \pi)$. Then, clearly, we have

$$P = P(2\pi/3 < \theta < \pi) = (\pi - 2\pi/3)/\pi = 1/3.$$

(c) Third model

Consider the mid-point of the chord with coordinates (X, Y). Suppose that (X, Y) has a two-dimensional uniform distribution over the circle. It is seen that

$$P = P[(X, Y) \in A],$$

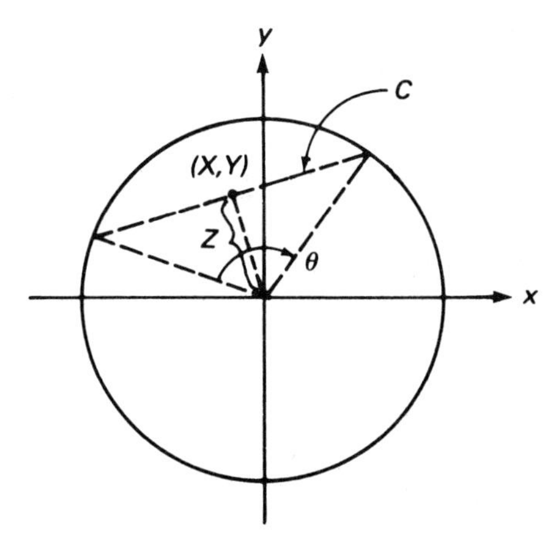

Fig. 4.7.

where A is the circular disc of radius $r/2$ (see Fig. 4.6). Hence we find (see Example 2 in §4.4)

$$P = (\pi r^2/4)/(\pi r^2) = 1/4.$$

Does this mean that we have obtained three different answers to the same question? Not at all: the reason is that the problem is not clearly stated. We have used our imagination too freely when building the models. It is impossible to decide whether any one of the three is correct until the method of drawing the chord is properly described. For example, let us suppose that two points are chosen at random and independently on the circumference of the circle and joined by a chord. Some reflection reveals that the second model is then obtained, and hence the correct answer is then 1/3. Therefore, Bertrand's paradox is no paradox; however, in the history of probability it bears this name. □

Example 7. Buffon's Needle Problem

A needle of length $2d$ is thrown at random on a plane filled by parallel lines distance $2a$ apart ($d < a$). Determine the probability P that the needle intersects one of the lines.

The following model for this experiment seems reasonable. Let X be the distance from the mid-point of the needle to the nearest line and let Y be the angle between the needle and the line (see Fig. 4.8). We assume that X and Y are independent and uniformly distributed over the intervals $(0, a)$ and $(0, \pi)$, respectively.

This means, vaguely speaking, that all distances are equiprobable, as are also all angles. The assumption of independence implies that the joint density function of X and Y is given by

$$f_{X,Y}(x, y) = \frac{1}{a}\cdot\frac{1}{\pi} \qquad (0 \le x \le a, 0 \le y \le \pi).$$

The needle hits a line if $X < d \sin Y$ (see the figure). Hence the probability we want is given by

$$P = \iint_A \frac{1}{a\pi}\, dx\, dy,$$

where A is the area defined by $x < d \sin y$, $0 \le x \le a$, $0 \le y \le \pi$. It is found that

$$P = \frac{1}{a\pi}\int_0^\pi \left[\int_0^{d \sin y} dx\right] dy = \frac{d}{a\pi}\int_0^\pi \sin y\, dy = \frac{2d}{a\pi}.$$

Hence π can be estimated by emptying a big box of matches over the American flag with a sweeping movement of the arm! (Compare the frequency interpretation of a probability in §2.3 and the theory of statistics.) □

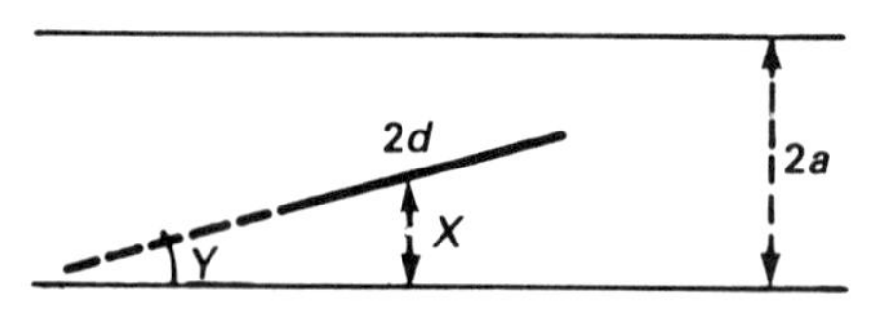

Fig. 4.8.

EXERCISES

401. In a certain part of a city, a number of families were surveyed at random regarding the number X of children and the number Y of rooms in their apartment. (Y does not include the kitchen.) Suppose that (X, Y) is a rv with the probability function $p_{X,Y}(j, k)$ given in the following table:

j \ k	1	2	3	4
0	0.11	0.09	0.07	0.01
1	0.07	0.12	0.12	0.02
2	0.02	0.05	0.17	0.05
3	0.00	0.02	0.04	0.02
4	0.00	0.00	0.01	0.01

(a) Find the probability that a randomly chosen family has at most one child and lives in an apartment with at most three rooms (and kitchen).
(b) If each family consists of two grown-up persons and children, what is the probability that a randomly chosen family lives in an overcrowded apartment? By "overcrowded" we mean that the number of persons/room (apart from the kitchen) exceeds two.
(c) Find the marginal probability functions $p_X(j)$ and $p_Y(k)$ of the number of children and the number of rooms, respectively. (§4.3)

402. The rv (X, Y) has density function

$$f_{X,Y}(x, y) = (c/x)e^{-x^2y} \qquad \text{if} \quad x \geq 1, y \geq 0.$$

Determine the constant c. (§4.4)

403. A rv (X, Y) has density function

$$f_{X,Y}(x, y) = \frac{1}{\pi^2(1 + x^2)(1 + y^2)}, \qquad -\infty < x < \infty, \quad -\infty < y < \infty.$$

Determine
(a) the distribution function $F_{X,Y}(x, y)$;
(b) the marginal density functions $f_X(x)$ and $f_Y(y)$. (§4.4)

404. The rv's X and Y are independent and have the following probability functions:

j	1	3	5	7	9
$p_X(j)$	0.10	0.20	0.40	0.20	0.10

k	2	4	6	8
$p_Y(k)$	0.10	0.20	0.30	0.40

(a) Find $P(X = 3, Y = 6)$.
(b) Find $P(X \leq 3, Y \leq 6)$. (§4.5)

405. The rv's X and Y are independent and assume the values j_1, j_2, j_3 and k_1, k_2, k_3, respectively. Complete the following incomplete table of the probability function $p_{X,Y}(j, k)$:

	k_1	k_2	k_3
j_1	0.03	0.15	?
j_2	0.04	?	?
j_3	0.03	?	?

(§4.5)

406. A person travels first by bus 1 and then by bus 2. The waiting times, X and Y, are independent and uniformly distributed over the intervals (0, 10) and (0, 8), respectively (unit: minutes). Compute the probability that the total waiting time is at least 16 minutes.
Hint: The rv (X, Y) is uniformly distributed over a rectangle. (§4.5)

407. Two persons decide to meet at a certain place on a certain day between 12:00 and 13:00. Each person is prepared to wait at most 20 minutes for the other. If both arrive independently and at random during the hour, what is the probability that they meet? (§4.5)

408. A continuous rv (X, Y) has density function

$$f_{X,Y}(x, y) = 2(1 + x + y)^{-3} \qquad \text{if} \quad x \geq 0, y \geq 0.$$

(a) Determine $f_X(x)$ and $f_Y(y)$.
(b) Determine $F_{X,Y}(x, y)$, $F_X(x)$ and $F_Y(y)$, and show that

$$F_{X,Y}(x, y) \neq F_X(x)F_Y(y) \qquad \text{if} \quad x > 0, y > 0.$$

(c) Are X and Y independent? (§4.5)

409. The rv (X, Y) has density function

$$f_{X,Y}(x, y) = \frac{1 + x + y + cxy}{c + 3} \cdot e^{-(x+y)} \qquad \text{if} \quad x \geq 0, y \geq 0.$$

(a) Determine $f_X(x)$.
(b) There exists exactly one value of c such that X and Y are independent. Which value? (§4.5)

*410. A symmetric die is thrown repeatedly. "One" is obtained for the first time in throw number X and "two" for the first time in throw number Y. Find the joint probability function of X and Y.

*411. The rv (X, Y) has density function

$$f_{X,Y}(x, y) = e^{-x^2 y} \qquad \text{if} \quad x \geq 1, y \geq 0.$$

(a) Show that $P(X^2 Y > 1) = e^{-1}$.
(b) Prove that X and Y are dependent. *Hint*: Form the ratio $f_{X,Y}(x, y)/f_X(x)$ and think (make no calculations).

*412. Two events E_1 and E_2 occur at times X and Y, respectively, after a certain initial time point. Here X and Y are independent rv's with density functions

$$f_X(x) = 2e^{-2x} \quad \text{if} \quad x \geq 0,$$

$$f_Y(y) = 4y^2e^{-2y} \quad \text{if} \quad y \geq 0.$$

Find the probability that E_1 occurs before E_2.

CHAPTER 5

Functions of Random Variables

5.1. Introduction

It is often of interest to consider a function $Y = g(X)$ of a rv X. Then Y itself is a rv with a distribution determined by $g(\cdot)$ and the distribution of X.

More generally, it is possible to consider a function of several rv's, for example, a sum $Z = X + Y$ of two rv's X and Y. (Still more generally, we can investigate several functions at the same time, but this situation will not be considered in this chapter.)

To handle the tools of the trade skilfully, readers are urged to study the examples given in the text and to work through the exercises.

In §5.2 we discuss functions of a single rv, in §5.3 sums of two or more rv's, in §5.4 the distribution of the largest or the smallest of two or more rv's, and in §5.5 the ratio of two rv's. To understand later chapters it is necessary to read §5.2 and §5.3.

5.2. A Single Function of a Random Variable

We shall consider the rv $Y = g(X)$ and begin with a special method for determining its distribution.

(a) A Special Method

If X has a discrete distribution, it is rather uncomplicated to find the distribution of Y. The function $g(X)$ is determined for all possible values of X, and the distribution is then evaluated by adding the probability function of X in an appropriate way. An example will show how to proceed.

Example 1. Errors of Measurement

The dimension of an object is measured. The error of measurement is assumed to be a rv X with the following distribution:

j	-3	-2	-1	0	1	2	3
$p_X(j)$	0.02	0.08	0.15	0.50	0.15	0.08	0.02

The probability function $p_Y(k)$ of the quadratic error $Y = X^2$ is given by

k	0	1	4	9
$p_Y(k)$	0.50	0.30	0.16	0.04

(If $j = -3$ or $j = 3$ we have $k = 9$ and hence we obtain $p_Y(9) = 0.02 + 0.02 = 0.04$, and so on.) □

More generally we obtain: If X assumes integer values and has probability function $p_X(j)$, then the rv $Y = g(X)$ has probability function

$$p_Y(k) = \sum_{j:g(j)=k} p_X(j). \tag{1}$$

(b) General Method

The method described above can be used only in the discrete case. A quite general method of proceeding is to first determine the distribution function $F_Y(y)$ of Y. Then the density function or the probability function can be found in the familiar way.

The procedure is easiest to apply if $g(x)$ is strictly increasing or strictly decreasing. We shall illustrate this situation by means of three examples.

Example 2. Linear Transformation

We assume that X is a continuous rv and take $Y = aX + b$.

(1) a positive

If a is positive we find

$$F_Y(y) = P(Y \le y) = P(aX + b \le y) = P\left(X \le \frac{y-b}{a}\right) = F_X\left(\frac{y-b}{a}\right).$$

Hence we obtain the distribution function of Y by replacing the argument x by

$(y - b)/a$ in the distribution function $F_X(x)$. The density function of Y is then obtained by differentiating the distribution function with respect to y:

$$f_Y(y) = \frac{1}{a} f_X\left(\frac{y-b}{a}\right).$$

(2) a negative

If a is negative we find instead

$$F_Y(y) = P(aX + b \le y) = P\left(X \ge \frac{y-b}{a}\right) = 1 - P\left(X < \frac{y-b}{a}\right).$$

Since X is a continuous rv we have $P\left(X < \frac{y-b}{a}\right) = P\left(X \le \frac{y-b}{a}\right)$ and hence we obtain

$$F_Y(y) = 1 - P\left(X \le \frac{y-b}{a}\right) = 1 - F_X\left(\frac{y-b}{a}\right).$$

Finally, a differentiation shows that

$$f_Y(y) = -\frac{1}{a} f_X\left(\frac{y-b}{a}\right).$$

As far as the density function is concerned, the two cases $a > 0$ and $a < 0$ can be summarized as follows:

$$f_Y(y) = \frac{1}{|a|} f_X\left(\frac{y-b}{a}\right).$$

In particular, if $a = -1$, $b = 0$, we get the distribution of the rv $Y = -X$:

$$F_Y(y) = 1 - F_X(-y),$$

$$f_Y(y) = f_X(-y).$$

□

Example 3. Logarithmic Transformation

Let us assume that X has a uniform distribution over the interval (0, 1). We then have (see §3.7)

$$F_X(x) = \begin{cases} 0 & x < 0, \\ x & 0 \le x \le 1, \\ 1 & x > 1. \end{cases}$$

The distribution function of the rv $Y = -m \ln X$ $(m > 0)$ is found to be

$$F_Y(y) = P(-m \ln X \le y) = P(X \ge e^{-y/m}) = 1 - P(X < e^{-y/m})$$

$$= 1 - P(X \le e^{-y/m}) = \begin{cases} 0 & \text{if } y < 0, \\ 1 - e^{-y/m} & \text{if } y \ge 0. \end{cases}$$

Hence we have that $Y \sim \text{Exp}(m)$.

□

Example 4. Square Root Transformation

Set $Y = \sqrt{X}$, where X is a continuous rv which assumes only positive values. We get

$$F_Y(y) = P(Y \le y) = P(\sqrt{X} \le y) = P(X \le y^2) = F_X(y^2).$$

A differentiation with respect to y shows that

$$f_Y(y) = 2yf_X(y^2). \qquad \square$$

For the benefit of friends of formulae we state a general result: If X is continuous and $g(x)$ is strictly increasing, we have

$$F_Y(y) = F_X[g^{-1}(y)], \tag{2}$$

where $g^{-1}(y)$ is the inverse of $g(x)$. The reader is encouraged to derive the formula as an exercise and to check it on the three examples given above.

We shall finish this section by demonstrating the procedure for finding the distribution of Y if $g(x)$ is neither strictly increasing nor strictly decreasing.

Example 5. Quadratic Transformation

Set $Y = X^2$ where X is a continuous rv. For $y > 0$ we see that

$$\begin{aligned} F_Y(y) &= P(Y \le y) = P(X^2 \le y) = P(-\sqrt{y} \le X \le \sqrt{y}) \\ &= P(X \le \sqrt{y}) - P(X < -\sqrt{y}). \end{aligned}$$

Using the fact that X is continuous, we have $P(X < -\sqrt{y}) = P(X \le -\sqrt{y})$ and hence we obtain

$$F_Y(y) = P(X \le \sqrt{y}) - P(X \le -\sqrt{y}) = F_X(\sqrt{y}) - F_X(-\sqrt{y}).$$

We now differentiate with respect to y:

$$f_Y(y) = \frac{1}{2\sqrt{y}}[f_X(\sqrt{y}) + f_X(-\sqrt{y})].$$

Let us consider the special case when X has a uniform distribution over the interval $(-1, 1)$ so that

$$f_X(x) = \begin{cases} 1/2 & \text{if } \quad -1 \le x \le 1, \\ 0 & \text{otherwise.} \end{cases}$$

Substituting this in the expression for $f_Y(y)$ we find

$$f_Y(y) = \begin{cases} \dfrac{1}{2\sqrt{y}} & \text{if } \quad 0 \le y \le 1, \\ 0 & \text{otherwise.} \end{cases} \qquad \square$$

5.3. Sums of Random Variables

Investigation of a sum of two or more rv's is a common problem. For example, consider n light bulbs with lifetimes $X_1, \ldots, X_n$. We may be interested in studying the total lifetime $X_1 + X_2 + \cdots + X_n$ of the bulbs.

We restrict ourselves to the case of two rv's and hence consider the rv $Z = X + Y$, where X and Y may be independent or dependent. Let us view Z as a function of the two-dimensional rv (X, Y); see Chapter 4. The "addition" of X and Y which we shall perform is often called a *convolution* of the distributions of X and Y.

(a) Convolution of Discrete Distributions

We assume first that X and Y are discrete and restrict ourselves as usual to the special case when they assume only integer values. This is then also true of Z, which has the probability function

$$p_Z(k) = P(Z = k) = P(X + Y = k).$$

We obtain the distribution of Z in the following way. The event $X + Y = k$ can occur in different ways. We may have $X = 0$, $Y = k$, or $X = 1$, $Y = k - 1$, and so on, up to and including $X = k$, $Y = 0$. Hence the probability of this event can be written as a sum (see Fig. 5.1)

$$P(X + Y = k) = \sum_{i+j=k} p_{X,Y}(i, j), \tag{3}$$

with addition over all nonnegative i and j having sum k. Hence we have

$$p_Z(k) = \sum_{i+j=k} p_{X,Y}(i, j). \tag{4}$$

Analogously, it is seen that the distribution function of Z is given by

$$F_Z(z) = \sum_{i+j \le z} p_{X,Y}(i, j). \tag{5}$$

In particular, if X and Y are independent, as is often the case, we have according to (10) in §4.5

$$p_Z(k) = \sum_{i+j=k} p_X(i)p_Y(j), \tag{6}$$

$$F_Z(z) = \sum_{i+j \le z} p_X(i)p_Y(j). \tag{7}$$

The expression (6) is called the *convolution formula for independent discrete*

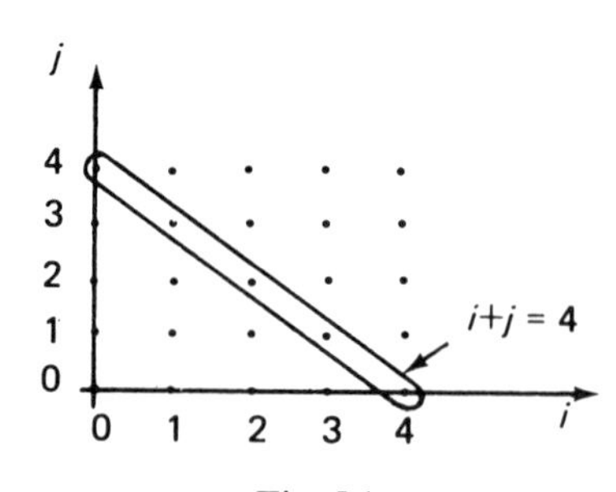

Fig. 5.1.

rv's. Alternatively, (6) can be written in the form

$$p_Z(k) = \sum_{i=0}^{k} p_X(i)p_Y(k-i). \tag{6'}$$

Example 6. Game of Chance

A person participates in a series of games of chance with a well-made die. In the first round he receives 1 dollar if "one" or "two" appears, 2 dollars if "three" or "four" appears and 3 dollars if "five" or "six" appears; the same rule is applied in the second round. Let X and Y denote the amount he gets in the two rounds. We want the distribution of the total amount $Z = X + Y$ that the person receives.

Clearly, X and Y have the probability functions

$$p_X(i) = 1/3 \quad (i = 1, 2, 3); \qquad p_Y(j) = 1/3 \quad (j = 1, 2, 3).$$

The sum $i + j$ for all combinations of i and j becomes:

i \ j	1	2	3
1	2	3	4
2	3	4	5
3	4	5	6

It is seen that the sum is 2 in one case, 3 in two cases, 4 in three cases, 5 in two cases and 6 in one case. Hence by (6) the probability function of Z is

$$p_Z(2) = 1/9; \quad p_Z(3) = 2/9; \quad p_Z(4) = 3/9; \quad p_Z(5) = 2/9; \quad p_Z(6) = 1/9.$$

The sum of the probabilities is 1 as it should be. □

(b) Convolution of Continuous Distributions

Suppose that X and Y are continuous rv's. The distribution function of Z is obtained by integrating the joint density function $f_{X,Y}(x, y)$ over the region

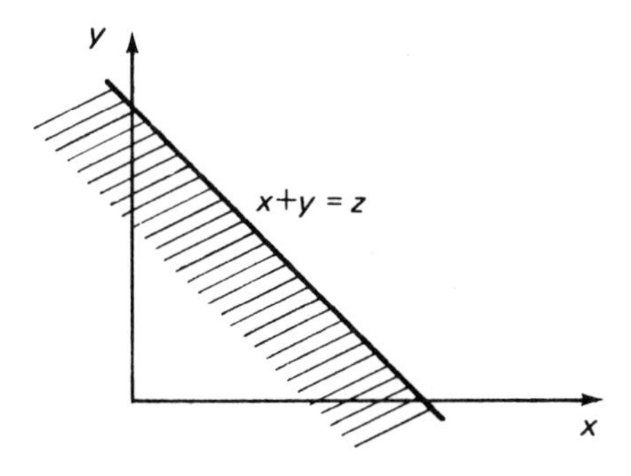

Fig. 5.2.

$x + y \leq z$ (see (8) in §4.4 and Fig. 5.2). Hence we get

$$F_Z(z) = \iint\limits_{x+y \leq z} f_{X,Y}(x, y)\, dx\, dy. \tag{8}$$

If X and Y are independent, we have according to (10) in §4.5

$$F_Z(z) = \iint\limits_{x+y \leq z} f_X(x) f_Y(y)\, dx\, dy. \tag{9}$$

Rewriting this, we see that

$$F_Z(z) = \int_{-\infty}^{\infty} f_X(x) \left[\int_{-\infty}^{z-x} f_Y(y)\, dy \right] dx = \int_{-\infty}^{\infty} f_X(x) F_Y(z - x)\, dx. \tag{10}$$

By differentiating this function under the integral sign with respect to z (we assume that this operation is permissible) we obtain the density function of Z:

$$f_Z(z) = \int_{-\infty}^{\infty} f_X(x) f_Y(z - x)\, dx. \tag{11}$$

This expression is called the *convolution formula for independent continuous* rv's.

In applications we can use either of the three formulae (9), (10), (11). Frequently, the density functions appearing in the integrands are zero in some regions, which requires great care when performing the integrations.

Example 7. Convolution of Exponential Distributions

The rv's X and Y are assumed to be independent and Exp(m). Hence we have

$$f_X(x) = \begin{cases} \dfrac{1}{m} e^{-x/m} & \text{if} \quad x \geq 0, \\ 0 & \text{otherwise,} \end{cases}$$

and analogously for Y.

Formula (9) shows that (see Fig. 5.3; since the integrand is 0 if x or y is negative,

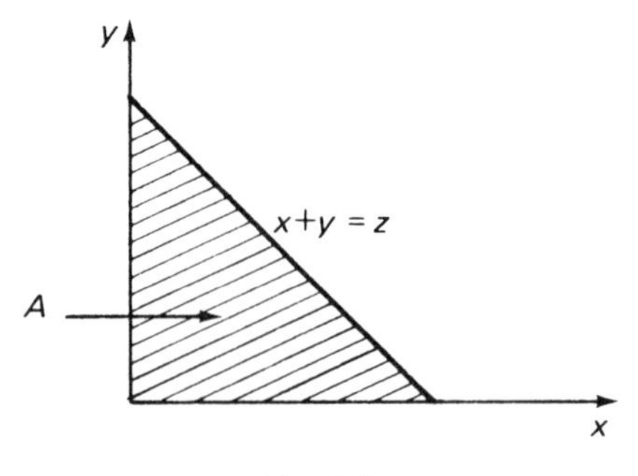

Fig. 5.3.

the area of integration can be restricted to the region A in the figure)

$$F_Z(z) = \iint_A (1/m)e^{-x/m}(1/m)e^{-y/m}\,dx\,dy$$

$$= \int_0^z (1/m)e^{-x/m}\left[\int_0^{z-x} (1/m)e^{-y/m}\,dy\right]dx$$

$$= (1/m)\int_0^z (e^{-x/m} - e^{-z/m})\,dx$$

$$= 1 - e^{-z/m} - (z/m)e^{-z/m}.$$

Differentiation with respect to z leads to

$$f_Z(z) = \frac{z}{m^2}e^{-z/m} \qquad (z \geq 0).$$

The problem which we have discussed here is often encountered in practice. Here is an example: If two electronic components have lifetimes that are independent and exponentially distributed, then the distribution of the sum of their lifetimes can be determined by means of the formulae given above. □

***Example 7. Convolution of Exponential Distributions (continued)**

The problem discussed above can be generalized by adding several rv's with the same distribution. Using induction, it can be proved that the sum $Z = X_1 + \cdots + X_n$ of n independent rv's which are Exp(m) has density function

$$f_Z(z) = \frac{1}{m^n(n-1)!}z^{n-1}e^{-z/m} \qquad (z \geq 0).$$

This means that Z has a gamma distribution with $p = n$, $a = m$; see §3.7. □

Example 8. Convolution of Uniform Distributions

Let X and Y be independent and uniformly distributed over the interval $(0, a)$. The density functions are given by

$$f_X(x) = 1/a \qquad (0 \leq x \leq a),$$

$$f_Y(y) = 1/a \qquad (0 \leq y \leq a),$$

and zero otherwise. We seek the density function of $Z = X + Y$.

First solution:

Let us use (11). The integrand is $1/a^2$ if $0 \leq x \leq a$, $0 \leq z - x \leq a$ and zero otherwise. This is illustrated in Fig. 5.4, where the shaded part shows where the integral is $1/a^2$.

As shown by the figure, there are two cases depending on the value of z:

(a) $0 \leq z \leq a$

We obtain from (11)

$$f_Z(z) = \int_0^z (1/a)^2\,dx = z/a^2.$$

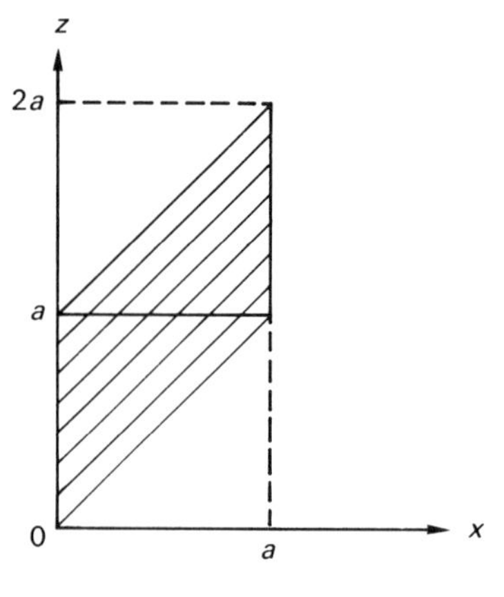

Fig. 5.4.

(b) $a \leq z \leq 2a$

We find

$$f_Z(z) = \int_{z-a}^{a} (1/a)^2 \, dx = (2a - z)/a^2.$$

The reader is recommended to verify that the integral over $f_Z(z)$ from 0 to $2a$ is 1 as it should be.

Second solution:

Let us use (9). We find, in the two cases described above, the following integrals:

$$F_Z(z) = \iint_A (1/a)^2 \, dx \, dy,$$

$$F_Z(z) = \iint_B (1/a)^2 \, dx \, dy.$$

Here A and B denote the shaded areas in Fig. 5.5(a) and (b), respectively. Evaluating the double integrals we have

$$F_Z(z) = \begin{cases} \dfrac{1}{2}\left(\dfrac{z}{a}\right)^2 & \text{if} \quad 0 \leq z \leq a, \\[2ex] 1 - \dfrac{1}{2}\left(\dfrac{2a - z}{a}\right)^2 & \text{if} \quad a < z \leq 2a. \end{cases}$$

A differentiation gives $f_Z(z)$. □

It is clear how, in principle, the distribution of the sum of more than two rv's can be evaluated. First, two variables are added, then one more is added,

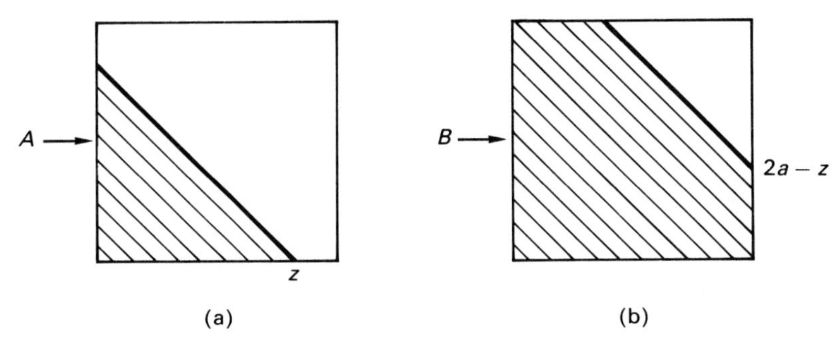

Fig. 5.5.

and so on. It is also realized that, by a small modification, the *difference* $X - Y$ between two rv's can be studied. Then one begins by determining the distribution of the rv $-Y$ (see the end of Example 2, p. 81) and then performs a convolution with the distribution of X.

5.4. Largest Value and Smallest Value

Problems often arise leading to the examination of the largest or the smallest of two or more rv's. We first discuss the case of two independent rv's X and Y.

(a) Larger of Two Values

Set $Z = \max(X, Y)$. Using the obvious fact that $Z \le z$ if and only if both $X \le z$ and $Y \le z$, we see that

$$F_Z(z) = P(Z \le z) = P(X \le z \text{ and } Y \le z) = F_X(z)F_Y(z). \tag{12}$$

(b) Smaller of Two Values

Set $Z = \min(X, Y)$. Since $Z > z$ if and only if both $X > z$ and $Y > z$, we get

$$\begin{aligned} F_Z(z) &= P(Z \le z) = 1 - P(Z > z) = 1 - P(X > z \text{ and } Y > z) \\ &= 1 - P(X > z)P(Y > z). \end{aligned}$$

But $P(X > z) = 1 - P(X \le z) = 1 - F_X(z)$ and analogously for Y. Hence we find that

$$F_Z(z) = 1 - [1 - F_X(z)][1 - F_Y(z)]. \tag{13}$$

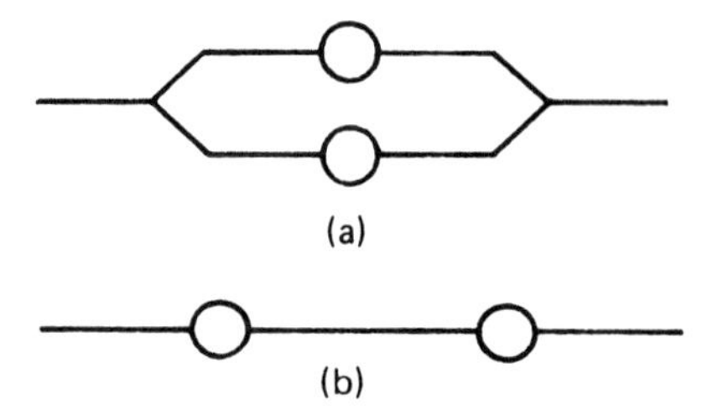

Fig. 5.6. (a) Parallel system. (b) Series system.

Example 9. Lifetime of an Electronic Instrument

An electronic instrument contains two identical components. Their lifetimes are independent and Exp(m). How long does the instrument function if it is switched on with new components and these are organized as, (a) a parallel system, and (b) a series system?

Let Z be the lifetime of the instrument.

(a) Parallel system

The instrument is functioning as long as at least one of the components is functioning. We then have $Z = \max(X, Y)$, and formula (12) shows that

$$F_Z(z) = (1 - e^{-z/m})^2 \qquad (z \geq 0).$$

The density function is obtained by a differentiation:

$$f_Z(z) = \frac{2}{m} e^{-z/m}(1 - e^{-z/m}) \qquad (z \geq 0).$$

(b) Series system

The instrument is functioning as long as both components are functioning. We then have $Z = \min(X, Y)$ and formula (13) gives

$$F_Z(z) = 1 - [1 - (1 - e^{-z/m})]^2 = 1 - e^{-2z/m} \qquad (z \geq 0).$$

By differentiating this function we see that the density function of the lifetime is given by

$$f_Z(z) = (2/m)e^{-2z/m},$$

that is, we have $Z \sim \text{Exp}(m/2)$. □

Example 10. Tensile Strength

A steel rod of given diameter and length is assumed to break when the tensile stress is X kp/cm^2, where X has a Weibull distribution (see §3.7) with distribution function

$$F_X(x) = \begin{cases} 0 & x < 0, \\ 1 - e^{-x^c/a} & x \geq 0. \end{cases}$$

Here a and c are constants which depend on the diameter, the length, the kind of steel, and other things. A rod of the same diameter but twice as long as the first is assumed to break when the tensile stress is Z kp/cm^2. Determine the distribution of Z. (For certain reasons, the parameter a is defined differently than in §3.7.)

We imagine that the rod is divided into two equally long parts. According to a theory proposed by Weibull, the rod breaks as soon as the weaker of the two parts breaks; furthermore, the parts are assumed to break independently. Let X and Y be the tensile strengths of the two parts. It follows from our assumptions that these rv's are independent, each having the distribution function given above. By the theory we have $Z = \min(X, Y)$. Using formula (13) we find after a simple reduction the distribution function

$$F_Z(z) = \begin{cases} 0 & z < 0, \\ 1 - e^{-2z^c/a} & z \geq 0. \end{cases}$$

The answer is interesting: The rod that is twice as long has a tensile strength which also has a Weibull distribution with the same parameter value c but with the parameter a replaced by $a/2$. By a slight extension of these considerations the following can be shown: According to Weibull's theory, a rod of length L has a tensile strength that has a Weibull distribution with parameters a_0/L and c. (Here a_0 is a new parameter which does not depend on the length of the rod.) □

(c) Largest and Smallest Value Among More Than Two Values

It is straightforward to generalize formulae (12) and (13) to more than two independent rv's. The result becomes particularly simple if we consider n independent rv's $X_1, \ldots, X_n$ with the same distribution function $F(x)$. If Z is the largest value, we have $F_Z(z) = [F(z)]^n$; if Z is the smallest value we have $F_Z(z) = 1 - [1 - F(z)]^n$.

5.5. Ratio of Random Variables

Let X and Y be two independent continuous rv's of which Y assumes only positive values. We wish to determine the distribution of the rv

$$Z = X/Y.$$

It follows that

$$F_Z(z) = P(X/Y \leq z) = P(X \leq Yz),$$

where we have used the assumption that Y is never negative. Further, we obtain

$$P(X \leq Yz) = \iint\limits_{x \leq yz} f_{X,Y}(x, y)\, dx\, dy = \iint\limits_{x \leq yz} f_X(x) f_Y(y)\, dx\, dy$$

$$= \int_0^\infty f_Y(y) \left[\int_{-\infty}^{yz} f_X(x)\, dx \right] dy = \int_0^\infty F_X(yz) f_Y(y)\, dy.$$

Hence the distribution function is

$$F_Z(z) = \int_0^\infty F_X(yz) f_Y(y)\, dy. \tag{14}$$

Whenever differentiation under the integral sign is permissible, we get the density function

$$f_Z(z) = \int_0^\infty y f_X(yz) f_Y(y)\, dy. \tag{15}$$

Example 11. Ratio of Exponentially Distributed rv's

Let X and Y be independent and Exp(m) so that

$$f_X(x) = \begin{cases} \dfrac{1}{m} e^{-x/m} & \text{if } x \geq 0, \\ 0 & \text{if } x < 0, \end{cases}$$

and analogously for Y. By (15) the quotient $Z = X/Y$ has for $z \geq 0$ the density function

$$f_Z(z) = \int_0^\infty y \frac{1}{m} e^{-yz/m} \frac{1}{m} e^{-y/m}\, dy = \int_0^\infty m^{-2} y e^{-y(1+z)/m}\, dy.$$

Substituting $y(1 + z) = t$ we get the answer

$$f_Z(z) = \begin{cases} \dfrac{1}{(1+z)^2} & \text{if } z \geq 0, \\ 0 & \text{if } z < 0. \end{cases}$$

□

Exercises

501. A person throws a die. If the outcome is X, he receives the amount $2X$. Determine the probability function of the amount Y he gets after one throw. (§5.2)

502. The rv X assumes the values $0, 1, \ldots, 9$ with equal probabilities $1/10$. Find the probability function of $Y = |X - 3|$. (§5.2)

503. The rv X is uniformly distributed over the interval $(-1, 1)$. Find the distribution of $Y = (X + 1)/2$. (§5.2)

504. Let the rv X have density function $f_X(x) = \dfrac{2}{\pi} \cdot \dfrac{1}{1 + x^2}$ if $x > 0$. Determine the density function of $Y = 1/X$. (§5.2)

505. In an investigation of lifetimes for a certain type of electronic component the components are, for practical reasons, inspected only once a day. If X denotes the actual lifetime of a unit, then $Y = [X]$ is the registered lifetime, where $[X]$ is the integer part of X. Suppose that X has an exponential distribution with density function

$$f_X(x) = \frac{1}{a} e^{-x/a} \quad \text{if} \quad x \geq 0.$$

Determine the distribution of Y. (§5.2)

506. An ant starts a walk from a certain point.
   (a) The ant moves distance X to the east, then distance Y north and distance Z west. The rv's X, Y and Z are independent and each assumes the value 0 with probability 1/2 and the value 1 with probability 1/2. After this walk the ant is at a point distant V from the starting-point. Find the probability function of V.
   (b) Assume that the ant also goes south a distance U, where U has the same distribution as the other rv's and is independent of them. What is the probability function of V now? (§5.3)

507. The rv's X and Y are independent and have the following probability functions:

i	0	1	2
$p_X(i)$	1/6	1/3	1/2

j	0	1	2
$p_Y(j)$	1/2	1/3	1/6

   (a) Which values can $Z = X + Y$ assume?
   (b) Determine, for each possible value k of Z, the number of terms in the convolution formula for determining $p_Z(k)$.
   (c) Find the probability function $p_Z(k)$. (§5.3)

508. The rv's X and Y are independent and have probability functions $p_X(i) = pq^i$ $(i = 0, 1, 2, \ldots)$ and $p_Y(j) = pq^j$ $(j = 0, 1, 2, \ldots)$, respectively. Find the probability function of $X + Y$. (§5.3)

509. A and B play a certain game five times. Their chances of winning a game are the same. After each game the winner receives one point. Let X and Y be the sums of the points won by A and B, respectively, after the five matches. Find the probability function of:
   (a) X;
   (b) $X - Y$;
   (c) the difference Z between the sum of the points won by the winner and that won by the loser. (§5.3)

510. The rv's X and Y are independent and have probability functions $p_X(j) = 1/n$ for $j = 1, 2, \ldots, n$ and $p_Y(k) = 1/n$ for $k = 1, 2, \ldots, n$. Find the probability function of the rv $|X - Y|$. (§5.3)

511. The rv's X and Y are independent and have density functions

$$f_X(x) = 2e^{-2x} \quad \text{if} \quad x \geq 0,$$

$$f_Y(y) = 3e^{-3y} \quad \text{if} \quad y \geq 0.$$

Determine the density function of $X + Y$.
Hint: When formula (11) is applied one obtains $f_Z(z) = \int_0^z f_X(x) f_Y(z - x)\, dx$ $(z \geq 0)$. (Note the area of integration.) Why is this so? (§5.3)

512. The rv's X and Y are independent and have density functions

$$f_X(x) = \frac{1}{\pi} \cdot \frac{1}{1 + x^2} \quad \text{if} \quad -\infty < x < \infty,$$

$$f_Y(y) = \frac{1}{2} \qquad \text{if} \quad -1 \le y \le 1.$$

Find the distribution of $X + Y$. (§5.3)

513. The rv's X and Y are independent and have density functions

$$f_X(x) = e^{-x} \qquad \text{if} \quad x \ge 0,$$

$$f_Y(y) = 1 \qquad \text{if} \quad 0 \le y \le 1.$$

Determine the density function of $X + Y$.
Hint: Distinguish between the cases $0 \le z \le 1$ and $z > 1$ when the convolution formula is applied. (§5.3)

514. The rv's X and Y are independent and uniformly distributed over the interval $(0, a)$. Find the distribution functions of Z_+ and of Z_-, where

$$Z_+ = \max(X, Y); \qquad Z_- = \min(X, Y). \quad (\text{§}5.4)$$

515. The rv's $X_1, X_2, \ldots, X_n$ are independent and have density functions

$$f_{X_i}(x) = \lambda_i e^{-\lambda_i x} \qquad \text{if} \quad x \ge 0.$$

What is the density function of $Z = \min(X_1, X_2, \ldots, X_n)$? (§5.4)

516. The time in minutes that it takes a cross-country runner to get round a certain track is a rv X with density function

$$f_X(x) = (125 - x)/450 \qquad \text{if} \quad 95 \le x \le 125.$$

What is the probability that, of eight different runners who run independently, after 100 minutes
(a) all have reached the goal;
(b) none has reached the goal? (§5.4)

517. The rv's X and Y are independent and have density functions $f_X(x) = e^{-x}$ $(x \ge 0)$ and $f_Y(y) = ye^{-y}$ $(y \ge 0)$. Find the density function of $Z = X/Y$. (§5.5)

*518. The rv's X and Y are independent and

$$f_X(x) = \frac{1}{a} e^{-x/a} \qquad \text{if} \quad x \ge 0,$$

$$p_Y(k) = pq^{k-1} \qquad \text{if} \quad k = 1, 2, 3, \ldots.$$

Determine the distribution function and the density function of the rv $Z = X/Y$.

*519. A person starts swimming in a dense mist from a straight shore line. He first swims 500 meters in a random direction. Then he stops and swims again in another random direction. What is the probability that he reaches the shore before he has swum a further distance of 500 meters?

*520. According to the time-table a train is to arrive at a railway station at 12:00, and to leave the station at 12:07. The delay in minutes, X, at arrival is uniformly distributed over the interval $(-2, 3)$. The stopping-time Y at the station is independent of X and is uniformly distributed over the interval $(3, 5)$. (Note that

the train never leaves the station before 12:07.) Find the distribution function of the delay with which the train leaves the station.
Hint: The rv (X, Y) varies over a region shaped like a rectangle. Draw a picture in this rectangle!

*521. The rv's X and Y are independent with the same normal distribution:

$$f_X(x) = \frac{1}{\sqrt{2\pi}} e^{-x^2/2} \quad (-\infty < x < \infty); \qquad f_Y(y) = \frac{1}{\sqrt{2\pi}} e^{-y^2/2} \quad (-\infty < y < \infty).$$

Prove that the quotient $Z = X/Y$ has density function

$$f_Z(z) = \frac{1}{\pi} \cdot \frac{1}{1 + z^2}, \qquad -\infty < z < \infty.$$

CHAPTER 6

Expectations

6.1. Introduction

The sections in this chapter have rather different contents but are bound together by a common theme: we discuss various aspects of the concept of expectation.

In §6.2, some fundamental notions are introduced and some important theorems are proved. In §6.3, we discuss expectations suitable as measures of location and measures of dispersion. In §6.4, measures of dependence of two rv's are presented.

6.2. Definition and Simple Properties

(a) Definition

We first take an example intended to explain the use of expectations.

Example 1. Throwing a Die

Let us return to an earlier example (Example 1 in §3.2). The situation is the following: A well-made die is thrown once. A person gets 1 dollar for "one", 2 dollars for "two" or "three", and 4 dollars for "four", "five" or "six". The amount he receives is a rv X that takes the values 1, 2 and 4.

As the expectation of X we take the expression

$$1 \cdot \frac{1}{6} + 2 \cdot \frac{1}{3} + 4 \cdot \frac{1}{2} = \frac{17}{6}.$$

Here each outcome of X has been multiplied by the corresponding probability, and then the products have been added.

To understand why we do so, we imagine that the following experiment is performed. Assume that the person throws the die many times, say 6,000 times. The three possible outcomes, 1, 2 and 4, will then, according to the frequency interpretation of a probability (see p. 11), occur about 1,000, 2,000 and 3,000 times, respectively. So the total amount that the person receives will be, approximately,

$$1 \cdot 1{,}000 + 2 \cdot 2{,}000 + 4 \cdot 3{,}000 = 17{,}000,$$

that is, about

$$\frac{1}{6{,}000}(1 \cdot 1{,}000 + 2 \cdot 2{,}000 + 4 \cdot 3{,}000) = \frac{17{,}000}{6{,}000}$$

in each throw. If we reduce the fractions, we obtain the expression given above as the definition of the expectation. Naturally, we might take some number other than 6,000, in fact, any number which is large enough.

We see that the term expectation is somewhat misleading. The person cannot expect to receive exactly 17/6 in any throw; but, as we have just explained, he may expect to obtain this on the average, if many throws are made. □

We are now prepared for the following general definition.

Definition. The *expectation* of the rv X is defined by

$$E(X) = \begin{cases} \sum_k k p_X(k) & \text{(discrete rv)}, \\ \int_{-\infty}^{\infty} x f_X(x)\,dx & \text{(continuous rv)}. \end{cases} \tag{1}$$

□

We assume that the sum is absolutely convergent. Instead of expectation we often use the term *mean*. The symbol $E(X)$ is often replaced by m, m_X or some other symbol.

Using the mechanical analogy mentioned earlier, we may regard the expectation as the centre of gravity of the mass distributed along the x-axis. Hence the expectation is a measure of location which shows where the mass is situated on the average.

When the two versions in (1) are compared, one should remember the relationship between discrete and continuous distributions. When a continuous distribution is approximated by a discrete one in the way described in §3.8, a sum $\sum k p_X(k)$ is obtained which is an approximation of the integral. Compare the definition of an integral as a limit of a sum.

Example 2. fft-Distribution

If X has an fft-distribution (see §3.5) we have $p_X(k) = q^{k-1}p$ $(k = 1, 2, \ldots)$, and hence (1) gives

$$E(X) = \sum_{k=1}^{\infty} k q^{k-1} p = p\frac{d}{dq}\left(\sum_{k=1}^{\infty} q^k\right) = p\frac{d}{dq}\left(\frac{q}{1-q}\right) = p\frac{1}{(1-q)^2} = \frac{1}{p}.$$

This is a result of practical importance. If a person is exposed to a risk of $p = 0.01$ several times, it takes on the average $1/0.01 = 100$ times until an accident occurs. □

Example 3. Uniform Distribution over (a, b)

If X is uniformly distributed over the interval (a, b), that is, if $f_X(x) = 1/(b - a)$ for $a \le x \le b$ and 0 otherwise, we get

$$E(X) = \int_a^b x \frac{1}{b-a}\, dx = \left[\frac{1}{b-a}\cdot\frac{x^2}{2}\right]_a^b = \frac{1}{2}\cdot\frac{b^2 - a^2}{b-a} = \frac{a+b}{2}.$$

Hence the mean lies at the mid-point of the interval. □

Example 4. Exponential Distribution

If $X \sim \text{Exp}(m)$, that is, if $f_X(x) = (1/m)e^{-x/m}$ $(x \ge 0)$, we find by a partial integration

$$E(X) = \int_0^\infty \frac{x}{m} e^{-x/m}\, dx = [xe^{-x/m}]_0^\infty + \int_0^\infty e^{-x/m}\, dx = [-me^{-x/m}]_0^\infty = m. \qquad \square$$

***Example 5. Cauchy Distribution**

If
$$f_X(x) = \frac{a}{\pi}\cdot\frac{1}{a^2 + x^2} \qquad (-\infty < x < \infty),$$

then X is said to have a Cauchy distribution. In the expression

$$\int_{-\infty}^{\infty} x f_X(x)\, dx = \int_{-\infty}^{0} x\frac{a}{\pi}\cdot\frac{1}{a^2 + x^2}\, dx + \int_0^\infty x\frac{a}{\pi}\cdot\frac{1}{a^2 + x^2}\, dx$$

the integrals on the right-hand side are divergent. Hence the mean does not exist. □

Remark. Alternative Determination of the Mean

In the first formula (1) we used the probability function when computing the mean of the discrete rv X. When X is defined on a discrete sample space with outcomes u_i $(i = 1, 2, \ldots)$ and the probabilities $P(u_i)$ are known, it is often more convenient to use the expression

$$E(X) = \sum_i X(u_i)P(u_i). \tag{2}$$

The probability function for X is then not needed.

That (2) gives the same result as the first formula (1) in this case is quickly realized: We get (1) from (2) by combining in (2) all u_i with the same value of X and adding the corresponding probabilities. □

Example 1. Throwing a Die (continued)

Formula (2) gives a sum of six terms corresponding to the six outcomes:

$$E(X) = 1\cdot\frac{1}{6} + 2\cdot\frac{1}{6} + 2\cdot\frac{1}{6} + 4\cdot\frac{1}{6} + 4\cdot\frac{1}{6} + 4\cdot\frac{1}{6} = 17/6,$$

hence the same value as before. □

(b) The Expectation of a Function of a rv

It is often interesting to study the expectation of a function

$$Y = g(X)$$

of a rv. It is, of course, possible to find $E(Y)$ by first deriving the distribution of Y from that of X, by means of the methods in Chapter 5, and then using our first definition of an expectation. However, it is generally easier to apply

Theorem 1. *If $Y = g(X)$ we have*

$$E(Y) = \begin{cases} \sum_k g(k)p_X(k) & (\textit{discrete}\ \text{rv}), \\ \int_{-\infty}^{\infty} g(x)f_X(x)\,dx & (\textit{continuous}\ \text{rv}). \end{cases} \tag{3}$$

□

The message that the theorem conveys is rather natural: For example, in the discrete case the expectation of $g(X)$ is obtained by multiplying for each possible value k of X the expression $g(k)$ by the corresponding probability and adding the products.

PROOF. A general proof is difficult (see, for example, B.V. Gnedenko, *Theory of Probability*, Chelsea, New York, 1962), and we restrict ourselves to the discrete case. We reorder the terms on the right-hand side of the first expression in (3) and obtain, by combining all terms with the same value of $g(k)$,

$$\begin{aligned} \sum_k \left[\sum_{g(j)=k} g(j)p_X(j) \right] &= \sum_k k \left[\sum_{g(j)=k} p_X(j) \right] \\ &= \sum_k k \left[\sum_{g(j)=k} P(X=j) \right] \\ &= \sum_k kP[g(X)=k] \\ &= \sum_k kp_Y(k) = E(Y). \end{aligned}$$

(That this reordering is allowed is due to the absolute convergence.) This proves the first expression (3). □

Remark.

Theorem 1 shows that E is a linear operator: It is easy to see that

$$E[g(X) + h(X)] = E[g(X)] + E[h(X)].$$

□

Theorem 1 has many important theoretical and practical applications. Here is a simple but important theoretical application: If Y is a linear function $Y = aX + b$, it follows that $E(aX + b) = aE(X) + b$. This confirms that the mean is a measure of location: If a constant b is added to the rv, then b is added to the mean.

Example 6. Lottery

In a certain Swedish state lottery the number of tickets sold is 1,310,000 at a cost of 20 Swedish Crowns each. The winning tickets are made up as follows: 1, 2, 6, 77, 166, 875, 1,296, 11,790 and 146,720 tickets with the winnings 500,000, 100,000, 50,000, 10,000, 5,000, 1,000, 250, 100 and 50 Swedish Crowns, respectively. There are also 26,200 prizes consisting of a new ticket. A person buys a ticket. Find his chance of obtaining a winning ticket and the expectation of his net gain.

The probability of obtaining a winning ticket is $(1 + 2 + 6 + \cdots + 146{,}720)/1{,}310{,}000 = 0.123$, or approximately 1 in 8; if the possibility of winning a new ticket is included, the probability increases to 1 in 7.

Let X be the winnings and $Y = X - 20$ the net gain. We find

$$E(X) = 500{,}000 \cdot \frac{1}{1{,}310{,}000} + 100{,}000 \cdot \frac{2}{1{,}310{,}000} + \cdots$$

$$+ 50 \cdot \frac{146{,}720}{1{,}310{,}000} + 20 \cdot \frac{26{,}200}{1{,}310{,}000} = 9.80.$$

Here the prize of a new ticket has been estimated as 20 Swedish Crowns. (Other estimates are possible.) The net gain is, on the average,

$$E(Y) = E(X - 20) = E(X) - 20 = 9.80 - 20.00 = -10.20.$$

Hence one suffers a loss, on the average. This is so in all lotteries! □

Example 7. Average Gain in a Factory

A factory produces certain parts of a prescribed length a. However, the buyer tolerates a length in the interval $(a - 1, a + 1)$; such parts yield the factory a gain of 1 dollar each. Shorter parts are useless, which causes a loss of 4 dollars. Parts longer than $a + 1$ must be cut off, which causes waste of material and reduces the gain to 0 dollars. The length X of a randomly chosen unit has, approximately, the density function

$$f_X(x) = e^{-2|x-a|} \qquad (-\infty < x < \infty).$$

(That this function can only be approximately true follows, among other things, from the fact that it is positive even if $x < 0$, which is, of course, impossible in practice; however, if e^{-2a} is small, the probability of a negative length is so small that it is negligible.)

Let Y be the gain from a unit. We then obtain

$$Y = \begin{cases} -4 & \text{if } \; X < a - 1, \\ 1 & \text{if } \; a - 1 \le X \le a + 1, \\ 0 & \text{if } \; X > a + 1. \end{cases}$$

Hence Y is a function of X. The average gain, that is, the expectation of Y, is then, according to the second expression (3),

$$E(Y) = \int_{-\infty}^{a-1} (-4)e^{-2|x-a|}\,dx + \int_{a-1}^{a+1} 1 \cdot e^{-2|x-a|}\,dx + \int_{a+1}^{\infty} 0 \cdot e^{-2|x-a|}\,dx.$$

Substituting $y = x - a$ and splitting the second integral into two parts, we obtain

$$E(Y) = -4\int_{-\infty}^{-1} e^{2y}\,dy + \int_{-1}^{0} e^{2y}\,dy + \int_{0}^{1} e^{-2y}\,dy$$

$$= -2e^{-2} + \frac{1}{2}(1 - e^{-2}) + \frac{1}{2}(1 - e^{-2})$$

$$= 1 - 3e^{-2} \approx 0.59. \qquad \square$$

(c) The Expectation of a Function of Several rv's

Theorem 1 can be extended to a function of several rv's. Let us restrict ourselves to a function of two rv's X and Y:

Theorem 1′. *If* $Z = g(X, Y)$ *then*

$$E(Z) = \begin{cases} \sum_{j,k} g(j, k)p_{X,Y}(j, k) & (\textit{discrete}\ \text{rv}), \\ \int_{-\infty}^{\infty}\int_{-\infty}^{\infty} g(x, y)f_{X,Y}(x, y)\,dx\,dy & (\textit{continuous}\ \text{rv}). \end{cases} \tag{4}$$

$\square$

Important applications of this theorem will be given in §6.4.

6.3. Measures of Location and Dispersion

In order to give a complete description of a rv we use the distribution function or the probability function/density function. When such detailed knowledge is not required, we may use other measures which describe some property of the distribution. Of particular value are measures of location and dispersion.

(a) Measures of Location

It has already been mentioned that the expectation $E(X)$ is a *measure of location* that shows where the mass is situated "on the average".

Other quantities may be used as measures of location. An often used measure is the *median* $x_{0.50}$ (see §3.6) which can also be denoted by $\tilde{m}$ (see Fig. 6.1). We use it, as in Chapter 3, only for continuous rv's.

The median is not always uniquely determined (see Fig. 6.2). When the distribution is symmetric, the median and the expectation coincide (if the median is unique and the expectation exists).

If the distribution is skew, the median and the expectation can be rather different. If the distribution has a long tail to the right as in Fig. 6.1, the median

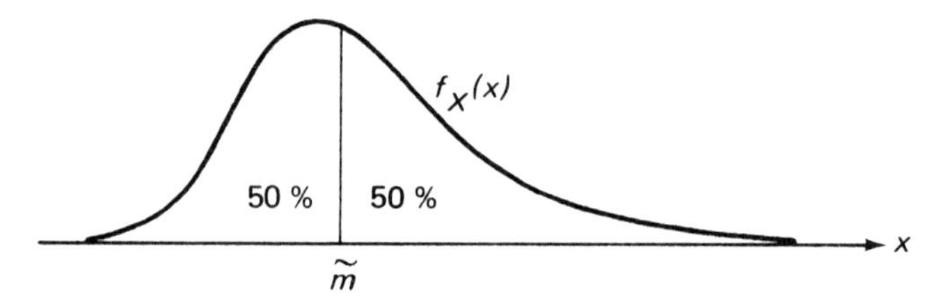

Fig. 6.1. Median.

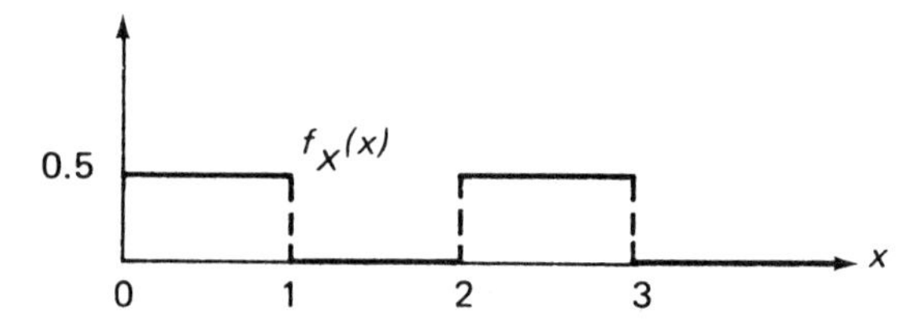

Fig. 6.2. Distribution without a unique median. (All values between 1 and 2 can, with equal right, be called medians.)

is smaller than the expectation. Note that the median does not change if, for example, in the 50% part to its right, the probability mass is "moved around" so that the tail increases in length (or becomes shorter in length). However, the expectation is affected by such changes.

(b) Measures of Dispersion

It is in general not enough to know only a measure of location of a rv. Two rv's may very well have exactly the same expectation but rather different distributions. For example, if a rv with a two-point distribution assumes the values -1 and 1 with the same probability, and another rv the values -5 and 5 with the same probability, it is obvious that both variables have expectation 0 but very different dispersions (see Fig. 6.3). Evidently, the distribution on the right in the figure is more dispersed than that on the left.

This simple example shows the need for a *measure of dispersion* of a rv. There are several such measures, the most common being the variance and

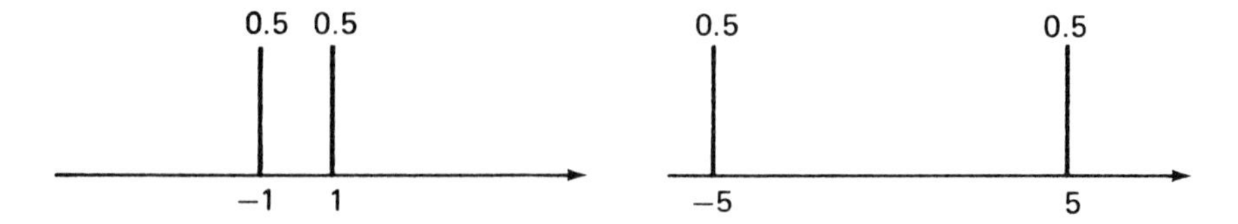

Fig. 6.3. Two-point distributions with the same expectation but different dispersions.

the standard deviation; a third one is the coefficient of variation. The three measures are closely related.

Definition. The *variance* $V(X)$ of the rv X is defined by

$$V(X) = E[(X - m)^2]. \qquad \square$$

(Instead of $V(X)$ it is common to write var(X).) Hence the variance is the expectation of the rv $Y = (X - m)^2$, where $m = E(X)$. According to Theorem 1 we have

$$V(X) = \begin{cases} \sum_k (k - m)^2 p_X(k) & \text{(discrete rv),} \\ \int_{-\infty}^{\infty} (x - m)^2 f_X(x)\, dx & \text{(continuous rv).} \end{cases} \tag{5}$$

From these expressions it emerges that the definition is sensible: If the distribution is concentrated around m, then $(X - m)^2$ assumes small values with large probability, and the variance is small. In particular, if the whole probability mass is concentrated at one single point, the variance is 0.

If the sum or the integral in (5) does not converge, we say that $V(X)$ does not exist.

Definition. The *standard deviation* $D(X)$ of the rv X is the square root of the variance, that is,

$$D(X) = \sqrt{V(X)}. \qquad \square$$

It is natural to take the square root because then $D(X)$ has the same dimension as the rv itself. For instance, if X is given in cm, then $D(X)$ also has the dimension cm, while $V(X)$ is of dimension cm^2.

Instead of using the symbol $D(X)$ for the standard deviation we often use the letter σ (sigma), that is,

$$D(X) = \sigma,$$

whence

$$V(X) = \sigma^2.$$

Definition. The ratio

$$R(X) = D(X)/E(X)$$

of the standard deviation to the expectation is called the *coefficient of variation.* $\square$

The coefficient of variation is used only when X assumes positive values. It is often expressed as a percentage. (But do not assume that it is always less than 100%. It can occasionally be larger.)

There are several reasons for using the coefficient of variation, two of which will be mentioned here. First, it does not depend on the unit in which X is expressed; that is, it is dimensionless. Second, we frequently observe, for example, when objects of different magnitudes are measured, that the errors of measurement increase with the size of the object in such a way that the standard deviation is roughly proportional to the size. When this is the case, the coefficient of variation is approximately constant, that is, independent of the size of the object.

When $V(X)$ is evaluated, it is often convenient to use, instead of the expressions (5), the following formula:

Theorem 2.

$$V(X) = E(X^2) - [E(X)]^2. \tag{6}$$

□

PROOF.

$$V(X) = E[(X - m)^2] = E[X^2 - 2mX + m^2] = E(X^2) - 2mE(X) + m^2$$
$$= E(X^2) - 2[E(X)]^2 + [E(X)]^2 = E(X^2) - [E(X)]^2.$$

□

Example 1. Throwing a Die (continued)

We know already that $E(X) = 17/6$ (see p. 96). Further, we obtain

$$E(X^2) = 1^2 \cdot \frac{1}{6} + 2^2 \cdot \frac{1}{3} + 4^2 \cdot \frac{1}{2} = 19/2.$$

Hence, according to (6),

$$V(X) = \frac{19}{2} - \left(\frac{17}{6}\right)^2 = 53/36; \qquad D(X) = \sqrt{53/36}.$$

In this simple case it is almost as easy to use the first expression (5) directly:

$$V(X) = \left(1 - \frac{17}{6}\right)^2 \cdot \frac{1}{6} + \left(2 - \frac{17}{6}\right)^2 \cdot \frac{1}{3} + \left(4 - \frac{17}{6}\right)^2 \cdot \frac{1}{2} = 53/36.$$

□

Example 3. Uniform Distribution over (a, b) (continued)

We know that $E(X) = (a + b)/2$ (see p. 98). Moreover, we find

$$E(X^2) = \int_{-\infty}^{\infty} x^2 f_X(x)\, dx = \int_a^b x^2 \frac{1}{b - a}\, dx = \frac{1}{3} \cdot \frac{b^3 - a^3}{b - a}.$$

By (6) and a rearrangement we obtain

$$V(X) = \frac{1}{3} \cdot \frac{b^3 - a^3}{b - a} - \left(\frac{a + b}{2}\right)^2 = \frac{(b - a)^2}{12},$$

$$D(X) = \frac{b - a}{\sqrt{12}}.$$

In particular, if X has a uniform distribution over the interval (0, 1), we have $a = 0$ and $b = 1$. The expectation is then 1/2 and the standard deviation $1/\sqrt{12}$. □

Example 4. Exponential Distribution (continued)

It is shown by two partial integrations that

$$E(X^2) = \int_0^\infty x^2 \frac{1}{m} e^{-x/m}\, dx = [-x^2 e^{-x/m}]_0^\infty + \int_0^\infty 2xe^{-x/m}\, dx$$

$$= [-2mxe^{-x/m}]_0^\infty + 2m \int_0^\infty e^{-x/m}\, dx = 2m^2.$$

On p. 98 we showed that $E(X) = m$. Hence we obtain by (6)

$$V(X) = 2m^2 - m^2 = m^2; \qquad D(X) = m.$$

So we have proved for an exponential distribution, that the standard deviation is equal to the mean; as a consequence, the coefficient of variation $R(X)$ is 1 or, if we prefer, 100%. □

The following rules of computation are very useful:

Theorem 3.

$$E(aX + b) = aE(X) + b,$$

$$V(aX + b) = a^2 V(X), \qquad (7)$$

$$D(aX + b) = |a| D(X).$$

□

Proof. The first rule was mentioned already on p. 99. The second rule is demonstrated as follows: If we put $E(X) = m$, the rv $aX + b$ has (by the first rule) expectation $am + b$. By the definition of a variance we see that $aX + b$ has variance

$$E[(aX + b - am - b)^2] = E[(aX - am)^2] = a^2 E[(X - m)^2] = a^2 V(X)$$

and the second rule is proved. The third rule is obtained from the second by taking the square root. □

Let us emphasize the following important consequences of the theorem. If a rv X is increased (decreased) by the addition (subtraction) of a constant b, then $E(X)$ also increases (decreases) by the same constant b, while $D(X)$ and $V(X)$ are *not* changed. See Fig. 6.4, which shows how such a change moves the distribution to the right or to the left, but does not affect its form; because of this, the measures of dispersion do not change. If a rv X is multiplied by a positive constant a, then $E(X)$ and $D(X)$ are multiplied by the same constant a, while $V(X)$ is multiplied by a^2.

We shall introduce a definition which will prove useful later:

Definition. If X is a rv with expectation m and standard deviation σ, we call $Y = (X - m)/\sigma$ a *standardized* rv. □

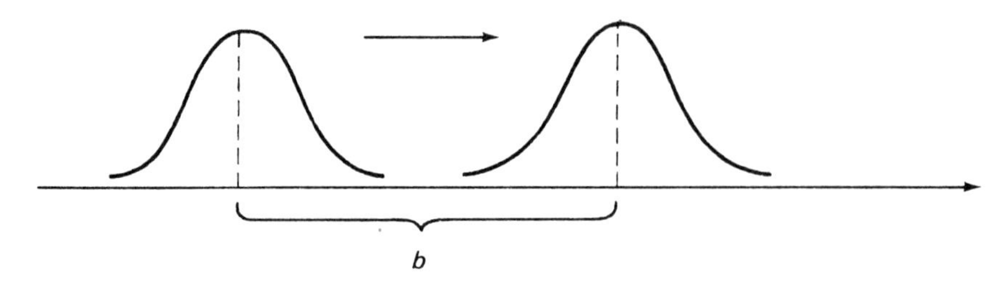

Fig. 6.4. If a rv X increases by a constant b, then $E(X)$ also increases with b, but $D(X)$ and $V(X)$ are left unchanged.

We shall determine the mean and the variance of Y. Theorem 3 shows that

$$E(Y) = E[(X - m)/\sigma] = \frac{1}{\sigma}E(X) - \frac{m}{\sigma} = 0,$$

$$V(Y) = \frac{1}{\sigma^2}V(X) = 1.$$

Hence a standardized rv has mean 0 and variance 1.

The terms mean and standard deviation of a rv make it possible to describe clearly what a systematic error is and what a random error is.

A person measures an object repeatedly and then records the result to so many digits that the result varies from measurement to measurement. If the instrument used by the person is correctly calibrated (and the person has no tendency to misread the instrument in a certain direction), no systematic error arises, only an error which we call random. The difference between measured value and true value is then regarded as a rv X with mean 0 and standard deviation σ. The rv X usually has a symmetric distribution, but this is not essential for the discussion (see Fig. 6.5a).

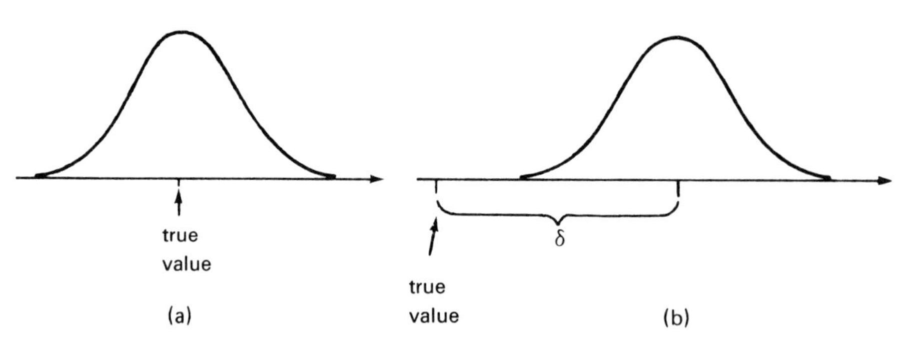

Fig. 6.5. Systematic error and random error. (a) Only random error. (b) Both systematic and random errors.

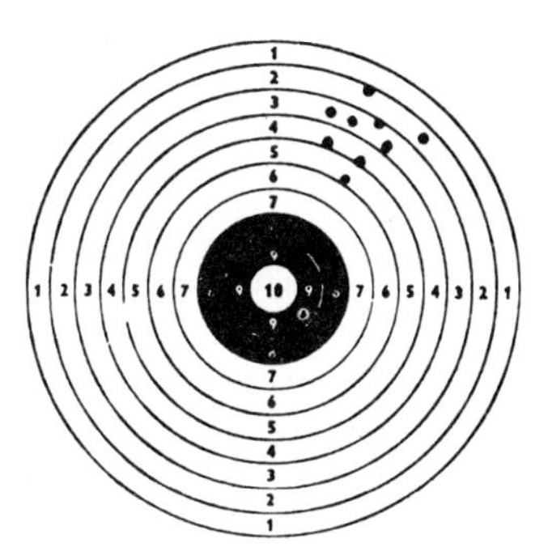

Fig. 6.6. Volley of shots with good precision but bad accuracy.

If the instrument is not correctly calibrated but reads consistently too high or too low, both systematic and random errors arise. If the systematic error is constant and equal to δ in the interval where the measurements lie, the difference between measured value and true value is a rv X with mean δ and standard deviation σ (see Fig. 6.5b).

We now introduce the following:

Definition. *Systematic error* = the difference between the mean of the measured value and the true value. *Random error* = the difference between the measured value and its mean. □

It is seen from this definition that the systematic error is a number, while the random error is a rv with mean 0. As a measure of the magnitude of the random error it is appropriate to use the standard deviation σ.

It is important to distinguish between *accuracy* and *precision*. By high accuracy we mean good agreement between measured value and true value; by high precision we mean small random error. (Example: A volley of shots which is well concentrated but hits far from the centre of the target has good precision but bad accuracy; see Fig. 6.6.)

Remark 1. About Fair Games and Related Concepts

Let A and B play the following game. A receives a dollars from B if a certain event H occurs, while A pays b dollars to B if H does not occur. (What H signifies has no importance; for example, we may have H = "A wins at a game of chess".) Let p be the probability that H occurs and set $q = 1 - p$.

The amount received by A is a rv X with a two-point distribution; it takes the values a and $-b$ with the probabilities p and q, respectively. The expectation is given by

$$E(X) = ap - bq.$$

If $E(X) > 0$, the game is said to be favourable to A; if $E(X) = 0$ it is said to be *fair*; in the latter case, we have

$$\frac{a}{b} = \frac{q}{p}.$$

Sometimes one encounters the term *odds* in this connection. That A's odds for H are 10:1 signifies that a game, in which A gets the amount 1 if H occurs and pays the amount 10 if H does not occur, is a fair game. Hence this means that in the relation given above we can take $a = 1$, $b = 10$, so that $p = 10/11$. (Other definitions of the term odds are sometimes used, but we restrict the discussion to the one just given.)

If the game is played at a casino with A = player, B = casino, H = event which A hopes will occur (for example, "rouge" or "noir"), then a is the amount paid to the player if he wins and b is the stake. The rules are always chosen so that $E(X)$ is negative and hence the play is unfavourable to A.

We return to the discussion of fair games in the Remark in §7.3. □

Remark 2. Subjective Probabilities and Odds

In the Remark in §2.3 we considered subjective probabilities. Such a probability measures an individual's private opinion concerning an uncertain event. It is interesting to ask how a numerical expression for such a probability can be assigned. There has been a lively discussion about this in the literature, a discussion which is still going on, more or less heatedly. The concept of odds (see the preceding Remark) can be used for this purpose. Let H be the event in which we are interested and let us ask the person to state the odds he is willing to take in a game where H is involved in the way described in Remark 1. If he chooses the odds 10:1, say, his subjective probability is taken to be $p = 10/11$. Note that p may change with time; before a horse race it is quite legitimate to modify one's odds in view of information received as the race approaches. □

6.4. Measures of Dependence

How should the dependence of two rv's be measured? The most complete picture is obtained by studying their joint probability function or density function, which furnishes all available information about the variables and hence also about their dependence (see Chapter 4). For example, if two continuous rv's have a density function with contour curves as in Fig. 6.7, it is apparent that there is a dependence between the variables: In Fig. 6.7(a) the variables have a tendency to deviate in the same direction from their means m_X and m_Y; that is, a large value of X has a tendency to be accompanied by a large value of Y, and vice versa. On the other hand, in Fig. 6.7(b), we see that X and Y tend to deviate in the opposite direction from their means.

This description of dependence, by reference to a figure, is rather vague, and we often need instead a measure in the form of a single number. We shall introduce two closely related measures of dependence, namely the covariance and the coefficient of correlation.

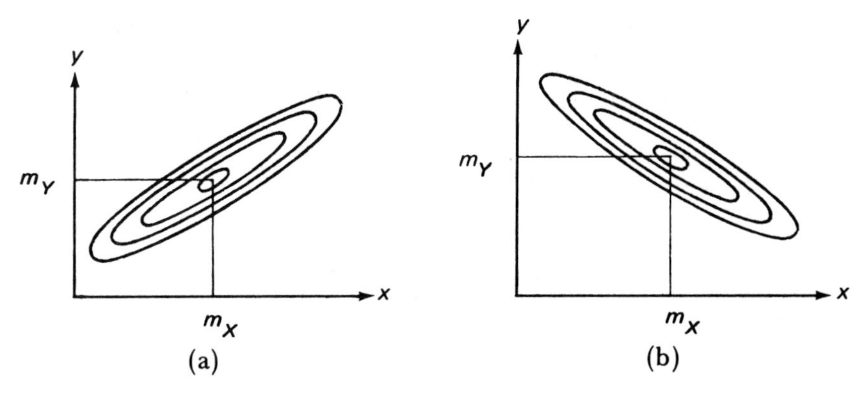

Fig. 6.7. Contour curves.

Definition. The *covariance* of X and Y is defined by

$$C(X, Y) = E[(X - m_X)(Y - m_Y)].$$ □

(Another common symbol for the covariance is cov(X, Y).) Theorem 1′ in §6.2 shows that the covariance can be obtained as follows:

$$C(X, Y) = \begin{cases} \sum_j \sum_k (j - m_X)(k - m_Y)p_{X,Y}(j, k) & \text{(discrete rv)}, \\ \int_{-\infty}^{\infty} \int_{-\infty}^{\infty} (x - m_X)(y - m_Y)f_{X,Y}(x, y)\, dx\, dy & \text{(continuous rv)}. \end{cases} \tag{8}$$

However, it is usually more convenient to use, instead,

Theorem 4.

$$C(X, Y) = E(XY) - E(X)E(Y). \tag{9}$$

□

Hence, to determine the covariance, we need the expectations of the variables and the expectation of their product.

PROOF. The expectation in the definition of the covariance may be written

$$E(XY - Xm_Y - m_X Y + m_X m_Y) = E(XY) - m_X m_Y - m_X m_Y + m_X m_Y,$$

which gives (9). □

One realizes after some contemplation that the covariance is positive if there is a tendency of the variables to deviate from their means in the same direction. For then the product $(X - m_X)(Y - m_Y)$ will more often be positive than negative. On the other hand, if the variables tend to deviate in the opposite direction from the means, the above product will more often be negative than positive. In the examples shown in Fig. 6.7(a) and 6.7(b) we therefore expect that $C(X, Y)$ is positive and negative, respectively.

We frankly admit that this commentary is vague. In order to find the covariance for a certain distribution you have to perform the detailed computation shown earlier!

Variables with covariance 0 have a special name:

Definition. If $C(X, Y) = 0$, the rv's X and Y are said to be *uncorrelated.* □

According to Theorem 4 we know that two rv's are uncorrelated if and only if

$$E(XY) = E(X)E(Y).$$

Now we have an important result:

Theorem 5. *If X and Y are independent, then they are also uncorrelated.* □

The proof of this theorem will be given later (see the beginning of §7.2). The converse statement is *not* true: uncorrelated rv's are not necessarily independent.

We now pass on to another measure of dependence, the coefficient of correlation. In fact, we introduce this measure only after some hesitation, for there is probably no concept in probability theory and statistical theory which has so often been misused and misinterpreted.

Definition. *The coefficient of correlation* of X and Y is defined by

$$\rho(X, Y) = \frac{C(X, Y)}{D(X)D(Y)}. \tag{10}$$

□

The relationship to the covariance is apparent. By dividing the covariance by $D(X)D(Y)$ we obtain a dimensionless quantity. We shall now prove that the coefficient of correlation always lies between -1 and 1. We start from the following obvious inequality:

$$E\left[\frac{X - m_X}{D(X)} \pm \frac{Y - m_Y}{D(Y)}\right]^2 \geq 0.$$

Expanding this squared expression and using the definitions of $\rho(X, Y)$, $V(X)$ and $V(Y)$ we arrive at the inequality

$$\frac{V(X)}{V(X)} + \frac{V(Y)}{V(Y)} \pm 2\rho \geq 0.$$

This is equivalent to

$$-1 \leq \rho \leq 1$$

as stated. The proof provides a further result. It shows that $\rho = \pm 1$ if and only if X and Y are linearly dependent, that is, $Y = aX + b$, where a and b are constants. This implies that, for any given value x of the rv X, the rv Y assumes the value $ax + b$; hence the whole probability mass is situated on a straight line. If $a > 0$, we have $\rho = 1$; if $a < 0$ we have $\rho = -1$.

Theorem 5 shows that if X and Y are independent, then $\rho = 0$.

Remark. Why is the Coefficient of Correlation So Often Misinterpreted?

The reader will, no doubt, want to know why the coefficient of correlation ρ (which mathematically speaking looks rather innocent) is so often misused and misinterpreted. This is a long story which we must make very short. The coefficient of correlation is often used in models designed for examining the connection between phenomena in the empirical world. Let us take the example X = height of the father and Y = intelligence quotient of the child. If we find that ρ for these two rv's is greater than zero, it may be tempting to draw the conclusion that it is the height of father that causes the intelligence of the child. The conclusion is too hasty, for some third factor may lie behind both variables and indirectly cause a connection between them. Let us, from the heap of possible explanations, pick out one (not meant to be taken too seriously): Intelligent women may have a liking for tall men! □

Example 8. Throwing a Die

Let a well-made die be thrown twice, and denote by X and Y the number of "ones" and "twos", respectively. We want to determine the covariance of X and Y.

We begin by computing the joint probability function of X and Y. Clearly, we have $P(X = 0, Y = 0) = P(\text{no throw gives "one" or "two"}) = (1 - 2/6)^2 = 16/36$ and $P(X = 1, Y = 0) = P(\text{"one" in throw 1, no "one" or "two" in throw 2, or vice versa}) = \frac{1}{6}\cdot\frac{4}{6} + \frac{4}{6}\cdot\frac{1}{6} = 8/36$. The other probabilities are found in a similar way, and the resulting probability function becomes (all numbers in the table should be divided by 36):

j \ k	0	1	2	$p_X(j)$
0	16	8	1	25
1	8	2	0	10
2	1	0	0	1
$p_Y(k)$	25	10	1	36

A simple computation shows that $E(X) = E(Y) = 1/3$. Moreover, the expectation of the product of the variables becomes

$$E(XY) = 0\cdot 0\cdot\frac{16}{36} + 0\cdot 1\cdot\frac{8}{36} + \cdots + 2\cdot 2\cdot\frac{0}{36} = \frac{2}{36} = \frac{1}{18}.$$

(We have dutifully indicated all terms, but only one of them is different from zero!) Formula (9) gives the covariance

$$C(X, Y) = E(XY) - E(X)E(Y) = \frac{1}{18} - \left(\frac{1}{3}\right)^2 = -\frac{1}{18}.$$ □

***Example 9. Lifetimes of Electronic Components**

This is a continuation of Example 4, p. 73, where we considered the density function

$$f_{X,Y}(x, y) = \begin{cases} \dfrac{1}{1+c}(xy + c)e^{-x-y} & \text{if} \quad x > 0, y > 0, \\ 0 & \text{otherwise.} \end{cases}$$

By dividing each integral into two parts and using the gamma function, p. 58, we find

$$E(X) = \int_0^\infty \int_0^\infty x \frac{1}{1+c}(xy + c)e^{-x-y}\, dx\, dy = (2 + c)/(1 + c),$$

$$E(X^2) = \int_0^\infty \int_0^\infty x^2 \frac{1}{1+c}(xy + c)e^{-x-y}\, dx\, dy = (6 + 2c)/(1 + c),$$

$$E(XY) = \int_0^\infty \int_0^\infty xy \frac{1}{1+c}(xy + c)e^{-x-y}\, dx\, dy = (4 + c)/(1 + c).$$

Moreover, we have $E(Y) = E(X)$ and $E(Y^2) = E(X^2)$. Using (6) and (9) we find, respectively,

$$V(X) = V(Y) = \frac{6 + 2c}{1 + c} - \left(\frac{2 + c}{1 + c}\right)^2 = (c^2 + 4c + 2)/(1 + c)^2,$$

$$C(X, Y) = E(XY) - E(X)E(Y) = \frac{4 + c}{1 + c} - \left(\frac{2 + c}{1 + c}\right)^2 = c/(1 + c)^2.$$

Finally, by formula (10) we obtain the coefficient of correlation

$$\rho(X, Y) = c/(c^2 + 4c + 2).$$

For example, if $c = 1$, we have $\rho(X, Y) = 1/7$. If $c = 0$, we find $\rho(X, Y) = 0$. This is in agreement with what we found on p. 73, for if X and Y are independent, they are also uncorrelated. □

Exercises

601. When playing a certain game with a die, the player moves an object the number of steps that the die shows, except when it shows 1; in this case, the object is moved six steps. Compute the expectation of the number of steps that the object is moved. (§6.2)

602. The rv X has density function

$$f_X(x) = 2x/a^2 \qquad (0 \le x \le a).$$

Find $E(X)$. (§6.2)

603. The discrete rv X has probability function $p_X(k) = Ck^{-2}$ $(k = 1, 2, \ldots)$. Show that $E(X)$ does not exist. (§6.2)

604. In a parking garage the total parking fee Y consists of 1 dollar plus a variable charge of 50 cents per hour. The parking time, X in hours, has density function $f_X(x) = e^{-x}$ $(x \ge 0)$. Compute $E(Y)$. (§6.2)

605. Determine $E(e^X)$ if $f_X(x) = 2e^{-2x}$ $(x \geq 0)$. (§6.2)

606. The rv X has distribution function $F_X(x) = 1 - (1 + x)^{-a}$ $(x \geq 0)$. Here a is a positive constant. Determine $E\left(\dfrac{1}{1+X}\right)$. (§6.2)

607. The rv X has density function $f_X(x) = 1/10\,(-5 \leq x \leq 5)$. Find $E[g(X)]$, where

$$g(x) = \begin{cases} -1 & \text{if } x < 0, \\ 2 & \text{if } x \geq 0. \end{cases} \qquad (\S 6.2)$$

608. Volume X of a liquid is poured into an empty vessel of volume a. (If $X \geq a$ there is, of course, some overflow.) Here X is a rv with density function $f_X(x) = (x + 1)^{-2}$ $(x \geq 0)$. Let Y be the final volume of liquid in the vessel. Determine $E(Y)$. (§6.2)

609. The rv X has expectation 81 and variance 81. Find its standard deviation and its coefficient of variation. (§6.3)

610. The rv X has density function $f_X(x) = 3x^{-4}$ $(x \geq 1)$. Determine its mean and variance. (§6.3)

611. The rv X has density function $f_X(x) = 2x$ $(0 \leq x \leq 1)$.
(a) Find the mean m and the standard deviation σ of X.
(b) Find $P(m - 2\sigma < X < m + \sigma)$.
(c) Find $P(m - \sigma < X < m + 2\sigma)$. (§6.3)

612. The rv X has a uniform distribution over the interval (0, 1). Determine the variance of X^2. (§6.3)

613. The acidity in a stream is measured regularly by means of a pH-meter. Each measurement contains a random error Y with standard deviation $\sigma = 0.05$ and, because of incorrect calibration of the instrument, a systematic error $d = 0.4$. Determine the mean and the standard deviation of the measured value if the true pH-value is 5.8. (§6.3)

614. A rv (X, Y) has the following probability function $p_{X,Y}(j, k)$:

k \ j	0	1	2
1	0.1	0.2	0.3
2	0.4	0	0

First determine $p_X(j)$ for $j = 0, 1, 2$ and $p_Y(k)$ for $k = 1, 2$, and then find $E(X)$ and $E(Y)$. Then give the values which the product XY can assume and the corresponding probabilities. Finally compute $E(XY)$ and $C(X, Y)$. (§6.4)

615. A purse contains two 10-cent coins and one 25-cent coin. Two coins are taken at random without replacement from the purse. Let X be the value of the first coin and Y the value of the second coin. Compute $E(X)$, $E(Y)$, $V(X)$, $V(Y)$, $E(XY)$, $C(X, Y)$ and $\rho(X, Y)$. (§6.4)

616. The two-dimensional rv (X, Y) assumes the values $(0, 1)$, $(1, 0)$, $(0, -1)$ and $(-1, 0)$, each with probability 1/4. (Draw a diagram!) Find $C(X, Y)$ and show that X and Y are uncorrelated. Also prove that X and Y are dependent. (§6.4)

617. The rv (X, Y) has density function

$$f_{X,Y}(x, y) = \frac{1}{2x}e^{-x} \quad \text{if} \quad x > 0, -x \leq y \leq x.$$

Determine $V(X)$, $V(Y)$ and $\rho(X, Y)$. Are X and Y uncorrelated? Are they independent? (§6.4)

*618. The velocity of a randomly chosen gas molecule in a certain volume of gas is a rv with density function

$$f_X(x) = \sqrt{\frac{2}{\pi}}\left(\frac{M}{RT}\right)^{3/2} x^2 e^{-Mx^2/(2RT)} \quad \text{if} \quad x \geq 0.$$

The meaning of the constants R, T and M has been commented upon in Exercise 312.

(a) Determine the mean of the velocity.

(b) Determine the mean of the kinetic energy $MX^2/(2N)$ of a randomly chosen molecule. Here N is Avogadro's number.

Hint:

$$\int_0^\infty x^k e^{-x^2}\, dx = \frac{1}{2}\Gamma\left(\frac{k+1}{2}\right).$$

*619. Certain electric units (example: light bulbs) may be checked using so-called group testing. In such testing, the units are arranged in a series system of k units per group, and the current is switched on. If the current passes through, all units in the group are functioning and only a single test is needed. However, if no current passes through the system, at least one unit is defective; then all units are tested individually, that is, $k + 1$ tests are performed in all.

Suppose that nk units are tested in this way by dividing them into n groups of k units in each. Suppose, further, that a randomly chosen unit is defective with probability p. If X is the total number of tests, then $Y = X/nk$ is the number of tests divided by the total number of units. Evidently, Y is of interest in assessing the economy of group testing: if $E(Y)$ is less than 1, group testing is cheaper on the average than individual testing of the units.

(a) Determine $E(Y)$ as a function of p.

(b) Suppose that $p = 0.05$. Determine k so that $E(Y)$ is a small as possible.

*620. In a factory, very long rods are produced. They are divided into lengths of 100 cm in two steps. First the rods are cut into parts of about 100 cm. Then the parts are treated in a machine working with great precision: the parts longer than 100 cm are cut down to a length of exactly 100 cm, while those too short are rejected. Hence a rod of length x causes a loss of material x if $x < 100$ and $x - 100$ if $x \geq 100$.

Suppose that the length of a part obtained at the first cutting is a rv X with density function

$$f_X(x) = \frac{1}{10} - \frac{1}{100}|x - m| \qquad \text{if} \quad |x - m| \leq 10.$$

Here $m = E(X)$ is a constant which may be varied in the interval (80, 120) by adjusting the first cutting machine. Let Y denote the loss of material from one of the parts produced in the first step. Give an expression for $E(Y)$. Determine the value of m for which $E(Y)$ is as small as possible.

CHAPTER 7

More About Expectations

7.1. Introduction

This chapter is a direct continuation of the preceding one, which we assume to be known to the reader by now.

In §7.2, we discuss products, sums and linear combinations of rv's; in §7.3, we discuss the law of large numbers and some other matters; and in §7.4, we discuss Gauss's approximation formulae, which are useful in many applications of probability theory. §7.5 contains some classical problems of probability.

7.2. Product, Sum and Linear Combination

(a) Product

Theorem 1. *If X and Y are independent* rv's, *then*

$$E(XY) = E(X)E(Y). \tag{1}$$

□

PROOF. By means of Theorem 1′ in §6.2 and the definition of independence in §4.5, we have, for example, in the discrete case,

$$\begin{aligned} E(XY) &= \sum_j \sum_k jk p_{X,Y}(j, k) = \sum_j \sum_k jk p_X(j) p_Y(k) \\ &= \sum_j j p_X(j) \sum_k k p_Y(k) = E(X)E(Y). \end{aligned}$$

□

From Theorem 1 we immediately obtain Theorem 5 in §6.4. Theorem 1 is easily extended to more than two independent rv's: For three rv's we have $E(XYZ) = E(X)E(Y)E(Z)$ and so on.

(b) Sum and Difference

Theorem 2. *For all* rv's *X and Y*

$$E(X + Y) = E(X) + E(Y). \tag{2}$$

If X and Y are independent, we also have

$$V(X + Y) = V(X) + V(Y), \tag{3}$$

$$D(X + Y) = \sqrt{D^2(X) + D^2(Y)}. \tag{4}$$

□

PROOF. We give the proof only in the discrete case; in the continuous case it is similar. We find successively

$$\begin{aligned} E(X + Y) &= \sum_j \sum_k (j + k)p_{X,Y}(j, k) \\ &= \sum_j \sum_k jp_{X,Y}(j, k) + \sum_j \sum_k kp_{X,Y}(j, k) \\ &= \sum_j jp_X(j) + \sum_k kp_Y(k) = E(X) + E(Y). \end{aligned}$$

Moreover, using the definition of a variance we obtain

$$\begin{aligned} V(X + Y) &= E[(X + Y - (m_X + m_Y))^2] \\ &= E[(X - m_X + Y - m_Y)^2] \\ &= E[(X - m_X)^2] + E[(Y - m_Y)^2] + 2E[(X - m_X)(Y - m_Y)] \\ &= V(X) + V(Y) + 2C(X, Y). \end{aligned}$$

Here we have used the definition of covariance on p. 109. Since the variables are independent, the covariance is zero (see Theorem 5 in §6.4), and thus relation (3) is proved. Relation (4) then immediately follows. □

Theorem 2 establishes some elegant properties of independent rv's: For a sum of independent rv's both expectations and variances are additive. The third relation in the theorem has a nice interpretation: If $D(X)$ and $D(Y)$ are the smaller sides of a right-angled triangle, then, according to Pythagoras' theorem, the hypotenuse is equal to $D(X + Y)$ (see Fig. 7.1).

If the rv's are correlated, that is, if their covariance $C(X, Y)$ is not 0, then relation (3) in Theorem 2 is replaced by

$$V(X + Y) = V(X) + V(Y) + 2C(X, Y).$$

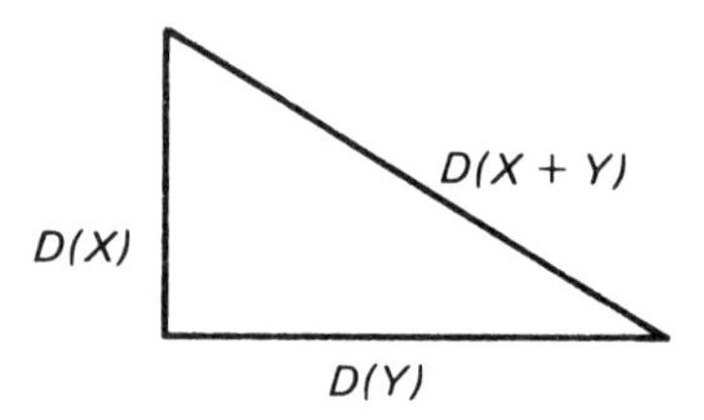

Fig. 7.1. "Pythagoras' theorem".

(This follows from the proof of Theorem 2.) Hence, depending on the sign of the covariance, the variance of a sum of dependent rv's is greater or smaller than when the variables are independent. If the variables are uncorrelated, the covariance in the relation disappears, and the same expression is valid as for independent variables.

We now consider the difference of two rv's.

Theorem 2′. *For all* rv's *X and Y*

$$E(X - Y) = E(X) - E(Y). \tag{5}$$

If X and Y are independent, we also have

$$V(X - Y) = V(X) + V(Y). \tag{6}$$

□

The proof is very similar to that of Theorem 2. Note that the variances are *added*, precisely as in Theorem 2. A mistake can lead to disaster!

If X and Y are correlated, (6) is replaced by

$$V(X - Y) = V(X) + V(Y) - 2C(X, Y).$$

(c) Linear Combination

To avoid too many theorems, we consider a general linear combination

$$c_1X_1 + c_2X_2 + \cdots + c_nX_n$$

of n rv's $X_1, X_2, \ldots, X_n$. The constants $c_1, c_2, \ldots, c_n$ can be positive or negative.

Theorem 3. *For all* rv's $X_1, X_2, \ldots, X_n$

$$\begin{aligned} E\left(\sum_{i=1}^{n} c_iX_i\right) &= \sum_{i=1}^{n} c_iE(X_i), \\ V\left(\sum_{i=1}^{n} c_iX_i\right) &= \sum_{i=1}^{n} c_i^2V(X_i) + 2\sum_{i<j} c_ic_jC(X_i, X_j). \end{aligned} \tag{7}$$

In particular, if $X_1, X_2, \ldots, X_n$ are independent, we have

$$V\left(\sum_{i=1}^{n} c_i X_i\right) = \sum_{i=1}^{n} c_i^2 V(X_i). \tag{8}$$

□

The proof is similar to that of Theorem 1, but there are more terms to keep in mind.

Corollary. *If $X_1, X_2, \ldots, X_n$ are* rv's *with the same expectation m, then*

$$E\left(\sum_{i=1}^{n} X_i\right) = nm. \tag{9}$$

If $X_1, X_2, \ldots, X_n$ are independent and have the same standard deviation σ, we have

$$V\left(\sum_{i=1}^{n} X_i\right) = n\sigma^2; \qquad D\left(\sum_{i=1}^{n} X_i\right) = \sigma\sqrt{n}. \tag{10}$$

□

Look at these expressions: When more and more independent rv's are added, the expectation of their sum is proportional to their number n, but the standard deviation is proportional to $\sqrt{n}$, and hence increases less rapidly.

7.3. Arithmetic Mean. Law of Large Numbers

(a) Arithmetic Mean

It is often of interest to examine the arithmetic mean of several independent rv's.

Theorem 4. *If $X_1, X_2, \ldots, X_n$ are independent* rv's, *each with expectation m and standard deviation σ, and*

$$\bar{X} = \sum_{i=1}^{n} X_i/n$$

is their arithmetic mean, then

$$E(\bar{X}) = m; \qquad V(\bar{X}) = \sigma^2/n; \qquad D(\bar{X}) = \sigma/\sqrt{n}. \tag{11}$$

□

Proof. Take $c_i = 1/n$ in Theorem 3. (The statement concerning $E(\bar{X})$ is also true for dependent rv's.) □

This theorem is worth thinking about for a while. That the expectation of $\bar{X}$ is equal to m is not at all remarkable; what is interesting is the form of the variance and that of the standard deviation. For instance, let us look at the latter. Evidently, as n increases, $D(\bar{X})$ decreases inversely proportional to the square root of n. Since the standard deviation is a measure of dispersion, a reasonable interpretation is that, as n increases, the distribution of $\bar{X}$ becomes concentrated more and more around m. This interpretation is correct, which we shall discuss again in connection with the so-called law of large numbers.

By combining the theorems proved so far in this chapter, it is possible to determine expectation, variance and standard deviation of many functions of rv's.

Example 1. Comparison of Measurements

When measuring an object, one person has obtained the measurements X_1 and X_2, another the measurements X_3, X_4 and X_5. All five values are assumed to be independent and to have the same distribution with mean m and standard deviation σ. Consider the difference

$$Z = U - V$$

of the arithmetic means $U = (X_1 + X_2)/2$ and $V = (X_3 + X_4 + X_5)/3$ of the two sets of measurements. Theorem 4 shows that U and V have means m and variances $\sigma^2/2$ and $\sigma^2/3$, respectively. Theorem 3 shows that Z has mean

$$E(Z) = m - m = 0$$

and variance

$$V(Z) = \sigma^2/2 + \sigma^2/3 = 5\sigma^2/6.$$

Hence the standard deviation is

$$D(Z) = \sigma\sqrt{5/6}.$$

□

(b) Law of Large Numbers

"The eyes follow the rolling die. What will the result be? Nobody knows in advance. If the manufacturer of the die has done a good job, all six faces have an equal chance of turning up .... We throw the die a thousand times, add the numbers that appear and divide by the number of throws. We repeat this long series another ten times or so. Remarkable! Each time we compute the arithmetic mean of the thousand throws, we get a number close to 3.5. If we had time and did a million throws, we would, with high probability, come still nearer to 3.5."

The quotation (a translation of a passage in a Swedish book by S. Fagerberg, *Dialog i det fria*) describes the *law of large numbers*, first

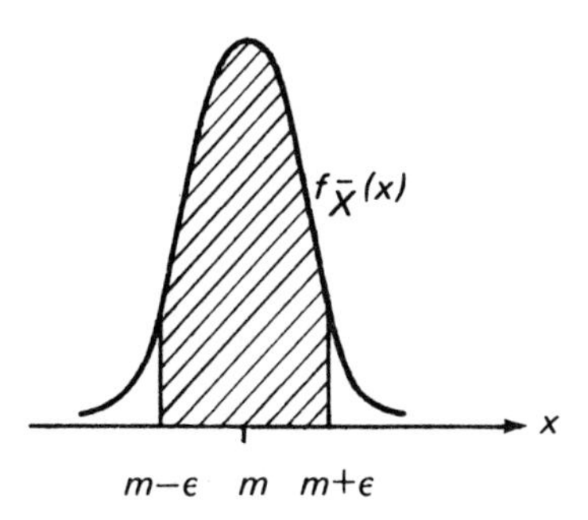

Fig. 7.2. The distribution of $\bar{X}$.

formulated by the Swiss mathematician James Bernoulli in the seventeenth century. This law states that the arithmetic mean of a large number of independent rv's with the same expectation m is close to m. A more precise statement is the following:

Theorem 5 ("Law of Large Numbers"). *Let* $X_1, X_2, \ldots$ *be independent* rv's, *each with expectation m and standard deviation* σ, *and set*

$$\bar{X}_n = \sum_{i=1}^{n} X_i/n.$$

However small $\varepsilon > 0$ *is taken,*

$$P(m - \varepsilon < \bar{X}_n < m + \varepsilon) \to 1$$

as n goes to infinity. □

In Fig. 7.2, the probability mentioned in the theorem has been indicated. The theorem states that the shaded area tends to 1 as n increases. Hence the probability mass lying outside $m - \varepsilon$ and $m + \varepsilon$ shrinks gradually, and the whole distribution becomes more and more concentrated about m.

In order to prove the theorem we need the following:

Lemma ("Chebyshev's Inequality"). *Let X be a* rv *with mean* $E(X) = m$ *and variance* $V(X) = \sigma^2$. *For each* $k > 0$

$$P(|X - m| \geq k\sigma) \leq 1/k^2. \tag{12}$$

□

Let us first illustrate the inequality and take $k = 2$. The inequality states that at most 1/4 of the mass of the distribution is situated outside the interval $(m - 2\sigma, m + 2\sigma)$.

Proof. The inequality is proved in the continuous case as follows (the proof in the discrete case is analogous):

$$\begin{aligned}\sigma^2 &= \int_{-\infty}^{\infty} (x-m)^2 f_X(x)\,dx \\ &= \int_{|x-m|\ge k\sigma} (x-m)^2 f_X(x)\,dx + \int_{|x-m|<k\sigma} (x-m)^2 f_X(x)\,dx \\ &\ge \int_{|x-m|\ge k\sigma} (x-m)^2 f_X(x)\,dx \ge k^2\sigma^2 \int_{|x-m|\ge k\sigma} f_X(x)\,dx \\ &= k^2\sigma^2 P(|X-m| \ge k\sigma).\end{aligned}$$

We now divide both sides of the inequality by $k^2\sigma^2$, and the lemma is proved. □

Proof of Theorem 5. We apply Chebyshev's inequality to the rv $\bar{X}_n$. The inequality then takes the form

$$P\left(|\bar{X}_n - m| \ge k\cdot\frac{\sigma}{\sqrt{n}}\right) \le 1/k^2,$$

where σ is the standard deviation of the rv's $X_1, \ldots, X_n$. Taking $k = \varepsilon\sqrt{n}/\sigma$, we get

$$P(|\bar{X}_n - m| \ge \varepsilon) \le \frac{\sigma^2}{n\varepsilon^2}.$$

Noting that the right-hand side tends to 0 as $n \to \infty$, the proof is complete. □

Remark. More About Fair Games (continued from Remark 1 in §6.3)

The law of large numbers has led to many speculations, some correct, some fallacious. We shall try to kill one fallacious idea. Suppose that A and B play a game of chance with a balanced coin and that A obtains 1 money unit from B (for example, 1 dollar or 100 dollars) if a head turns up, and B obtains 1 unit from A if a tail turns up. Moreover, let X_i be the amount that A obtains at the ith toss. Clearly, X_i is a rv, which takes the values 1 and -1 with equal probabilities. We have $m = 0$, and the game is then said to be *fair*. The law of large numbers tells us that, if A and B go on playing long enough, then the arithmetic mean of the amounts received by A will lie very near 0. Thus it may seem safe to play a fair game. However, this is by no means true!

(1) It is assumed at the beginning that A and B are in the position of being able to play for an unlimited time. In practice, each player has limited capital and hence may be ruined during the course of the play. In fact, the probability of ruin is very large if one plays against a very rich adversary; see Example 25 in Chapter 2.

(2) The theorem states only that the *arithmetic mean* of the amounts that A gets is, with high probability, near zero for large n, not that the *sum* of the amounts is near zero. And it is, of course, the sum which is decisive for the player! As a matter of fact,

it can be proved that the probability that $|\sum X_i| > a$, for any given $a > 0$, tends to *one*, as n goes to infinity. Hence, after many rounds of a fair game, the player will probably either have earned much money or lost much money; it is quite improbable that he has about the same amount as when he started.

We understand after these considerations that it is risky to rely on the law of large numbers without examining the chances of winning more closely. More about these fascinating problems, which have relevance not only for players of games of chance but also for insurance companies and other enterprises, may be found in W. Feller, *An Introduction to Probability Theory and Its Applications*, Vol. I (3rd ed., Wiley, New York, 1968). □

7.4. Gauss's Approximation Formulae

As we have seen before in this chapter, it is rather easy to determine means and variances of simple functions of rv's, for example, linear combinations, especially when the variables are independent. For many functions, however, an exact determination of these quantities is difficult and sometimes almost impossible from a computational point of view. Then an approximate method due to Gauss is often of great value. Whenever nonlinear expressions are studied, as is often the case in science, Gauss's method may be useful.

We begin with a function $g(X)$ of a single rv X.

Theorem 6 ("Gauss's Approximation Formulae for One rv").

$$E[g(X)] \approx g[E(X)],$$
$$V[g(X)] \approx V(X)[g'(E(X))]^2. \tag{13}$$

□

Proof. Expand $g(X)$ in a Taylor series about $E(X) = m$:

$$g(X) = g(m) + (X - m)g'(m) + R.$$

Suppose that the error term R may be neglected: hence let it be zero. On taking the expectation of the remaining terms, we get the first formula of (13). Moreover, the second expression of (7) in §6.3 leads to

$$V[g(X)] \approx [g'(m)]^2 V(X - m) = V(X)[g'(m)]^2,$$

and so also the second formula of (13) is proved. □

The reader is *warned* not to apply the formulae indiscriminately, for they can be very inaccurate. A condition for good approximation is that $g(x)$ is approximately linear within the region where the main part of the mass of the distribution of X is situated (see Fig. 7.3). If $g(x)$ is linear, so that $g(X) = aX + b$, the formulae hold exactly.

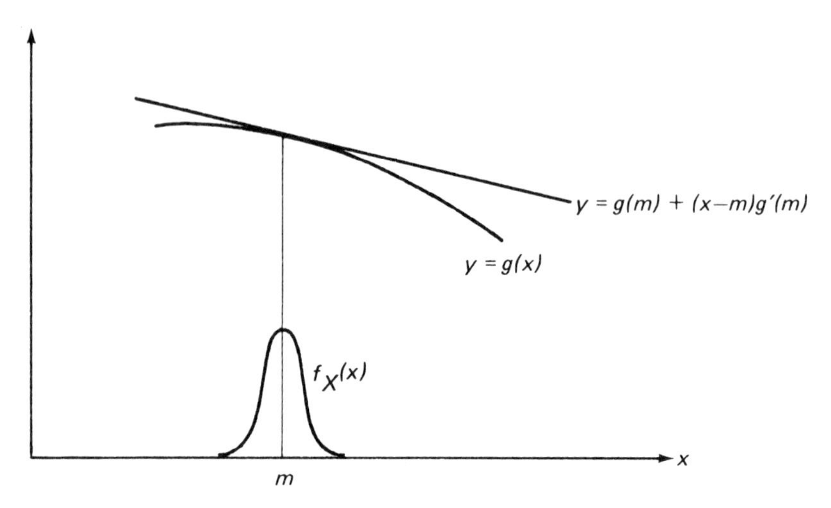

Fig. 7.3. Gauss's approximation formulae are good if the distribution of X is concentrated around m and if also $g(x)$ is approximately linear near m.

Example 2. Reciprocal of X

If $g(X) = 1/X$, Gauss's approximation formulae lead to the approximate expressions

$$E(1/X) \approx 1/E(X),$$

$$V(1/X) \approx V(X)/[E(X)]^4.$$

The accuracy is usually good if $E(X)$ is distinct from zero and $V(X)$ is small. However, a positive mass at $X = 0$ completely spoils both formulae, for then neither mean nor variance of $1/X$ exists.

Suppose now that X assumes only positive values. Dividing the square root of the second equation by the first equation then gives (see the definition of coefficient of variation on p. 103)

$$R(1/X) \approx R(X).$$

Hence the coefficients of variation of X and $1/X$ are about equal. The agreement is better the smaller $R(X)$ is. □

Example 3. Logarithmic Transformation

Let X be a positive rv with large mean m and small variance σ^2. We are interested in studying the mean and variance of the rv $\ln X$. Gauss's approximation formulae can then be used, for under the given conditions the function $\ln x$ is approximately linear in the region near $x = m$ and the probability mass of the rv $\ln X$ is concentrated around this point. The approximation formulae show that

$$E(\ln X) \approx \ln m,$$

$$V(\ln X) \approx \sigma^2/m^2.$$

By the definition of the coefficient of variation we also obtain

$$D(\ln X) \approx R(X). \qquad \square$$

We shall now generalize Gauss's approximation formulae to a function of n rv's $X_1, X_2, \ldots, X_n$ with expectations $m_1, m_2, \ldots, m_n$.

Theorem 6′ ("Gauss's Approximation Formulae for Several rv's").

$$E[g(X_1, \ldots, X_n)] \approx g(m_1, \ldots, m_n),$$

$$V[g(X_1, \ldots, X_n)] \approx \sum_{i=1}^{n} V(X_i)\left(\frac{\partial g}{\partial m_i}\right)^2 + 2\sum_{i<j} C(X_i, X_j)\frac{\partial g}{\partial m_i}\cdot\frac{\partial g}{\partial m_j}. \tag{14}$$

(Here $\partial g/\partial m_i$ signifies that g is differentiated partially with respect to the i*th variable in the function g; then $m_1, m_2, \ldots, m_n$ are inserted in the derivative.)*

$\square$

The proof is based on a multivariate Taylor series. The usefulness is limited as before. The accuracy will be acceptable if $g(X_1, \ldots, X_n)$ can be described approximately by a plane in the region where the main part of the mass of the multidimensional rv $(X_1, \ldots, X_n)$ is situated. (Note that the probability mass is now distributed over n-dimensional space.)

It is important to notice that, if $g(X_1, \ldots, X_n) = \sum_1^n c_i X_i$, the formulae are exact.

When $X_1, X_2, \ldots, X_n$ are independent, the second formula (14) becomes particularly simple, for then all covariances are zero:

$$V[g(X_1, \ldots, X_n)] \approx \sum_{i=1}^{n} V(X_i)\left(\frac{\partial g}{\partial m_i}\right)^2. \tag{15}$$

Example 4. The Variance of a Product

The product XY of two independent rv's has by (15) the approximate variance

$$V(XY) \approx V(X)[E(Y)]^2 + V(Y)[E(X)]^2.$$

Dividing by $[E(XY)]^2 = [E(X)E(Y)]^2$ we obtain the simple and useful formula

$$R^2(XY) \approx R^2(X) + R^2(Y).$$

In this case, it is easy to judge how accurate these two approximations are: It is a good exercise to show that the *exact* expressions are, respectively,

$$V(XY) = V(X)[E(Y)]^2 + V(Y)[E(X)]^2 + V(X)V(Y),$$

$$R^2(XY) = R^2(X) + R^2(Y) + R^2(X)R^2(Y).$$

It can be seen from these expressions that the approximating formulae are good if both coefficients of variation are small (or, at least, if one of them is small).

Let us give a simple practical illustration: We want to find the area of a rectangle. When the sides are measured, errors arise such that the coefficient of variation is 2%.

This means that the measured lengths X and Y have $R(X) = R(Y) = 2\%$. We assume that the measurements are independent. The computed area XY has the approximate coefficient of variation

$$R(XY) \approx \sqrt{0.02^2 + 0.02^2} = 2.8\%. \qquad \square$$

Example 5. Parallel System

We have a parallel system of three resistances of R_1, R_2 and R_3 ohm. The voltage U gives the current

$$I = U\left(\frac{1}{R_1} + \frac{1}{R_2} + \frac{1}{R_3}\right).$$

Because of errors of measurement, U, R_1, R_2, R_3 are rv's which are assumed to have means 120, 10, 15, 20, respectively, and standard deviations 15, 1, 1, 2, respectively; the four rv's are assumed to be independent.

We find

$$\frac{\partial I}{\partial U} = \sum 1/R_i, \qquad \frac{\partial I}{\partial R_i} = -U/R_i^2,$$

and the approximation formulae (14) give

$$E(I) \approx 120(1/10 + 1/15 + 1/20) = 26,$$

$$V(I) \approx 15^2(1/10 + 1/15 + 1/20)^2 + 1^2(-120/10^2)^2$$
$$+ 1^2(-120/15^2)^2 + 2^2(-120/20^2)^2 = 2.61.$$

Thus the mean and standard deviation of I are, approximately, 26 and $\sqrt{2.61} = 1.6$, respectively. □

Remark. Some Cautionary Examples

Expectations do not always behave as one might think initially. Remember that we have (except in the trivial case of a one-point distribution)

$$E(1/X) \neq 1/E(X),$$

$$E(X^2) \neq [E(X)]^2,$$

$$V(XY) \neq V(X)V(Y).$$

Resist the temptation to put equality signs in these relations! As a matter of fact we have $>$ in the first formula whenever X assumes only positive values and $>$ in the other formulae, except when we have a one-point distribution. On the other hand, the first formula reminds us of a sometimes useful approximation (see Example 2). □

7.5. Some Classical Problems of Probability[1]

Example 6. Petersburg Paradox

A person is invited by a rich friend to participate in the following game. A fair coin is tossed until a head appears for the first time. If this happens on the kth toss, the player

[1] Special section, which may be omitted.

receives 2^k dollars ($k = 1, 2, \ldots$). Obtain the expectation of the amount that the player receives.

Let Y be this amount. The number of tosses has an fft-distribution with the probability function $p_X(k) = (1/2)^k$. (Take $p = 1/2$ in the expression for the probability function in §3.5.) We find

$$E(Y) = E(2^X) = 2\frac{1}{2} + 2^2\frac{1}{2^2} + 2^3\frac{1}{2^3} + \cdots.$$

The sum diverges and we see that $E(Y)$ is infinite. Hence the game seems to be exceptionally favourable to the player, since his rich friend apparently loses "an infinite amount of money". This deserves a closer examination.

First, suppose that the game is played only once. It is seen that $Y \leq 2^6$ with the probability

$$P(X \leq 6) = 1/2 + 1/2^2 + \cdots + 1/2^6 = 1 - 1/2^6 = 63/64.$$

This tells us there is a high probability that the friend escapes with paying at most $2^6 = 64$ dollars. Similarly, it is seen that the chance is 1,023/1,024, that is, nearly one, that the friend has to pay at most $2^{10} = 1{,}024$. To let the person play once thus seems rather safe for the rich friend.

It is worse for the friend if the player makes use of the offer many times. Then it is no longer possible to disregard the fact that the mean is infinite, for the mean shows, as we know, what will be paid on the average. The friend may perhaps demand a stake S which makes the game fair (see Remark 1 in §6.3). Then $E(Y) - S$ should be zero, and, paradoxically enough, the stake must be infinite!

Such a game is impossible to play, for example, in Monte Carlo, and must be modified in some way. For example, it may be prescribed that the amount 2^k is paid to the player only if the number k of tosses is at most equal to a given number N. Then the expectation of the payment becomes

$$2\frac{1}{2} + 2^2\frac{1}{2^2} + \cdots + 2^N\frac{1}{2^N} = N,$$

and it is possible to make the game fair.

The Petersburg paradox presumably got its name because the mathematician Daniel Bernoulli discussed it about 1730 in proceedings published by the Academy in Petersburg (the present Leningrad). □

Example 7. The Prize in the Food Package (continued)

The situation in Example 27 in §2.8 is changed in the following way. A person buys one packet at a time until he gets a complete set of prizes, that is, at least one of each type. Find the expectation $E(X)$ of the number of packages which he has bought.

We use a trick. Write

$$X = Y_1 + Y_2 + \cdots + Y_N,$$

where Y_j is the number of packages the person must buy in order to increase his collection from $j - 1$ to j different types. We have

$$E(X) = E(Y_1) + E(Y_2) + \cdots + E(Y_N).$$

Suppose that the person has $j - 1$ types so that $N - j + 1$ are missing. The number, Y_j, which he must buy in order to get one more type, has a fft-distribution:

$$p_{Y_j}(k) = (1 - p_j)^{k-1}p_j \qquad (k = 1, 2, \ldots),$$

where $p_j = (N - j + 1)/N$. By Example 2 in §6.2, we have

$$E(Y_j) = 1/p_j = N/(N - j + 1),$$

and hence

$$E(X) = N\left(\frac{1}{N} + \frac{1}{N-1} + \cdots + 1\right).$$

If N is large, this is approximately equal to $N \ln N$.

Example 8. The Problem of Rencontre (continued)

We return to the problem of rencontre in Example 26 in §2.8. Let us evaluate the mean and variance of the number of rencontres.

The number of rencontres, X, may be written as a sum

$$X = U_1 + U_2 + \cdots + U_N,$$

where $U_i = 1$ if a rencontre is obtained at the ith draw, and $U_i = 0$ otherwise.

We begin with the mean. It is seen that $P(U_i = 1) = (N - 1)!/N!$, for there are $N!$ possible orderings, of which $(N - 1)!$ give a rencontre in the ith draw. Hence we obtain $E(U_i) = 1 \cdot (1/N) + 0 \cdot (1 - 1/N) = 1/N$ and $E(X) = E(U_1) + \cdots + E(U_N) = N(1/N) = 1$. Note that we have this simple result in spite of the dependence of the U_i.

We now turn to the variance, which is more difficult to obtain. We then need the variances and the covariances of the rv's U_i. To this end, we have

$$E(U_i^2) = 1^2 \cdot (1/N) + 0^2 \cdot (1 - 1/N) = 1/N,$$

$$V(U_i) = E(U_i^2) - [E(U_i)]^2 = 1/N - (1/N)^2 = (1/N)(1 - 1/N).$$

Moreover, we see that the product $U_i U_j$ is 1 if both U_i and U_j are 1, and 0 otherwise. Hence, we obtain for $i \neq j$

$$E(U_i U_j) = P(U_i = 1, U_j = 1) = (N - 2)!/N! = 1/[N(N - 1)].$$

This leads to

$$C(U_i, U_j) = 1/[N(N - 1)] - (1/N)^2 = 1/[N^2(N - 1)].$$

Theorem 3 states that

$$\begin{aligned} V(X) &= \sum_{i=1}^{N} V(U_i) + 2 \sum_{i<j} C(U_i, U_j) \\ &= NV(U_i) + N(N - 1)C(U_i, U_j). \end{aligned}$$

Substitution gives

$$V(X) = N \cdot (1/N)(1 - 1/N) + N(N - 1)/[N^2(N - 1)] = 1.$$

Hence both the mean and the variance of the number of rencontres are 1. □

Exercises

701. Let X, Y and Z be independent rv's such that

$$E(X + Y) = 1; \quad E(X - Z) = -4; \quad E(Y - Z) = -3,$$

$$V(X) = 4; \quad V(Y) = 3; \quad V(Z) = 12.$$

(a) Find $E(X)$, $E(Y)$ and $E(Z)$.
(b) Which of the following statements are true?
b_1: $V(X + Y) = 25$.
b_2: $V(X - Y) = 7$.
b_3: $V(X \pm Y \pm Z) = 19$ for all combinations of + and −. (§7.2)

702. The rv's X_1, X_2, X_3, X_4 are independent with means $2m, 2m, 3m, 2m$, respectively, and common standard deviation σ. Form the linear combination

$$Y = \tfrac{4}{3}X_1 - \tfrac{1}{2}X_2 - \tfrac{1}{3}X_3 - \tfrac{1}{3}X_4.$$

Find $E(Y)$ and $D(Y)$. (§7.2)

703. A person buys a ticket in a lottery every month. His gain each time is a rv with expectation -0.5 and standard deviation $\sqrt{15}$. Determine the expectation and the variance of his total gain during 12 months. (§7.2)

704. Prove that if $V(X) = 1$ and $V(Y) = 16$, then $9 \leq V(X + Y) \leq 25$. (§7.2)

705. A rv X is such that

$$E(X) = 0; \qquad V(X) = 1; \qquad E(X^3) = 2; \qquad V(X^2) = 4.$$

Find $V(X + X^2)$. (§7.2)

706. The weights (unit: kilo) of three randomly selected school children in a school are regarded as independent rv's X_1, X_2, X_3 with mean 36 and standard deviation 3. Find the expectation and the standard deviation of the arithmetic mean $\bar{X} = (X_1 + X_2 + X_3)/3$ of the three weights. (§7.3)

707. When determining the melting-point of cooking fat, the resulting measurement may be regarded as a rv X with standard deviation 2. How many measurements are needed to make the standard deviation of the arithmetic mean of the measurements at most 0.4? (§7.3)

708. The height of a radio-transmitting tower was determined by measuring first the distance (unit: meter) from a certain point P in the horizontal plane to the base of the tower and then the angle (unit: radian) between the horizontal plane and the sighting-line from P to the top of the tower. Because of errors of measurements these values can be regarded as independent rv's X and Z, respectively, with $E(X) = 550$, $E(Z) = 0.48$, $D(X) = 10$, $D(Z) = 0.01$. An estimate Y of the height is then computed by means of the formula $Y = X \tan Z$. Compute approximate values of the mean and standard deviation of Y. (§7.4)

709. The percentage of albumin is determined in four food samples, the approximate percentages being 0.1, 0.3, 3.0, 30.0. The coefficient of variation of a measurement is 10%, independently of the percentage of albumin. Set $Y = \log_{10} X$, where X is one of the measurements. Find approximate values of mean and standard deviation for each sample separately. (§7.4)

710. The yield (unit: percentage) was studied when oxidizing ammonia in a converter. The percentage X of NH_3 in the in-going gas and the percentage Y of NO in the out-going gas were measured, and then the yield Z was computed from the formula

$$Z = Y(100/X - 1.25).$$

By performing several double determinations the following approximate standard deviations were found:

$$\sigma_X = 7.7 \cdot 10^{-2}; \qquad \sigma_Y = 9.6 \cdot 10^{-2}.$$

In one of the experiments the values $X = 12.0$ and $Y = 13.5$ were obtained. Determine the standard deviation of Z in this experiment. Lacking better information you are allowed to use the estimates $m_X = 12.0$ and $m_Y = 13.5$ and, moreover, may assume that X and Y are independent. (§7.4)

*711. The rv's $X_1, X_2, \ldots$ are independent and have the same distribution. Set

$$\bar{X}_n = \frac{1}{n} \sum_{j=1}^{n} X_j \qquad (n = 1, 2, \ldots).$$

Show that

$$C(\bar{X}_n, \bar{X}_n - \bar{X}_k) = 0 \qquad \text{if} \quad k < n.$$

*712. On each side of a square of side 1 a point is chosen at random. Determine the mean and standard deviation of the area of the quadrilateral with vertices at the four points.

*713. Let X have a uniform distribution in the interval $(a, a + 1)$, where $a > 0$. Set $Y = 1/X$.

(a) Find the mean and variance of X.

(b) Find the approximation $V_{\text{appr}}(Y)$ of $V(Y)$ given by Gauss's approximation formula.

(c) Find the exact value of $V(Y)$.

(d) Compute numerically the ratio $V_{\text{appr}}(Y)/V(Y)$ for $a = 0.1$, $a = 1$ and $a = 10$.

CHAPTER 8

The Normal Distribution

8.1. Introduction

The normal distribution has already been presented as an example of a continuous distribution (see §3.7). In this chapter, we shall study this important distribution in more detail.

In §8.2 the importance of the normal distribution is outlined, §8.3 treats the standard normal distribution, and §8.4 the general normal distribution. In §8.5, we study the distribution of sums and other functions of normally distributed rv's. In §8.6, the reader is acquainted with the central limit theorem, and in §8.7 the lognormal distribution is mentioned briefly.

8.2. Some General Facts About the Normal Distribution

The normal distribution is often used for describing the variability of different phenomena. A considerable part of statistical theory is based upon this distribution. However, it is not correct to believe that the normal distribution is ubiquitous and that a deviation from this distribution means something "abnormal". There are unlimited possibilities for finding theoretical density functions which may agree quite well with collected data. The reason why the normal distribution is employed wherever possible is that it possesses several nice mathematical properties which simplify its use.

Instead of normal distribution or normal curve we sometimes encounter the terms Gaussian distribution, Gaussian curve, and error distribution.

We again give the density function and also the distribution function:

$$f_X(x) = \frac{1}{\sigma\sqrt{2\pi}} e^{-(x-m)^2/2\sigma^2},$$
$$F_X(x) = \frac{1}{\sigma\sqrt{2\pi}} \int_{-\infty}^{x} e^{-(t-m)^2/2\sigma^2}\, dt \qquad (-\infty < x < \infty). \qquad (1)$$

We also remind the reader of the code name $X \sim N(m, \sigma^2)$.

8.3. Standard Normal Distribution

If $m = 0$ and $\sigma = 1$ we obtain the important special case $X \sim N(0, 1)$. We then call X a *standard normal* rv. The density function of this rv is usually denoted by $\varphi(x)$ and the distribution function by $\Phi(x)$. Hence, we have

$$\varphi(x) = \frac{1}{\sqrt{2\pi}} e^{-x^2/2},$$
$$\Phi(x) = \int_{-\infty}^{x} \varphi(t)\, dt = \frac{1}{\sqrt{2\pi}} \int_{-\infty}^{x} e^{-t^2/2}\, dt. \qquad (2)$$

The shapes of these functions can be seen in Fig. 8.1(a), (b).

A table of the function $\Phi(x)$ is found in Table 1 at the end of the book. We give a sample, where $\varphi(x)$ is also included:

x	0.0	0.5	1.0	1.5	2.0	2.5	3.0
$\varphi(x)$	0.399	0.352	0.242	0.130	0.054	0.018	0.0044
$\Phi(x)$	0.500	0.691	0.841	0.933	0.977	0.994	0.9987

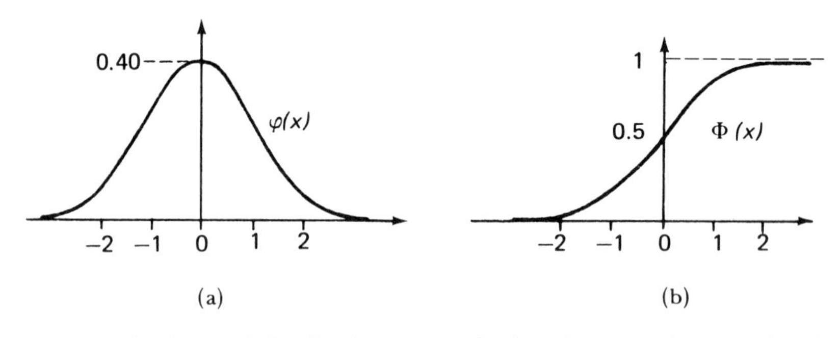

Fig. 8.1. Standard normal distribution. (a) Density function $\varphi(x)$ of $N(0, 1)$. (b) Distribution function $\Phi(x)$ of $N(0, 1)$.

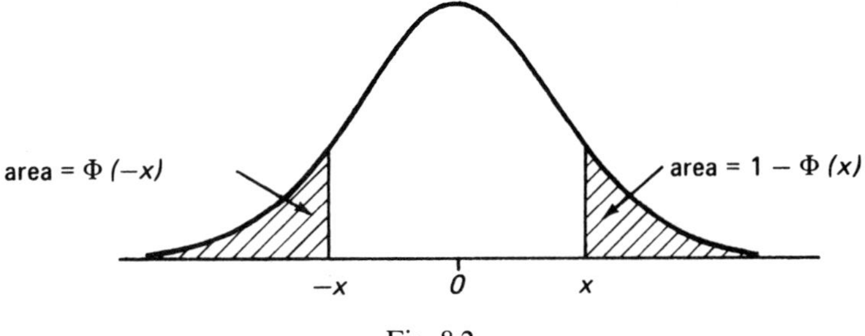

Fig. 8.2.

For negative x one uses the fact that the density function is an even function so that $\varphi(-x) = \varphi(x)$. Moreover, observing that $\Phi(x)$ is the area under the density function to the left of x, we realize that (see Fig. 8.2)

$$\Phi(-x) = 1 - \Phi(x). \tag{3}$$

Example.

For $x = -0.5$ we find $\varphi(-0.5) = \varphi(0.5) = 0.352$ and $\Phi(-0.5) = 1 - \Phi(0.5) = 1 - 0.691 = 0.309$. □

The probability that X lies between two values a and b is (see Theorem 2 in §3.3)

$$P(a < X \leq b) = \Phi(b) - \Phi(a). \tag{4}$$

Since the distribution is continuous it is possible to replace $\leq$ by $<$, or conversely, without changing the probability. For example, we have

$$P(-1 < X < 1) = \Phi(1) - \Phi(-1) = 0.841 - (1 - 0.841) = 0.682,$$
$$P(-2 < X < 2) = \Phi(2) - \Phi(-2) = 0.977 - (1 - 0.977) = 0.954.$$

We introduce the following notation, which will be used over and over again in the sequel: *The α-quantile of a standard normal distribution is denoted by* λ_α (see the definition of a quantile in §3.6). Hence the area under the density function to the *right* of λ_α is equal to α (see Fig. 8.3), that is,

$$P(X > \lambda_\alpha) = \alpha. \tag{5}$$

In Table 2 at the end of the book the quantiles λ_α are tabulated; we reproduce most of the values below:

α	0.0005	0.001	0.005	0.01	0.025	0.05	0.10
λ_α	3.29	3.09	2.58	2.33	1.96	1.64	1.28

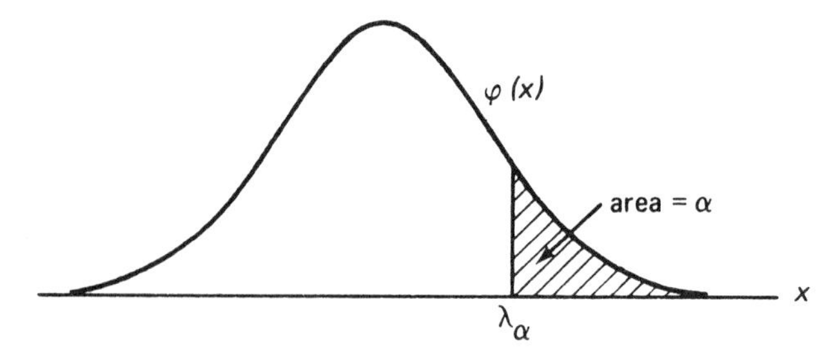

Fig. 8.3. Definition of λ_α.

In view of the symmetry of the density function about 0, it is seen that

$$\lambda_{1-\alpha} = -\lambda_\alpha.$$

For example, we have $\lambda_{0.99} = -\lambda_{0.01} = -2.33$.

One often wants to compute the probability that X is situated between the symmetrical values $\lambda_{1-\alpha/2} = -\lambda_{\alpha/2}$ and $\lambda_{\alpha/2}$. Note that we have replaced α by $\alpha/2$. It is found that (see Fig. 8.4)

$$P(-\lambda_{\alpha/2} < X < \lambda_{\alpha/2}) = 1 - \alpha \qquad \text{if} \quad X \sim N(0, 1). \tag{6}$$

Thus we have proved: The area under the density function between $-\lambda_{\alpha/2}$ and $\lambda_{\alpha/2}$ equals $1 - \alpha$. In particular, we have

$$P(-1.96 < X < 1.96) = 0.95,$$

$$P(-2.58 < X < 2.58) = 0.99,$$

$$P(-3.29 < X < 3.29) = 0.999.$$

Finally, we shall prove that if $X \sim N(0, 1)$ then

$$E(X) = 0; \qquad V(X) = 1.$$

(It is because the mean is 0 and the variance is 1 that the notation $N(0, 1)$ has

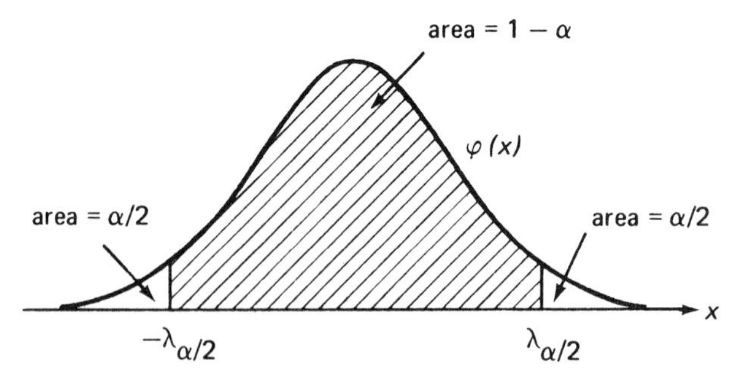

Fig. 8.4.

been introduced; see the general definition of a standardized rv in §6.3.) In fact, using the definition of a mean, we find

$$E(X) = \int_{-\infty}^{\infty} x \cdot \frac{1}{\sqrt{2\pi}} e^{-x^2/2}\,dx = \int_{-\infty}^{0} + \int_{0}^{\infty} = I_1 + I_2.$$

It is easy to see that I_2 exists. (As a matter of fact, $I_2 = 1/\sqrt{2\pi}$.) By substituting $x = -y$ in I_1, we see that $I_1 = -I_2$, and thus the mean is 0. Moreover, a partial integration leads to

$$\begin{aligned} E(X^2) &= \int_{-\infty}^{\infty} x^2 \frac{1}{\sqrt{2\pi}} e^{-x^2/2}\,dx = \int_{-\infty}^{\infty} x\left(\frac{1}{\sqrt{2\pi}} x e^{-x^2/2}\right) dx \\ &= \left[-x \frac{1}{\sqrt{2\pi}} e^{-x^2/2}\right]_{-\infty}^{\infty} + \int_{-\infty}^{\infty} \frac{1}{\sqrt{2\pi}} e^{-x^2/2}\,dx \\ &= \int_{-\infty}^{\infty} \frac{1}{\sqrt{2\pi}} e^{-x^2/2}\,dx = 1. \end{aligned}$$

Hence by Theorem 2 in §6.3, we conclude that $V(X) = E(X^2) - [E(X)]^2 = 1$.

8.4. General Normal Distribution

We shall now study the general normal distribution, that is, we shall consider $X \sim N(m, \sigma^2)$. The density function and the distribution function in (1) can be expressed by means of the functions $\varphi(x)$ and $\Phi(x)$ introduced in the previous section:

$$f_X(x) = \frac{1}{\sigma}\varphi\left(\frac{x - m}{\sigma}\right); \qquad F_X(x) = \Phi\left(\frac{x - m}{\sigma}\right). \tag{7}$$

The correctness of the first expression is recognized directly; the second expression is obtained by the substitution $u = (t - m)/\sigma$. This is convenient: In order to study a general normal distribution we need only the tables of $\varphi(x)$ and $\Phi(x)$ mentioned before.

When working with a general normal distribution one often takes recourse to

Theorem 1. *If $X \sim N(m, \sigma^2)$ and $Y = (X - m)/\sigma$, then $Y \sim N(0, 1)$.* □

PROOF.

$$\begin{aligned} P(Y \le x) &= P\left(\frac{X - m}{\sigma} \le x\right) = P(X \le m + \sigma x) \\ &= \frac{1}{\sigma\sqrt{2\pi}} \int_{-\infty}^{m+\sigma x} e^{-(t-m)^2/2\sigma^2}\,dt. \end{aligned}$$

The substitution $u = (t - m)/\sigma$ gives

$$P(Y \le x) = \frac{1}{\sqrt{2\pi}} \int_{-\infty}^{x} e^{-u^2/2}\, du = \Phi(x).$$

Thus we have proved that Y has the distribution function $\Phi(x)$, that is, that $Y \sim N(0, 1)$. It is also simple to prove the theorem using the second expression (7). □

Theorem 2. *If* $X \sim N(m, \sigma^2)$ *we have*

$$E(X) = m; \qquad V(X) = \sigma^2; \qquad D(X) = \sigma. \qquad \square$$

This theorem is of basic importance and tells us the following: The parameters m and σ^2 in the code name $N(m, \sigma^2)$ are the mean and the variance of X, respectively. This is in agreement with the corresponding statement for a standard normal distribution.

PROOF. Theorem 1 gives $X = m + \sigma Y$ and hence by Theorem 3 in §6.3 we have

$$E(X) = m + \sigma E(Y); \qquad V(X) = \sigma^2 V(Y).$$

But we know from the previous section that Y has mean 0 and variance 1, whence the theorem follows. □

Remark.

We demonstrated earlier that any standardized rv $Y = (X - m)/\sigma$ has mean 0 and variance 1 (see §6.3). Theorem 1 shows that a much stronger result holds for the normal distribution: The standardized rv Y belongs to the same family of distributions as the nonstandardized rv X. □

Theorem 1 is a special case of

Theorem 1′. *If* $Y = aX + b$, *where* $X \sim N(m, \sigma^2)$, *we have*

$$Y \sim N(am + b, a^2\sigma^2). \qquad \square$$

The proof is left to the reader as an exercise. (Discuss the two cases $a > 0$ and $a < 0$ separately.) Note that the expressions for the mean and the variance agree with the general results in Theorem 3 in §6.3.

The shape of the density function of X depends on the parameters m and σ. As shown by Fig. 8.5, this function is symmetric about m, and the smaller σ, the more concentrated is the function around m. By assigning different values to m and σ, a whole family of symmetric distributions is obtained.

We now use Theorem 1 to obtain the probability that X lies between two values a and b, or equivalently to determine the area under the density function

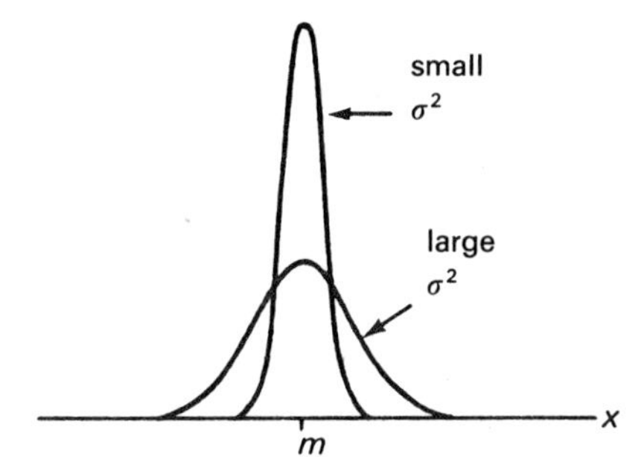

Fig. 8.5. The density function of $N(m, \sigma^2)$ for different σ^2.

between a and b. The theorem gives

$$\begin{aligned} P(a < X < b) &= P\left(\frac{a-m}{\sigma} < \frac{X-m}{\sigma} < \frac{b-m}{\sigma}\right) \\ &= P\left(\frac{a-m}{\sigma} < Y < \frac{b-m}{\sigma}\right) \\ &= \Phi\left(\frac{b-m}{\sigma}\right) - \Phi\left(\frac{a-m}{\sigma}\right). \end{aligned} \tag{8}$$

More directly, the expression may be obtained by means of the second expression (7). Hence, if a table of $\Phi(x)$ is available, any such probability can be computed. In particular, for $a = -\infty$ and $b = \infty$, respectively, we obtain (since $\Phi(-\infty) = 0$, $\Phi(\infty) = 1$)

$$\begin{aligned} P(X < b) &= \Phi\left(\frac{b-m}{\sigma}\right), \\ P(X > a) &= 1 - \Phi\left(\frac{a-m}{\sigma}\right). \end{aligned} \tag{9}$$

If instead $a = m + k_1\sigma$ and $b = m + k_2\sigma$ we get the important relation

$$P(m + k_1\sigma < X < m + k_2\sigma) = \Phi(k_2) - \Phi(k_1). \tag{10}$$

In particular, if $k_1 = -k$, $k_2 = k$, we have

$$P(m - k\sigma < X < m + k\sigma) = \Phi(k) - \Phi(-k). \tag{11}$$

For $k = 1, 2, 3$ this gives

$$P(m - \sigma < X < m + \sigma) = 0.682,$$

$$P(m - 2\sigma < X < m + 2\sigma) = 0.954,$$

$$P(m - 3\sigma < X < m + 3\sigma) = 0.997.$$

These relations should be studied carefully. The first one states that the area under the density function between $m - \sigma$ and $m + \sigma$ is about 0.68; that

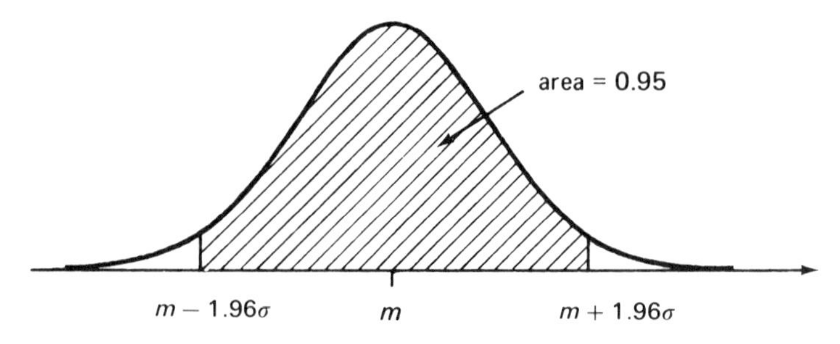

Fig. 8.6. The area between $m - 1.96\sigma$ and $m + 1.96\sigma$ for the density function of $N(m, \sigma^2)$ is 0.95.

is, the probability is 0.68 that the variable differs from the mean by at most one standard deviation. This should be kept in mind, for there sometimes arises a serious misunderstanding: It is believed that the probability of obtaining a value outside this interval is very small. This is not so; the probability is $1 - 0.68 = 0.32$, which means that it is approximately equal to 1/3! The second relation tells us that the probability is approximately 0.95 that the variable differs from the mean by less than two standard deviations. This in turn implies that the probability is about 0.05 that the variable differs from the mean by more than two standard deviations.

Let us give two more formulae, which correspond to (5) and (6):

$$P(X > m + \lambda_\alpha \sigma) = \alpha, \tag{12}$$

$$P(m - \lambda_{\alpha/2}\sigma < X < m + \lambda_{\alpha/2}\sigma) = 1 - \alpha. \tag{13}$$

In particular, the second formula shows that

$$P(m - 1.96\sigma < X < m + 1.96\sigma) = 0.95,$$

$$P(m - 2.58\sigma < X < m + 2.58\sigma) = 0.99,$$

$$P(m - 3.29\sigma < X < m + 3.29\sigma) = 0.999.$$

See Fig. 8.6, where the first of these relations is illustrated.

The reader will perhaps find all these statements about probabilities and corresponding areas rather monotonous, but their importance will emerge later.

8.5. Sums and Linear Combinations of Normally Distributed Random Variables

The normal distribution has an important property which simplifies its application quite remarkably: Sums of two or more independent normally distributed variables, the difference between two such variables, and the arithmetic mean of several such variables are also normally distributed; and this is true even if the variables have different means and different variances. We start with a special case:

Theorem 3. *If $X \sim N(m_X, \sigma_X^2)$ and $Y \sim N(m_Y, \sigma_Y^2)$, and X and Y are independent, then*

$$\begin{aligned} X + Y &\sim N(m_X + m_Y, \sigma_X^2 + \sigma_Y^2), \\ X - Y &\sim N(m_X - m_Y, \sigma_X^2 + \sigma_Y^2). \end{aligned} \tag{14}$$

□

Sketch of the Proof. The convolution formula (11) in §5.3 gives

$$f_{X+Y}(z) = \int_{-\infty}^{\infty} \frac{1}{\sigma_X\sqrt{2\pi}} e^{-(x-m_X)^2/2\sigma_X^2} \cdot \frac{1}{\sigma_Y\sqrt{2\pi}} e^{-(z-x-m_Y)^2/2\sigma_Y^2}\, dx.$$

A rather long calculation shows that this equals

$$\frac{1}{\sqrt{2\pi(\sigma_X^2 + \sigma_Y^2)}} e^{-(z-m_X-m_Y)^2/2(\sigma_X^2+\sigma_Y^2)} \cdot \int_{-\infty}^{\infty} \frac{1}{B\sqrt{2\pi}} e^{-(x-A)^2/2B^2}\, dx,$$

where A and B are certain expressions which we do not give in detail. Using the fact that the integral is 1 (see the first formula (1) in §8.2) the expression simplifies to the density function of a normally distributed rv with $m = m_X + m_Y$, $\sigma^2 = \sigma_X^2 + \sigma_Y^2$, and so the first part of the theorem is demonstrated. The second part follows easily from $X - Y = X + (-Y)$; then apply Theorem 1′ in §8.4 with $a = -1$, $b = 0$. □

Much more general than Theorem 3 is:

Theorem 3′. *If $X_1, X_2, \ldots, X_n$ are independent and $N(m_1, \sigma_1^2)$, $N(m_2, \sigma_2^2)$, $\ldots$, $N(m_n, \sigma_n^2)$, respectively, then for any constants $c_1, c_2, \ldots, c_n$*

$$\sum_1^n c_i X_i \sim N\left(\sum_1^n c_i m_i, \sum_1^n c_i^2 \sigma_i^2\right). \tag{15}$$

□

The theorem should be compared with Theorem 3 in §7.2. In that theorem, we did not assume normal distributions and could then give the mean and standard deviation of $\sum c_i X_i$ (which agree with those given in Theorem 3′), but could not state the form of the distribution. Now a much more detailed statement can be given.

Corollary 1. *If $X_1, X_2, \ldots, X_n$ are independent and $N(m, \sigma^2)$ and $\bar{X} = \sum_1^n X_i/n$ is their arithmetic mean, then*

$$\bar{X} \sim N(m, \sigma^2/n). \tag{16}$$

□

Corollary 2. *If $X_1, X_2, \ldots, X_{n_1}$ are $N(m_1, \sigma_1^2)$ and $Y_1, Y_2, \ldots, Y_{n_2}$ are $N(m_2, \sigma_2^2)$, and all variables are independent, then*

$$\bar{X} - \bar{Y} \sim N(m_1 - m_2, \sigma_1^2/n_1 + \sigma_2^2/n_2). \tag{17}$$

□

Both these results are of great importance.

Example 1. Measurements

Certain measurements are assumed to be normally distributed with mean 28.0 and standard deviation 0.25, or more briefly, $N(28.0, 0.25^2)$.

(a) Find the probability that a measurement lies between 27.5 and 28.5

Formula (8) in §8.4 gives the probability

$$\Phi\left(\frac{28.5 - 28.0}{0.25}\right) - \Phi\left(\frac{27.5 - 28.0}{0.25}\right) = \Phi(2) - \Phi(-2) = \Phi(2) - [1 - \Phi(2)] = 0.954.$$

(b) Find the probability that the arithmetic mean of four independent measurements lies between 27.5 and 28.5

By Corollary 1 the distribution of the arithmetic mean is $N(28.0, 0.25^2/4)$, that is, $N(28.0, 0.125^2)$. Hence by formula (8) the desired probability is given by

$$\Phi\left(\frac{28.5 - 28.0}{0.125}\right) - \Phi\left(\frac{27.5 - 28.0}{0.125}\right) = \Phi(4) - \Phi(-4) = 0.99994.$$

The answer should be compared with that in (a), which showed that the probability that a single measurement lies between the same limits is about 0.95. Hence the arithmetic mean of several measurements tends to lie closer to the mean. This is shown still more clearly if we consider the probability of obtaining values *outside* the given interval. Of single values, about 5%, or one value out of 20, fall outside (27.5, 28.5); of arithmetic means of four values only 0.006% will fall outside, that is about one $\bar{X}$-value out of 16,000.

(c) Find the probability that, if two persons take four measurements each, the difference (regardless of the sign) between their means exceeds 0.3

Corollary 2 shows that (take $n_1 = n_2 = 4$, $m_1 = m_2 = 28.0$, $\sigma_1 = \sigma_2 = 0.25$)

$$\bar{X} - \bar{Y} \sim N(0, 0.25^2/4 + 0.25^2/4) = N(0, 0.177^2).$$

The complement of the desired event thus has probability

$$P(-0.3 < \bar{X} - \bar{Y} < 0.3) = \Phi\left(\frac{0.3 - 0}{0.177}\right) - \Phi\left(\frac{-0.3 - 0}{0.177}\right)$$
$$= 0.954 - 0.046 = 0.908,$$

whence the answer is $1 - 0.908 = 0.092$. □

Example 2. Lengths of Tiles

The length (in cm) of certain tiles is approximately $N(25, 0.4)$.

(a) Twenty randomly chosen tiles are aligned lengthwise. What is the probability that their total length exceeds 505 cm?

Call the lengths $X_1, X_2, \ldots, X_{20}$. According to Theorem 3′ we have $\sum X_i \sim N(500, 20 \cdot 0.4)$. This gives

$$P\left(\sum X_i > 505\right) = 1 - \Phi\left(\frac{505 - 500}{\sqrt{20 \cdot 0.4}}\right) = 1 - \Phi(1.77) = 0.038.$$

(b) Two rows are composed of 40 tiles each. What is the probability that the difference between the lengths of the rows is less than 10 cm?

Call the lengths $X_1, X_2, \ldots, X_{40}$ and $Y_1, Y_2, \ldots, Y_{40}$. We find

$$\sum_1^{40} X_i \sim N(1{,}000, 40 \cdot 0.4), \qquad \sum_1^{40} Y_i \sim N(1{,}000, 40 \cdot 0.4),$$

and thus $\sum X_i - \sum Y_i \sim N(0, 80 \cdot 0.4)$. Hence

$$P\left(\left|\sum X_i - \sum Y_i\right| < 10\right) = \Phi\left(\frac{10}{\sqrt{80 \cdot 0.4}}\right) - \Phi\left(\frac{-10}{\sqrt{80 \cdot 0.4}}\right)$$
$$= \Phi(1.77) - \Phi(-1.77) = 0.924. \qquad \square$$

8.6. The Central Limit Theorem

In this chapter the rv's have hitherto been assumed to have exactly normal distributions. We shall now briefly consider one of the most remarkable and important results in probability theory. The normal distribution turns out to be much more applicable than one may think at first sight. In fact, it can be proved that a sum of independent identically distributed rv's with an arbitrary distribution is as a rule *approximately* normally distributed, if only the number of components in the sum is large enough.

As an illustration of this property of the normal distribution we shall use an earlier example.

Example 3. Sum of rv's with Uniform Distribution over 1, 2, 3

In Example 6 in §5.3, we considered the distribution of the sum of two independent rv's taking the values 1, 2, 3 with equal probabilities. Let us now extend the calculations to a sum of three rv's and then to four rv's. The result is reproduced graphically in Fig. 8.7, where the cases of one and two rv's have also been included for comparison. It is seen that the distribution changes rapidly, and seems to assume more and more the shape of a bell characteristic of the normal distribution. □

A further illustration showing how a convolution of several distributions leads to an approximately normal distribution will be given in Fig. 9.3.

We shall now give a mathematical formulation of what has been said and illustrated above:

Theorem 4 ("Central Limit Theorem"). *If $X_1, X_2, \ldots$ is an infinite sequence of independent identically distributed* rv's, *each with expectation m and standard*

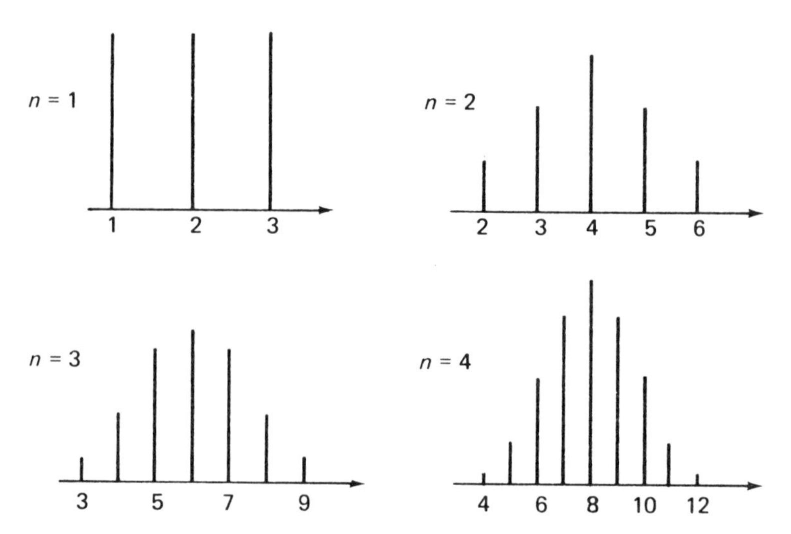

Fig. 8.7. Convolution of n independent rv's, each of which takes the values 1, 2, 3 with equal probabilities.

deviation $\sigma > 0$, *and if we set*

$$Y_n = X_1 + \cdots + X_n,$$

then we have

$$P(a < (Y_n - nm)/(\sigma\sqrt{n}) < b) \to \Phi(b) - \Phi(a) \qquad \textit{as} \quad n \to \infty. \tag{18}$$

□

Note that $E(Y_n) = nm$, $D(Y_n) = \sigma\sqrt{n}$. Hence, for any given n, we see that $(Y_n - nm)/(\sigma\sqrt{n})$ is a standardized rv (see §6.3). Hence it has the same mean $(= 0)$ and variance $(= 1)$ as a standard normal rv. The theorem contains a result which is much sharper: As n tends to infinity, the whole distribution of the given standardized variable will approach a standard normal distribution.

Theorem 4 can be proved in various ways; see, e.g. H. Cramér, *Mathematical Methods of Statistics* (Almqvist and Wiksell, Uppsala, 1945; Princeton University Press, Princeton, NJ, 1946).

We shall introduce a definition which may seem complicated at first, but which is useful when working with rv's satisfying the central limit theorem.

Definition. If Z_n, $n = 1, 2, \ldots$, is an infinite sequence of rv's and it is possible to find numbers A_n and B_n, $n = 1, 2, \ldots$, such that

$$P(a < (Z_n - A_n)/\sqrt{B_n} < b) \to \Phi(b) - \Phi(a) \qquad \text{when} \quad n \to \infty,$$

then Z_n is said to be *asymptotically normally distributed* with parameters A_n and B_n, or more briefly

$$Z_n \sim AsN(A_n, B_n).$$

□

Using this definition, the message of Theorem 4 is that $Y_n \sim AsN(nm, n\sigma^2)$.

Theorem 4 has an interesting consequence: For large n we may treat the rv Y_n as approximately normally distributed with the given parameters. This is a very important result.

We shall state a special consequence of Theorem 4:

Corollary. *If $X_1, X_2, \ldots,$ is an infinite sequence of independent identically distributed* rv's *with mean m and standard deviation σ, then $\bar{X}_n \sim AsN(m, \sigma^2/n)$.* □

The corollary is of basic importance for statistical applications of probability theory, in view of the fact that arithmetic means of identically distributed rv's often appear in such applications. It is then of great value to know that such a mean is approximately normally distributed if only the number of components is large enough.

However, some difficulty can arise in this connection. How large must n be in order to allow the normal approximation to be used? This question has no simple answer. In Example 3 the condition $n \geq 6$ gives enough accuracy for many purposes. It should be noted that, in general, the shape of the distribution of the variables X_i is important for the accuracy. In fact, the skewness of this distribution plays a decisive role. The more skew the distribution, the more components are needed for a good approximation.

Example 4. Round-Off Errors

Add 48 amounts, each of which has been rounded to whole dollars. The round-off errors $X_1, X_2, \ldots, X_{48}$ are assumed to be independently and uniformly distributed over the interval $(-0.5, 0.5)$. By taking $a = -0.5$, $b = 0.5$ in Example 3, p. 104, it is seen that these rv's have mean 0 and variance 1/12. According to the central limit theorem, the total round-off error $Y = X_1 + \cdots + X_{48}$ is approximately $N(0, 48/12)$, that is, $N(0, 2^2)$. Formula (18) shows that, for instance,

$$P(|Y| \leq 6) \approx \Phi\left(\frac{6-0}{2}\right) - \Phi\left(\frac{-6-0}{2}\right) = \Phi(3) - \Phi(-3) = 0.997.$$

Hence it is very improbable that the total round-off error exceeds 6 dollars.

This example is instructive because it reveals that the practice (common in certain circles) of computing maximum errors is inappropriate. The maximum positive error of the above sum is, as we see, 24 dollars, but this computation is far too pessimistic and not very interesting. Indeed, we have shown that even an error amounting to 6 dollars or more is improbable. □

Under certain general conditions, the central limit theorem also holds for sums of independent nonidentically distributed rv's. $X_1, X_2, \ldots.$ Set $Y_n = X_1 + \cdots + X_n$, $m_n = E(Y_n)$, $\sigma_n^2 = V(Y_n)$. Then we have

$$\begin{aligned} m_n &= E(X_1) + \cdots + E(X_n), \\ \sigma_n^2 &= V(X_1) + \cdots + V(X_n). \end{aligned} \tag{19}$$

The central limit theorem states that $Y_n \sim AsN(m_n, \sigma_n^2)$.

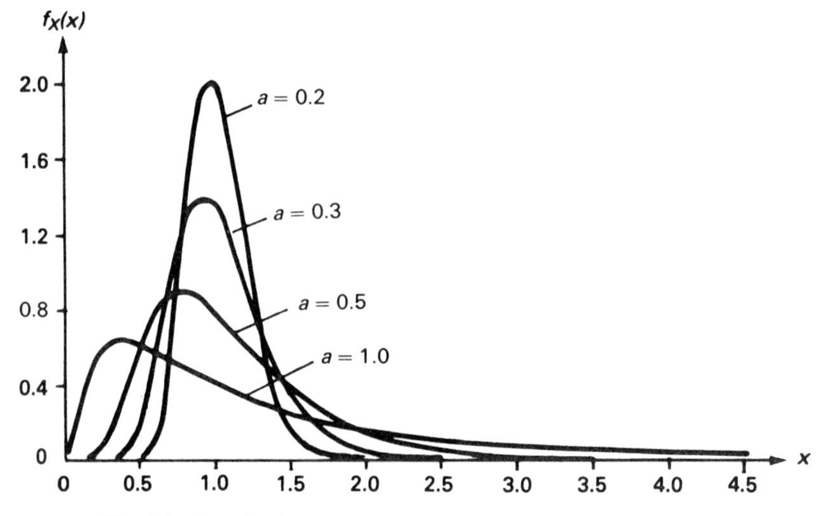

Fig. 8.8. Density functions of lognormal distributions ($b = 0$).

8.7. Lognormal Distribution

Definition. If the rv X has density function

$$f_X(x) = \begin{cases} 1/[xa\sqrt{2\pi}]e^{-(\ln x - b)^2/2a^2} & x \geq 0, \\ 0 & x < 0, \end{cases}$$

where $-\infty < b < \infty$ and $a > 0$, then X is said to have a *lognormal distribution*. □

The definition entails that $\ln X \sim N(b, a^2)$. (Perform the transformation $Y = \ln X$ in the expression for $f_X(x)$.) Therefore, it is not necessary to memorize the form of the density; it is easier to remember that taking the logarithm results in an ordinary normal distribution.

Figure 8.8 depicts the density function for $b = 0$ and some values of a. Clearly, the lognormal distribution contains a family of more or less skew distributions of positive rv's. This family is quite useful, for skew distributions are not uncommon in practice.

Exercises

801. The rv X is $N(0, 1)$. Compute $P(0.21 < X < 0.29)$, $P(-0.21 < X < 0.29)$, and $P(-0.29 < X < -0.21)$. (§8.3)

802. The rv X is $N(0, 1)$. Determine x such that:
 (a) $P(X > x) = 0.001$;
 (b) $P(X > x) = 0.999$;
 (c) $P(|X| < x) = 0.95$;
 (d) $P(X < -x) = 0.10$. (§8.3)

803. The rv X is $N(0, 1)$ and $Y = 3X + 2$. Find $E(Y)$ and $D(Y)$. (§8.3)

804. The rv X is $N(-1, 0.01^2)$. Compute $P(X < 0.99)$, $P(X < -0.99)$, $P(X > -0.99)$ and $P(-1.3 < X \leq -1.03)$. (§8.4)

805. The rv X is $N(20, 3^2)$. Find x such that $P(X \leq x) = 0.01$. (§8.4)

806. Consider a rv X which is $N(180, 5^2)$. Compute the probability that $X \geq 170$ and the probability that $170 \leq X \leq 200$. (§8.4)

807. When packaging margarine the weight X of a packet (unit: kilo) is a rv which is, approximately, $N(0.5, 0.003^2)$. Determine the probability that a packet weighs at least 0.495. Also determine d such that, in the long run,
   (a) 50%
   (b) 95%
   (c) 99%
   of all packets of margarine have a weight between $0.5 - d$ and $0.5 + d$. (§8.4)

808. It is known from long experience that the diameters of balls in a certain ball-bearing may be considered as normally distributed. For the purpose of rapidly determining the parameters in this normal distribution, one counts the number of balls that can pass circular holes of different magnitudes. On one occasion, it is found that holes with diameters 4.90 and 5.00 mm are passed by 23% of the balls and 59% of the balls, respectively. Determine the mean and standard deviation of the diameters of the balls. (§8.4)

809. On a lathe, metal cylinders are produced to have a certain diameter which is $N(12.4, \sigma^2)$. What is the largest possible value of σ if at most 5% of the cylinders are permitted to have a diameter outside the interval (12.0, 12.8)? (§8.4)

810. Let X and Y be independent and $N(1, 1)$ and $N(-1, 2^2)$, respectively. Find the distribution of $X + Y$ and that of $X - Y$. (§8.5)

811. Let X and Y be independent and $N(150, 3^2)$ and $N(100, 4^2)$, respectively.
   (a) Find the distributions of $X + Y$, $X - Y$ and $(X + Y)/2$.
   (b) Determine the probability that $X + Y < 242.6$, the probability that $|X - Y| < 40$, and the probability that $|(X + Y)/2 - 125| > 5$. (§8.5)

812. A's and B's monthly expenses may be regarded as independent rv's $X \sim N(720, 30^2)$ and $Y \sim N(640, 25^2)$, respectively (unit: dollar).
   (a) Find the distribution of $X + Y$ and that of $X - Y$.
   (b) Find the probability that their total expenses for a month exceed 1,400 dollars.
   (c) Find the probability that B's expenses exceed those of A. (§8.5)

813. (Continuation of the preceding excercise)
   (a) Determine the distribution of A's total yearly expenses. The different monthly expenses are assumed to be independent.
   (b) What is the probability that A's total expenses during a year are less than 8,900 dollars?
   (c) Compute the probability that B's total yearly expenses exceed those of A. Compare the answer of (c) with the answer of (c) in Exercise 812. (§8.5)

814. The rv's X_1 and X_2 are independent and $N(1, 2^2)$. Find the distribution of $\bar{X} = (X_1 + X_2)/2$. (§8.5)

815. $X_1, X_2, \ldots, X_n$ are independent and $N(m, 0.2^2)$.
   (a) Find the distribution of $\bar{X} - m$.
   (b) Find $P(|\bar{X} - m| > 0.2/\sqrt{n})$.
   (c) Find $P(|\bar{X} - m| > 0.1)$ if $n = 16$.
   (d) We want that $P(|\bar{X} - m| > 0.01)$ is less than 0.001. How large must n be? (§8.5)

816. Consider a large batch of pills. The weight (unit: gram) of a randomly chosen pill can with good accuracy be regarded as a rv $X \sim N(m, 0.02^2)$, where m is the mean weight of the pills in the batch. To check the weight a number of pills are chosen and weighed. Suppose that $m = 0.65$.
   (a) Determine the probability that the weight of a pill lies outside the interval (0.60, 0.70).
   (b) Determine the probability that the arithmetic mean $\bar{X}$ of the weights of 30 pills lies outside the interval (0.64, 0.66).
   (c) We want the probability to be at most 0.01 that $\bar{X}$ is outside the interval (0.64, 0.66). How many pills should be weighed? (§8.5)

817. $X_1, X_2, \ldots$ are independent errors of measurement with mean 0.1 and standard deviation 8. Put $\bar{X}_n = (X_1 + \cdots + X_n)/n$ and determine approximately $P(|\bar{X}_n| > 0.08)$ for $n = 16$, 1,600 and 160,000. (§8.6)

818. The weight (unit: gram) of a randomly chosen pill of a certain type is a rv with mean 0.65 and standard deviation 0.02.
   (a) Find the mean and standard deviation of the total weight of 100 pills (whose weights are supposed to be independent).
   (b) Use the central limit theorem to determine approximately the probability that 100 pills weigh at most 65.3 grams. (§8.6)

819. Let $X_1, X_2, \ldots, X_{100}$ be independent rv's with density function $f(x) = e^{-x}$ $(x \geq 0)$ and set $Y = X_1 + \cdots + X_{100}$. Evaluate approximately $P(Y > 110)$. (§8.6)

820. Determine the median of a lognormal distribution. (§8.7)

*821. In a factory, items are manufactured that are put together two at a time to a certain product. Each item is correct with probability 0.9 and defective with probability 0.1. The product gives a profit of 1 dollar if it is correct, a profit of 0 dollar if exactly one of the two items is defective, and a loss of 10 dollars if both items are defective. Compute approximately the probability that the profit of 1,000 sold products is at least 800 dollars.

*822. (Continuation of Exercise 818.) A box of pills should contain 100 pills. When filling a box it is convenient to pour pills on to the scale of a balance and stop when the weight exceeds 65 grams. Use the central limit theorem to find the probability that a box will contain at least 100 pills.
   Hint: The box contains at least 100 pills when and only when the weight of 99 pills is less than 65 grams.

CHAPTER 9

The Binomial and Related Distributions

9.1. Introduction

The binomial distribution was introduced in §3.5, where we also mentioned the hypergeometric distribution, the Poisson distribution and the multinomial distribution. All these distributions are related to the binomial distribution. In the present chapter, we will discuss the binomial distribution in §9.2, the hypergeometric distribution in §9.3, the Poisson distribution in §9.4, and the multinomial distribution in §9.5.

9.2. The Binomial Distribution

As has been previously stated, a rv X has a binomial distribution if

$$p_X(k) = \binom{n}{k} p^k q^{n-k} \qquad (k = 0, 1, \ldots, n), \tag{1}$$

where $0 \leq p \leq 1$ and $q = 1 - p$. We remind the reader of the code name $X \sim \text{Bin}(n, p)$.

(a) Occurrence

The binomial distribution occurs in certain types of random trials: At each of n independent trials it is noted if a certain event A occurs or not. For example, A may be "a six is obtained in a throw of a die" or "a student passes an examination". The probability that A occurs is p in each trial, and hence the probability of the complementary event A^* is $q = 1 - p$.

Let X be the number of times that A occurs in the n trials; that is, X is the *frequency* of A. Then X has the probability function in (1). We sum up this in:

Theorem 1. *Assume that the probability of occurrence of an event A in a single trial is p. If n independent trials are performed and X is the frequency of A, then* $X \sim \text{Bin}(n, p)$. □

PROOF. We have already proved the theorem in a special case, namely, when drawing with replacement from an urn with two kinds of objects (see §2.4). The same method of proof is used now. Consider first all possible sequences of outcomes such that A occurs exactly k times. One such sequence is

$$\underbrace{AA A \ldots A}_{k}\underbrace{A^* A^* A^* \ldots A^*}_{n-k},$$

another being

$$\underbrace{AAA \ldots A}_{k-1}\underbrace{A^* A A^* A^* A^* \ldots A^*}_{n-k+1}.$$

The total number of such sequences is equal to the number of ways that k A's can be placed into n positions, that is $\binom{n}{k}$ (see §2.7). The probability of obtaining the first sequence above is

$$\underbrace{pp \cdot \ldots \cdot p}_{k} \cdot \underbrace{qq \cdot \ldots \cdot q}_{n-k} = p^k q^{n-k}.$$

All other possible sequences occur with the same probability. To determine the desired probability $P(X = k)$ we need only sum over all the $\binom{n}{k}$ sequences, for they are mutually exclusive.

Hence (1) follows, proving the theorem. □

The terminology of independent trials is useful in many applications. However, it is also instructive to formulate the problem in another way. Consider a "point" (particle, for example) that moves stepwise in the first quadrant of a rectilinear coordinate system starting at the origin. At each stage, the point moves $+1$ horizontally and then a distance U vertically equal to $+1$ or 0 with probabilities p and $q = 1 - p$, respectively. The steps at the different stages are independent. Evidently, the point will perform a *random walk* as shown in Fig. 9.1. (In this figure the vertical steps are 1 0 0 1 0 1 1 0 0 0 1 0.)

Many questions concerning the behaviour of the point can be formulated, for example: What is the probability that the point reaches the position (n, k)

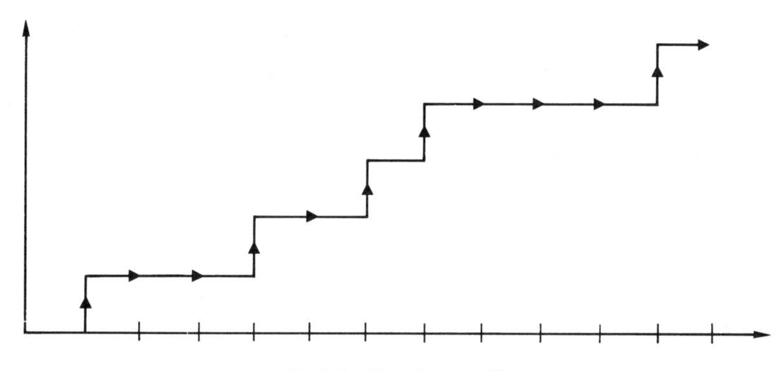

Fig. 9.1. Random walk.

in n steps? After some reflection the reader will realize that this is exactly the same problem as discussed earlier.

The theory of random walks is very comprehensive and important in several fields, for example, in statistical mechanics. In fact, Einstein showed, in a classical paper from the beginning of this century, that Brownian motion can be described mathematically as a random walk in three dimensions.

The binomial distribution also appears as the distribution of the sum of independent identically distributed rv's with a two-point distribution. As a matter of fact, this was proved already in Theorem 1: If a rv has a two-point distribution it takes on values a and b with probabilities p and q. Associate with each of the n independent trials a rv that assumes the value 1 if the event A occurs and 0 if not. Call these rv's $U_1, \ldots, U_n$. Clearly, the number of such variables equal to 1 is just the number of times X when A occurs, that is,

$$X = U_1 + U_2 + \cdots + U_n. \tag{2}$$

But the distribution of X is known from Theorem 1. Hence we have shown that the sum of the U's has the same distribution as X, that is, a binomial distribution.

We note in passing that, as is immediately apparent, $U_i \sim \mathrm{Bin}(1, p)$. Hence (2) shows that, if the rv's U_i, $i = 1, \ldots, n$, are independent and $\mathrm{Bin}(1, p)$, then we have $\sum_{i=1}^{n} U_i \sim \mathrm{Bin}(n, p)$.

(b) Exact Properties

The probability function and the distribution function,

$$\left.\begin{aligned} p_X(k) &= \binom{n}{k} p^k q^{n-k}, \\ F_X(k) &= P(X \le k) = \sum_{j=0}^{k} \binom{n}{j} p^j q^{n-j}, \end{aligned}\right. \qquad (k = 0, 1, \ldots, n) \tag{1'}$$

can easily be calculated numerically for small values of n. A table is found in Table 8 at the end of the book; larger tables are given in the references at the end of the book.

Example 1. Binomial Distribution for $n = 5$

For $n = 5$ and $p = 0.1, 0.5, 0.9$ the probability functions are as follows:

	p		
k	0.1	0.5	0.9
0	0.590	0.031	0.000
1	0.328	0.156	0.001
2	0.073	0.313	0.008
3	0.008	0.313	0.073
4	0.001	0.156	0.328
5	0.000	0.031	0.590

In Fig. 9.2 these probability functions are shown graphically. As in these figures it is generally true that the distribution is concentrated on the left when p is small, on the right when p is large, and is symmetrical for $p = 1/2$. □

It is important to know the mean, variance and standard deviation of a binomially distributed rv. They are given in

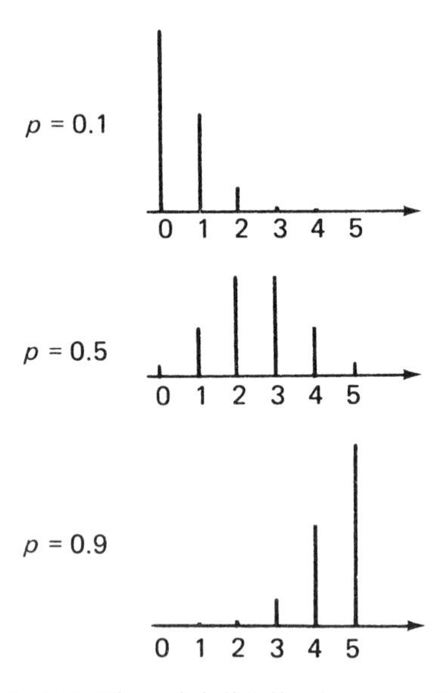

Fig. 9.2. Binomial distributions, $n = 5$.

Theorem 2. *If X is* Bin(n, p) *we have*

$$E(X) = np; \qquad V(X) = npq; \qquad D(X) = \sqrt{npq}. \tag{3}$$

□

PROOF. We give two proofs, the second of which is the nicer.

(a) Let us evaluate the mean directly from formula (1) in §6.2. Since

$$k\binom{n}{k} = k \cdot \frac{n(n-1)\cdot\ldots\cdot(n-k+1)}{k!} = n\binom{n-1}{k-1},$$

we obtain

$$E(X) = \sum_{k=0}^{n} kp_X(k) = \sum_{k=0}^{n} k\binom{n}{k}p^k q^{n-k} = np\sum_{k=1}^{n}\binom{n-1}{k-1}p^{k-1}q^{n-k}$$

$$= np\sum_{j=0}^{n-1}\binom{n-1}{j}p^j q^{n-1-j} = np(q+p)^{n-1} = np.$$

Likewise, we have

$$E(X^2) = E[X(X-1)] + E(X) = n(n-1)p^2 + np.$$

Hence

$$V(X) = E(X^2) - [E(X)]^2 = n(n-1)p^2 + np - (np)^2 = np - np^2 = npq,$$

$$D(X) = \sqrt{V(X)} = \sqrt{npq}.$$

(b) From the representation (2) of a binomially distributed rv as a sum it follows that

$$E(X) = E(U_1) + \cdots + E(U_n),$$

$$V(X) = V(U_1) + \cdots + V(U_n).$$

Each U_i is 1 and 0 with probabilities p and q, respectively, and hence we have

$$E(U_i) = 1 \cdot p + 0 \cdot q = p; \qquad E(U_i^2) = 1^2 \cdot p + 0^2 \cdot q = p,$$

$$V(U_i) = E(U_i^2) - [E(U_i)]^2 = p - p^2 = p(1-p) = pq.$$

When we insert this in the expressions for $E(X)$ and $V(X)$ above, we obtain the results in the theorem. □

There is a convolution theorem for the binomial distribution:

Theorem 3. *If $X \sim$* Bin(n_1, p) *and $Y \sim$* Bin(n_2, p), *where X and Y are independent, then $X + Y \sim$* Bin$(n_1 + n_2, p)$. □

The theorem is intuitively obvious: Consider a sequence of independent trials which are performed in two rounds with n_1 and n_2 trials.

(c) Approximate Properties

For large n the exact expressions for the probability function and the distribution function are awkward to handle numerically. Fortunately, good approximations can be obtained in this case.

(1) Normal Approximation. For large n the binomial distribution can be approximated by a normal distribution. In fact, X is then approximately normally distributed with mean and variance given by (3); that is, we have approximately $X \sim N(np, npq)$. By formula (4) in §8.2 we now have

$$P(a < X \leq b) \approx \Phi\left(\frac{b - np}{\sqrt{npq}}\right) - \Phi\left(\frac{a - np}{\sqrt{npq}}\right). \tag{4}$$

The accuracy is generally somewhat improved if, on the right-hand side, b is replaced by $b + 1/2$ and a by $a + 1/2$ so that

$$P(a < X \leq b) \approx \Phi\left(\frac{b + 1/2 - np}{\sqrt{npq}}\right) - \Phi\left(\frac{a + 1/2 - np}{\sqrt{npq}}\right). \tag{4'}$$

We say that we have then used a *continuity correction.* Formula (4′) gives fair accuracy if npq is at least 10 or so.

Figure 9.3 explains formula (4′) to some extent. The probability function of the binomial distribution is illustrated in this figure by rectangles of breadth 1. The total area of the rectangles for $a + 1$, $a + 2$, ..., b is obtained approximately by integrating the given normal density function in the figure from $a + 1/2$ to $b + 1/2$.

Rigorous proofs for the arguments behind formulae (4) and (4′) are given in many textbooks of greater scope, for example, in W. Feller, *An Introduction to Probability Theory and Its Applications*, Vol. I (3rd ed., Wiley, New York, 1968, Ch. VII). However, some hints on the background of the approximations can be given. According to formula (2), the binomially distributed rv X can be represented as a sum of n rv's with

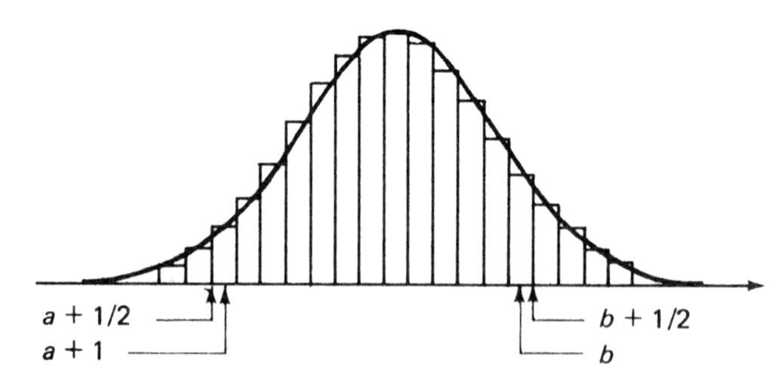

Fig. 9.3. Approximation of binomial distribution by normal distribution when $n = 100$, $p = 0.20$, $a = 12$, $b = 25$.

a two-point distribution. Combine this with the central limit theorem, p. 141, which states that the sum of many independent identically distributed rv's is approximately normally distributed. This shows that the binomial distribution, being the sum of n variables each with a two-point distribution, is approximately normally distributed and that the approximation improves with increasing n.

Example 2. Tossing a Coin

Let us determine the probability that, when a well-made coin is tossed 200 times, the number of heads lies between 95 and 105, the limits included. By (1) the exact expression is

$$P = \sum_{j=95}^{105} \binom{200}{j} \left(\frac{1}{2}\right)^j \left(\frac{1}{2}\right)^{200-j}.$$

This sum is awkward to calculate, since available tables are not large enough. If we use a normal approximation with continuity correction, that is, formula (4′), we find ($a = 94$; $b = 105$; $np = 200 \cdot 1/2 = 100$; $npq = 200 \cdot 1/2 \cdot 1/2 = 50$)

$$P \approx \Phi\left(\frac{105.5 - 100}{\sqrt{50}}\right) - \Phi\left(\frac{94.5 - 100}{\sqrt{50}}\right) = 0.56331.$$

The correct value is 0.56325. The accuracy is not always as good as this. □

(2) Poisson Approximation. When p is small the binomial distribution can be approximated by the Poisson distribution (see §3.5 and §9.4 later in this chapter). Then X has approximately a Poisson distribution with parameter $m = np$; that is, we have approximately

$$p_X(k) = e^{-np}(np)^k/k! \qquad (k = 0, 1, \ldots). \tag{5}$$

For practical purposes the approximation is usually good enough if p does not exceed 0.1.

Example 3. Birthdays

Calculate the probability that exactly 2 out of 100 people have their birthday on the first of January. Let us assume the probability of birth on that day is 1/365 (which is not quite true, since the birth rate varies over the year). If the 100 births are considered independent, Theorem 1, p. 148, shows that the desired probability is

$$\binom{100}{2} \left(\frac{1}{365}\right)^2 \left(\frac{364}{365}\right)^{100-2}.$$

Since $p < 0.1$ we may use the Poisson approximation which gives ($np = 100/365 = 0.27$)

$$e^{-0.27} \cdot 0.27^2/2! = 0.028.$$

If, instead, we ask for the probability that at least two of the persons were born on the given day, the same approximation gives

$$\sum_{j=2}^{\infty} e^{-0.27} 0.27^j/j! = 0.031.$$

Here, we have used Table 7 of the Poisson distribution. Alternatively, we may write the sum in the form

$$1 - \sum_{j=0}^{1} e^{-0.27} \cdot 0.27^j/j! = 1 - e^{-0.27} - 0.27e^{-0.27} = 0.031. \qquad \square$$

(d) Relative Frequencies

Suppose that in each of n independent trials we observe whether a certain event A occurs. We know already that the frequency X satisfies $X \sim \text{Bin}(n, p)$. It is often of interest to consider the *relative frequency* of A, that is, the ratio $Y = X/n$ of the frequency to the total number of trials. Of course, this ratio lies between 0 and 1.

The distribution of Y is easily obtained from that of X given in formula (1). Using the rules in Theorem 3 of §6.3 we obtain from (3) the following important expressions for mean, etc. of Y:

$$E(Y) = p; \qquad V(Y) = pq/n; \qquad D(Y) = \sqrt{pq/n}. \tag{6}$$

It is seen that the standard deviation $D(Y)$ is small when n is large. Hence Y then varies only little around its mean p. This is in accordance with common sense: For example, if a good die is thrown many times, the relative frequency of ones, say, ought to be near $p = 1/6$. We discussed this situation already on p. 10, and are now satisfied to see that the theory works, for it gives sensible results.

Let us now suppose that n increases to infinity. Then the following famous theorem can be formulated:

Theorem 4 ("Bernoulli's Theorem"). *For any given* $\varepsilon > 0$ *we have*

$$P(|Y - p| > \varepsilon) \to 0 \qquad \text{as} \quad n \to \infty. \qquad \square$$

The theorem tells us that the probability mass outside the interval $(p - \varepsilon, p + \varepsilon)$ tends to zero when n tends to infinity, and this is true however small ε is chosen. A most remarkable and elegant result!

James Bernoulli, who proved this result in the seventeenth century, is said to have toiled on it for 20 years. We children of the twentieth century write down the proof in a trice, using Chebyshev's inequality (which was unknown to Bernoulli). It is handled precisely as in the proof of the law of large numbers in §7.3.

The reader will easily see that Bernoulli's theorem is a special case of the law of large numbers. The representation (2) of a binomially distributed rv and a moment's thought is all that is needed.

Clearly, the approximations given earlier can be also used for relative frequencies. For example, for large n we have approximately

$$Y = \frac{X}{n} \sim N(p, pq/n).$$

This approximation has several important uses, such as the following: Let $Y_1 = X_1/n_1$ and $Y_2 = X_2/n_2$ be two independent relative frequencies, where $X_1 \sim \text{Bin}(n_1, p_1)$ and $X_2 \sim \text{Bin}(n_2, p_2)$. We then have approximately (see Theorem 3 in §8.5) that

$$Y_1 - Y_2 \sim N\left(p_1 - p_2, \frac{p_1 q_1}{n_1} + \frac{p_2 q_2}{n_2}\right). \tag{7}$$

In fact, it is possible to extend Theorem 4 in §8.6, and show that asymptotically normally distributed rv's satisfying the conditions of Theorem 4 may be handled in the same way as exactly normally distributed rv's.

9.3. The Hypergeometric Distribution

(a) Occurrence

As we saw in §2.4, the hypergeometric distribution occurs in connection with urn models. We repeat the description here, using somewhat different terminology and different notation.

In a population of N elements, Np have the characteristic A and the remaining Nq do not, where $q = 1 - p$. Hence p is the relative frequency of "A-elements" in the population. We draw n elements at random *without replacement* and count the number X of A-elements in our sample. Then X has a hypergeometric distribution. We give a more complete description in:

Theorem 5. *Let the relative frequency of A-elements in a population of N elements be p. If n elements are drawn at random without replacement, the number X of A-elements has the hypergeometric distribution*:

$$p_X(k) = \binom{Np}{k}\binom{Nq}{n-k}\bigg/\binom{N}{n} \tag{8}$$

(*k takes on all integer values such that* $0 \leq k \leq Np$, $0 \leq n - k \leq Nq$). □

The proof of this theorem was given earlier (see §2.4).

Remark 1.

We recall the fact that the result is different if the sampling is performed with replacement. The problem is then of the type discussed in §9.2. The event A there referred to "A-element is obtained" and the probability that A occurs in a certain drawing is p. Theorem 1 in §9.2 then shows that $X \sim \text{Bin}(n, p)$. □

Remark 2.

If, in Theorem 5, one lets N tend to infinity, a binomial distribution is obtained in the limit. (The proof of this is not entirely straightforward.) This result is quite reasonable.

If N is large compared to the number n of elements taken from the population, the population changes very little as the elements are drawn one by one; that is, the drawings may be regarded approximately as independent trials with constant probability of obtaining an A-element. Then we have the situation which led in §9.2 to a binomial distribution. □

The model discussed here is very useful. The population may be, for example, a group of individuals; an A-element may then signify a person with a certain opinion or some other characteristic. In another common situation the population is a batch of units; an A-element may then signify a defective unit. Like the binomial distribution, the hypergeometric distribution therefore has great practical importance, as will be exemplified later.

(b) Exact Properties

The probability function and the distribution function are, respectively,

$$p_X(k) = \binom{Np}{k}\binom{Nq}{n-k}\bigg/\binom{N}{n},$$
$$F_X(k) = P(X \le k) = \sum_{j=0}^{k} \binom{Np}{j}\binom{Nq}{n-j}\bigg/\binom{N}{n}. \qquad (8')$$

Tables of these functions are available (see the references at the end of the book).

We state without proof:

Theorem 6. *For a hypergeometric distribution we have*

$$E(X) = np; \qquad V(X) = \frac{N-n}{N-1}npq; \qquad D(X) = \sqrt{\frac{N-n}{N-1}}\cdot\sqrt{npq}. \qquad (9)$$

□

The mean is the same as for the binomial distribution, but the variance is different in so far as there is a factor $(N-n)/(N-1)$ in the hypergeometric case (see Theorem 2 in §9.2). We introduce the notation

$$d_n^2 = \frac{N-n}{N-1}; \qquad d_n = \sqrt{\frac{N-n}{N-1}}, \qquad (10)$$

and call d_n^2 the *finite population correction.* The reason for the name is given in Remark 2 above: If N is large compared to n, the rv X is approximately binomially distributed and d_n is near 1. When $n > 1$, we have $d_n < 1$ which means that the variance of the hypergeometric distribution is less than that of the binomial distribution.

It is instructive to see how d_n depends on the relative number of elements, n/N, taken from the population. Apart from the factor $\sqrt{N/(N-1)}$, which is near 1 if N is not very small, we have

n/N	0.02	0.04	0.06	0.08	0.10	0.20	0.30	0.40	0.50
d_n	0.99	0.98	0.97	0.96	0.95	0.89	0.84	0.77	0.71

If the quotient n/N is small (say less than about 0.1), we see that d_n is rather close to 1. Hence the error is then small if d_n is taken equal to 1, which means that the variance of the binomial distribution is in fact used (see Remark 2).

(c) Approximate Properties

As in the case of the binomial distribution it is often difficult and time-consuming to use the exact expressions for the probability function and the distribution function. It is then of great value to have good approximations available:

(1) Binomial Approximation. If the quotient n/N is small (less than about 0.1), X can be approximated by a binomially distributed rv with parameters n and p, that is, we have

$$p_X(k) \approx \binom{n}{k} p^k q^{n-k}.$$

A reason for this approximation has already been given in Remark 2 (see also the discussion of the finite population correction above).

(2) Normal Approximation. If n is so large that the variance $[(N-n)/(N-1)]npq$ is at least about 10, the normal approximation can be used. The procedure is the same as in the binomial case, with the modification that, instead of npq, the above exact expression for the variance is used. A continuity correction generally improves the accuracy.

Example 4. Market Research

Four hundred out of 1,000 housewives like a certain type of soap, while the rest dislike it. A random sample of 150 housewives taken from this population are questioned. What is the probability that at most 50 like the soap?

Let X be the number of positive answers. Clearly, X has a hypergeometric distribution. Taking $p = 400/1{,}000 = 0.4$ in Theorem 6, we obtain

$$E(X) = 150 \cdot 0.4 = 60; \qquad V(X) = \frac{1{,}000 - 150}{1{,}000 - 1} 150 \cdot 0.4 \cdot 0.6 = 30.6.$$

Formula (4) in §9.2, with modification of the variance and with continuity correction, gives

$$P(X \leq 50) \approx \Phi\left(\frac{50.5 - 60}{\sqrt{30.6}}\right) = \Phi(-1.71) = 1 - \Phi(1.71) = 0.044. \qquad \square$$

(3) Poisson Approximation. If p is small and N large compared to n, the Poisson distribution furnishes a good approximation:

$$p_X(k) \approx e^{-np}(np)^k/k! \qquad (k = 0, 1, \ldots).$$

The accuracy is usually good enough if $p + n/N \leq 0.1$.

(d) More About Relative Frequencies

In subsection (d) of §9.2 we studied the relative frequency $Y = X/n$ in the case of n independent trials. The discussion can easily be adapted to the situation we now consider, that is, to sampling without replacement from a population of N elements. We skip the details and give only formulae corresponding to (6):

$$E(Y) = p; \qquad V(Y) = d_n^2 pq/n; \qquad D(Y) = d_n\sqrt{pq/n}. \tag{11}$$

Apart from the factor d_n, the formulae are as before.

9.4. The Poisson Distribution

The Poisson distribution, which was first introduced in §3.5, is an interesting and important distribution. The probability function is given by

$$p_X(k) = e^{-m}m^k/k! \qquad (k = 0, 1, \ldots). \tag{12}$$

We recall the code name $X \sim \text{Po}(m)$.

(a) Occurrence

The Poisson distribution occurs under general conditions when studying events that happen at random in time or space. Let us consider events A in time. For example, A may be calls made to a telephone exchange or customers arriving at a shop; see Fig. 9.4. Let X be the number of events occurring in an interval of length t. We are interested in the distribution of X.

By events occurring at random in time we mean events that may happen at any moment and occur independently. The assumption of independence implies that, if several events have occurred during a time period, this does not affect the number of events taking place during a later period.

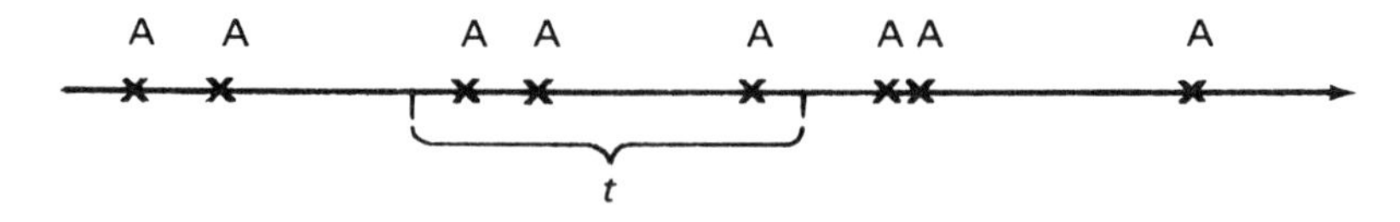

Fig. 9.4.

If these somewhat vaguely described conditions are satisfied, then X will have a Poisson distribution. Now also assume that the events arise with constant intensity so that, on the average, λ events take place during a period of unit length; hence λt events, on the average, during a time period of length t. Then it is true that $X \sim \text{Po}(\lambda t)$.

Proofs of what has been said above can be found in larger textbooks, for example, in W. Feller, *An Introduction to Probability Theory and Its Applications*, Vol. I (3rd ed., Wiley, New York, 1968).

Example 5. Radioactive Decay

When studying radioactive decay of a sufficiently large substance, the conditions given above are satisfied. The number of particles disintegrating during the time interval $(T, T + t)$ is not affected by the number of particles having disintegrated before T. We may therefore expect that X has a Poisson distribution. In this case λ is the decay intensity, which is the mean number of disintegrations taking place during a time period of unit length. For example, if $\lambda = 20$ per minute and $t = 0.5$ minutes, then $\lambda t = 20 \cdot 0.5 = 10$ and $X \sim \text{Po}(10)$. □

The Poisson distribution also arises as an approximation to the binomial and hypergeometric distributions (see the earlier sections of this chapter).

(b) Exact Properties

The distribution function is tabulated in Table 7 at the end of the book; for more extensive tables, see the references.

Example 6. Poisson Distribution for $m = 0.5$, 1.0 and 2.0

For the above values of m the probability functions are as follows:

	m		
k	0.5	1.0	2.0
0	0.607	0.368	0.135
1	0.303	0.368	0.271
2	0.076	0.184	0.271
3	0.013	0.061	0.180
4	0.001	0.015	0.090
5	0.000	0.003	0.036
6	0.000	0.001	0.012
7	0.000	0.000	0.003
8	0.000	0.000	0.001

In Fig. 9.5 these probability functions are illustrated. It is seen that the distribution becomes more symmetric for larger values of m; this can be confirmed by a more detailed study.

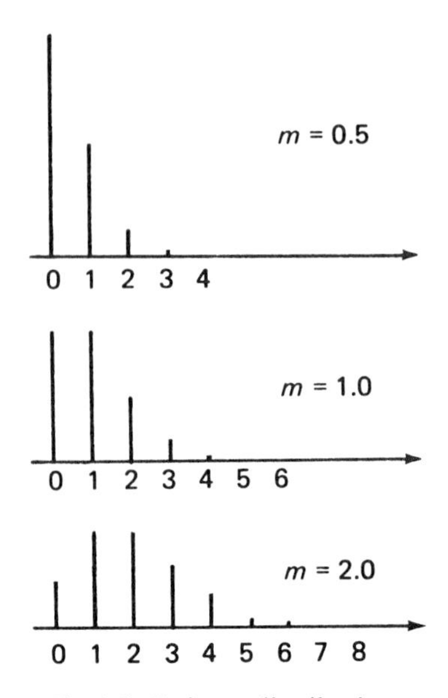

Fig. 9.5. Poisson distributions.

It is essential to know:

Theorem 7. *If X is $\text{Po}(m)$ we have*

$$E(X) = m; \qquad V(X) = m; \qquad D(X) = \sqrt{m}. \tag{13}$$

□

We note that the variance is equal to the mean, which is an interesting and important property of the distribution.

PROOF. The expressions can be derived in the usual manner starting from the probability function. For example, we find

$$E(X) = \sum_{k=0}^{\infty} ke^{-m}m^k/k! = me^{-m} \sum_{k=1}^{\infty} m^{k-1}/(k-1)!$$

$$= me^{-m} \sum_{j=0}^{\infty} m^j/j! = me^{-m}e^m = m.$$

□

The Poisson distribution has an important additive property:

Theorem 8. *If $X \sim \text{Po}(m_1)$ and $Y \sim \text{Po}(m_2)$, where X and Y are independent, we have $X + Y \sim \text{Po}(m_1 + m_2)$.* □

PROOF. The convolution formula (6′) in §5.3 and a rearrangement leads to

$$p_{X+Y}(k) = \sum_{i=0}^{k} e^{-m_1}\frac{m_1^i}{i!} \cdot e^{-m_2}\frac{m_2^{k-i}}{(k-i)!} = e^{-(m_1+m_2)}\frac{(m_1+m_2)^k}{k!}K,$$

where

$$K = \sum_{i=0}^{k} \binom{k}{i} \left(\frac{m_1}{m_1 + m_2} \right)^i \left(\frac{m_2}{m_1 + m_2} \right)^{k-i}$$

$$= \sum_{i=0}^{k} \binom{k}{i} \alpha^i (1 - \alpha)^{k-i}.$$

Here $\alpha = m_1/(m_1 + m_2)$. By the binomial theorem the last sum is 1, and the theorem is proved. □

Example 7. Defects in Woollen Fabrics

In a certain type of woollen fabric there are, on the average, 0.1 weaving defects per meter. The number of weaving defects in a piece of cloth has a Poisson distribution and is independent of the number of defects in other pieces. Two pieces, of lengths 30 and 40 meters, are sent to a customer. Find the distribution of the total number of defects in the two pieces.

Let X and Y be the number of defects and Z their sum. We have $X \sim \text{Po}(30 \cdot 0.1)$, $Y \sim \text{Po}(40 \cdot 0.1)$. By Theorem 8 we have $Z \sim \text{Po}(7)$. □

(c) Approximate Property

Since the Poisson distribution contains only one parameter, it is easier to handle than the binomial and the hypergeometric distributions, and approximations for it are not as often needed as for these distributions. However, the following approximation is useful: For large m we have, approximately, $X \sim N(m, m)$. The accuracy is, in general, satisfactory if m is at least 15. A continuity correction generally improves the accuracy (see p. 152).

Example 8. Poisson Distribution for $m = 100$

We want to evaluate $P(90 \leq X \leq 110)$ for $X \sim \text{Po}(100)$. Using a continuity correction the normal approximation gives the answer

$$P(90 \leq X \leq 110) \approx \Phi\left(\frac{110.5 - 100}{\sqrt{100}}\right) - \Phi\left(\frac{89.5 - 100}{\sqrt{100}}\right)$$

$$= \Phi(1.05) - \Phi(-1.05) = 0.706.$$ □

9.5. The Multinomial Distribution

In §4.3 the multinomial distribution was mentioned as an example of a multidimensional discrete distribution. We shall now examine it in a little more detail.

We repeat the definition: $(X_1, \ldots, X_r)$ has a multinomial distribution if

$$p_{X_1,\ldots,X_r}(k_1, \ldots, k_r) = \frac{n!}{k_1! \ldots k_r!} p_1^{k_1} \cdot \ldots \cdot p_r^{k_r}, \tag{14}$$

where $\sum p_i = 1$. In this expression $k_1, \ldots, k_r$ assume all nonnegative integer values with sum n.

The importance of this distribution in practice follows from a generalization of Theorem 1 in §9.2, which we do not prove:

Theorem 1′. *A trial can result in r different alternatives $A_1, \ldots, A_r$ with probabilities $p_1, \ldots, p_r$, respectively, where $\sum p_i = 1$. If n independent trials are performed and X_i is the frequency of A_i, then the* rv $(X_1, \ldots, X_r)$ *has a multinomial distribution with probability function* (14). □

If the trials have only two outcomes, A_1 and A_2, (14) reduces to

$$p_{X_1,X_2}(k_1, k_2) = \frac{n!}{k_1!\,k_2!} p_1^{k_1} p_2^{k_2}.$$

Since $k_1 + k_2 = n$, $p_1 + p_2 = 1$, the event $X_1 = k_1$, $X_2 = k_2$ can just as well be described by $X_1 = k_1$; hence we find

$$p_{X_1}(k_1) = \binom{n}{k_1} p_1^{k_1}(1 - p_1)^{n-k_1},$$

that is, $X_1 \sim \text{Bin}(n, p_1)$. Hence the multinomial distribution is a generalization of the binomial distribution.

It is also clear that, if in the general case one is interested only in the ith frequency, then $X_i \sim \text{Bin}(n, p_i)$.

Example 9. Roulette

Consider a roulette wheel with 38 possible outcomes 1–36, 0 and 00, each outcome having probability 1/38. The probability that, in 20 spins, 0, 00 and some number 1–36 appear 1, 2 and 17 times, respectively, is by (14)

$$\frac{20!}{1!\,2!\,17!}\left(\frac{1}{38}\right)^1\left(\frac{1}{38}\right)^2\left(\frac{36}{38}\right)^{17} = 0.025.$$

□

Example 10. Classification of Product

When manufacturing certain items the length X is a rv such that approximately $X \sim N(2.5, 0.1^2)$, in a suitable unit of measurement. Items shorter than 2.4 are classified as "too short" and those longer than 2.6 as "too long". Thirty randomly chosen items are sent to a customer. Obtain the probability that the customer gets 28 satisfactory items, 1 "too short" and 1 "too long".

Let A_1, A_2, A_3 denote the events than an item is "too short", "satisfactory" and "too long", and p_1, p_2, p_3 the corresponding probabilities. Using the normal distribution we obtain

$$p_1 = P(X < 2.4) = \Phi\left(\frac{2.4 - 2.5}{0.1}\right) = \Phi(-1) = 1 - \Phi(1) = 1 - 0.84 = 0.16,$$

$$p_3 = P(X > 2.6) = 1 - \Phi\left(\frac{2.6 - 2.5}{0.1}\right) = 1 - \Phi(1) = 0.16,$$

$$p_2 = 1 - p_1 - p_3 = 0.68.$$

Formula (14) now gives for $n = 30$, $k_1 = 1$, $k_2 = 28$, $k_3 = 1$, the probability

$$\frac{30!}{1!\,28!\,1!} 0.16 \cdot 0.68^{28} \cdot 0.16 = 0.00046. \qquad \square$$

Finally, we shall list expectations and some other moments for the multinomial distribution. Since the marginal distributions are binomial we have

$$E(X_i) = np_i; \qquad V(X_i) = np_i(1 - p_i); \qquad D(X_i) = \sqrt{np_i(1 - p_i)}.$$

Using the definition of covariance in §6.4 it is possible to prove (with some difficulty) that X_i and X_j have covariance

$$C(X_i, X_j) = -np_ip_j.$$

Formula (10) in §6.4 now shows that the coefficient of correlation is

$$\rho(X_i, X_j) = -\sqrt{\frac{p_ip_j}{(1 - p_i)(1 - p_j)}}.$$

Here is an example of a negative coefficient of correlation. It is quite natural that the coefficient is negative: Since the sum $\sum_1^r X_i$ is equal to n, that is, constant, a large value of X_i tends to be accompanied by a small value of X_j.

Exercises

901. Fifteen persons each toss two well-made coins. Find the distribution of X = number of persons with the same outcome on both tosses. (§9.2)

902. From a standard deck of cards the men $M_1, M_2, \ldots, M_5$ and their wives $W_1, W_2, \ldots, W_5$ draw one card each. After each drawing the card is returned to the deck. If a black card is drawn, the person wins 1 dollar.
    (a) Find the probability that a given family gets some payment.
    (b) Find the distribution of the number of families that get some payment.
    (c) Find the probability that exactly three families get some payment. (§9.2)

903. Eight persons each throw a die twice. Let X be the number of persons who get "one" as many times as "two".
    (a) Find the distribution of X.
    (b) Find the probability that exactly four of the eight persons get "one" as many times as "two". (§9.2)

904. Find $P(X \leq 3)$ if $X \sim \text{Bin}(7, 3/4)$. (§9.2)

905. Find $P(4 < X < 8)$ and $P(X = 6)$ if $X \sim \text{Bin}(16, 0.40)$. (§9.2)

906. A certain trial results in "success" with probability 0.80 and "failure" with probability 0.20. Twelve independent trials are performed.
   (a) Find the distribution of X = number of successes.
   (b) Find the distribution of Y = number of failures.
   (c) Find $P(2 \leq Y \leq 4)$.
   (d) Find the probability that the number of successes exceeds 7 but does not exceed 10. (§9.2)

907. A seed bag bears the inscription "percentage of germination 75%". If 15 seeds are sown, find the probability that between 65% and 90% of the seeds germinate. (§9.2)

908. A die is thrown 288 times. Let X be the number of times a one or six is obtained.
   (a) Give the distribution of X.
   (b) Find $E(X)$ and $V(X)$.
   (c) Can the normal approximation be used?
   (d) Compute approximately the probabilities $P(X \leq 100)$ and $P(X > 80)$. (§9.2)

909. Forty-eight persons each toss five coins. Let X be the number of persons who get the same outcome on all five tosses.
   (a) Find the distribution of X.
   (b) Find $E(X)$ and $V(X)$.
   (c) Can the normal approximation be used? The Poisson approximation?
   (d) Compute approximately the probabilities $P(X \leq 3)$ and $P(X > 4)$. (§9.2)

910. The probability that a student passes a certain examination is 0.70, independently of what happens to the other students. Find the probability that at least 75% pass:
   (a) if 4 students are examined;
   (b) if 16 students are examined;
   (c) if 400 students are examined. (§9.2)

911. A unit consists of two components A and B. During the production, an A-component becomes defective with probability 0.05 and a B-component with probability 0.10. All components are independent in this respect. At a certain stage n_A A-components and n_B B-components were produced.
   (a) Let X_A and X_B be the number of satisfactory A-components and B-components, respectively. State the distributions of X_A and X_B.
   (b) Find the probability that, if $n_A = 1{,}200$, $n_B = 1{,}300$, there are more satisfactory A-components than B-components. (§9.2)

912. In a factory certain items are produced. Each item turns out defective with probability 0.005, independently of what happens to other items. After production the items are packed (without any checking) in cartons of 100 each. A carton is considered unacceptable if it contains more than three defective items. Compute the probability that, of 10,000 cartons, more than 25 are unacceptable. (§9.2)

913. A population consists of N persons, Np of which are males. A sample of n persons is chosen at random, X of which are males. Find $E(X)$ and $V(X)$ in the two

cases: drawing with replacement and drawing without replacement. Perform a numerical calculation when $N = 200$, $n = 100$, $p = 0.60$, and compare the variances in the two cases. Repeat the calculations for $N = 1{,}000$. (§9.3)

914. The rv X has a hypergeometric distribution with parameters N, n and p. Which of the following statements are true?
(a) $X \sim$ approximately $\mathrm{Bin}(n, p)$ if $N = 100{,}000$, $n = 1{,}000$.
(b) $X \sim$ approximately $N(np, d_n^2 npq)$ if $d_n^2 npq = 78$.
(c) $X \sim$ approximately $N(np, npq)$ if $npq = 345$ and $n/N = 0.05$.
(d) $X \sim$ approximately $\mathrm{Po}(np)$ if $N = 2{,}000$, $n = 50$, $p = 0.04$. (§9.3)

915. A population consists of N persons, 20% of which are males. Four persons are chosen at random without replacement. Find the probability that at least one male appears in the sample if:
(a) $N = 10$;
(b) $N = 500$. (§9.3)

916. When checking the quality of a batch of 1,000 items, 50 are chosen at random and checked. The batch is accepted if at most two of the units in the sample are defective. Suppose that there are 12 defective items in the whole batch. Find the probability that the batch is accepted. (§9.3)

917. The rv X has a Poisson distribution with coefficient of variation $R(X) = 0.50$. Find $P(X = 0)$. (§9.4)

918. If $X \sim \mathrm{Po}(7.5)$, find $P(X \leq 4)$, $P(6 \leq X \leq 11)$, $P(X \geq 10)$ and $P(X = 8)$. (§9.4)

919. A nuclear physicist studying a radioactive substance performs the following experiment. During time t he counts the number of particles disintegrating in that period. The decay intensity is λ, and so the count is $\mathrm{Po}(\lambda t)$. Find the probability that the count exceeds the expectation by more than 10%, if
(a) $\lambda t = 8$;
(b) $\lambda t = 14$;
(c) $\lambda t = 100$. (§9.4)

920. After exposure of a photographic emulsion to radioactivity for some time, traces of particles in the emulsion are studied. The number of traces is assumed to have a Poisson distribution. As a result of many experiments performed under constant conditions, the probability that a plate has no traces at all is taken to be 0.07. What is the most common number of traces in a plate? (§9.4)

921. The number of white blood corpuscles in 1 mm^3 of blood from an individual has a Poisson distribution with an expectation λ that depends on the individual. Consider an individual with $\lambda = 6{,}000$.
(a) Find the probability that a drop of blood of volume 1 mm^3 from this individual contains less than 5,000 white corpuscles.
(b) 1 ml of blood from the person is diluted a thousand times to a volume of 1 liter. Then a drop of volume 1 mm^3 is taken out. The number of white corpuscles still has a Poisson distribution. Find the probability that the drop contains fewer than 5 white corpuscles. (§9.4)

922. Nine persons each toss two coins. Compute the probability that two of them get two heads, four get one head, and three get no head. (§9.5)

923. The rv $(X_1, \ldots, X_5)$ has the following distribution:

$$p_{X_1,\ldots,X_5}(k_1, \ldots, k_5) = \frac{n!}{k_1! \ldots k_5!}\left(\frac{1}{3}\right)^{k_1}\left(\frac{1}{4}\right)^{k_2}\left(\frac{1}{6}\right)^{k_3+k_4}\left(\frac{1}{12}\right)^{k_5}$$

for $k_1 + \cdots + k_5 = n$. Find the distribution of (Y_1, Y_2, Y_3), where $Y_1 = X_1 + X_4$, $Y_2 = X_3$, and $Y_3 = X_2 + X_5$. (§9.5)

*924. Consider two biased coins that both come up head with probability p. The coins are tossed simultaneously until either (head, tail) or (tail, head) has appeared n times in all. (Hence tosses with (head, head) or (tail, tail) are excluded from consideration.) Find the distribution of the number of tosses which result in (head, tail).

*925. The following problem is often called Banach's match box problem. (Banach was a Polish mathematician.) A person has two match boxes, each containing n matches. Each time he needs a match he chooses a box at random and takes a match. Sooner or later he will find a box empty. Let X be the number of matches that are then left in the other box. Find the distribution of X.
Hint: Box 1 is eventually found empty and box 2 then contains k matches if and only if: of the first $2n - k$ matches drawn, n are taken from box 1, $n - k$ from box 2, and box 1 is chosen next.

*926. Assume that the number of eggs laid by an insect is $\text{Po}(\lambda)$ and that the probability is p that a certain egg is hatched. Determine the distribution of the number of hatched eggs.

CHAPTER 10

Introduction to Statistical Theory

10.1. Introduction

This chapter contains an introduction to statistical theory and gives examples from different fields showing the scope of this theory. Several terms and definitions are also introduced.

In §10.2 we treat statistical investigations from a general point of view, in §10.3 we give four examples of sampling investigations and a scheme for such investigations, in §10.4 we present the main problems of statistical theory, and in §10.5 we give some glimpses of the history of statistical theory.

10.2. Statistical Investigations

A *statistical investigation* usually consists of four parts: planning, collection of data, analysis and presentation.

Planning consists of all sorts of preparations. We return to this important activity in Chapter 16.

Collection of data is a general term that may mean, for example, an experimenter reading an instrument or an interviewer asking people about their opinions.

The *analysis* can assume very different forms. In simple cases it may consist only of condensation of data in a table or a diagram. For this purpose we use *descriptive statistics*; see Chapter 11. For a more detailed analysis of sampling investigations (see below) we use some form of *statistical analysis*. Such an analysis rests upon statistical theory, which we discuss from Chapter 12 on.

The *presentation* may consist of graphical illustrations, a summary of results and conclusions and practical recommendations.

We shall now present various concepts important in different types of statistical studies.

(a) Pervasiveness of Statistical Investigations

Statistical investigations enter into almost all scientific activities (although the term is not always used), for example, physics, chemistry, technology, social sciences and nowadays, increasingly, the humanities. Examples of the last occur in linguistic research and in the study of literature. All experimental activities may be said to consist of such investigations. Also, in other parts of public life, statistical investigations are common, for example, in commerce, industry and government.

(b) Population and Element

A statistical investigation consists in studying some aspects of a *population* (sometimes several populations at the same time). This term, which we have used before, concerns a set of *elements* (for example, objects or persons). We often give the word element a more abstract meaning and let it denote some property of the object or person. A population is then a *set of data* or, as we often say, a *set of observations*.

There is a good rule for anybody preparing a statistical investigation: *Formulate clearly both what an element refers to and what the population of elements is.* If this rule is not followed, which is unfortunately sometimes the case, there is a risk of misinterpreting the observed results and hence a risk of erroneous recommendations and decisions.

Example. Control of Medicine

A US government agency wishes to check whether the content of a certain important ingredient in a drug is within prescribed limits. As an element it would be possible to use, for example, the bottle bought by the customer. As a population it would be possible to take all bottles sold in the USA during a certain year, or all bottles manufactured by pharmaceutical industries in the country. □

Example. Household Survey

English households are to be investigated with respect to their food consumption. As an element one takes the household and as a population all households in England. But what is a household? What about people living together in a house and having one single meal together each day? Do immigrants who are not yet British citizens belong to the population? Which products should be covered by the survey? Which period should it cover? □

Fig. 10.1. Smiling-face scale.

(c) Discrete and Continuous Variation

The elements in a statistical investigation can be very different depending on the property studied. An element can be a number, but sometimes it is impossible to use numbers. An example of the former situation is the weight of an object; the data are then a sequence of weights. An example of the latter situation is the colour of the eyes of a person; then the data are a series of appropriately chosen symbols, for example, GR, GR, BL, BL, BR, ..., where GR = green, BL = blue, BR = brown.

Hence it is not true, as is often stated, that only elements in the form of numbers are studied in statistical theory. Sometimes rather subjective phenomena are studied that cannot be easily quantified. See, for example, the "smiling-face" scale in Fig. 10.1. It can be used when tasting food. The children participating in the food panel circle the face corresponding most closely to how they feel about the food.

Sometimes each element is represented by a pair of numbers, a triple of numbers, and so on. Examples: the diameter at chest height and the volume of a tree; income, expenses and capital of a person. Analogously, the element may consist of a pair, a triple, and so on, of other symbols. An example: sex and hair colour of a person.

We use the term *discrete variation* when the number of possible different elements is finite or denumerably infinite. Elements that are not given in the form of numbers are always of this type, as are also elements obtained as a result of a *count* and hence consisting of integers 0, 1, ...

We use the term *continuous variation* to denote the situation when an element can assume any value in some interval. Measurements usually are of this type.

(d) Finite and Infinite Populations

A population is often *finite*; that is, it consists of a finite number of elements. Examples: the trees in a forest on January 1, 1986, US citizens on January 1, 1986, repeated measurements of a physical constant made at a certain laboratory during 1986. Sometimes it is informative to speak of an *infinite* population. Examples: an infinite sequence of measurements of a physical constant, all possible samples of 20 grams that can be taken from several tons of finely crushed iron ore.

(e) Complete Investigation and Sampling Investigation

In a *complete investigation* the whole population is studied. Such investigations are important in many areas, for example, when official statistics of births, deaths, and so on, are produced. It is also important in industry (determination of total production; inspection of each unit when shipping valuable products, such as anti-aircraft guns or railway cars). It often happens that a complete investigation is too expensive or time-consuming to perform, or even impossible in principle (for example, when a product is submitted to destructive testing). Complete investigations do not require any statistical theory and will hardly be discussed in the sequel.

In a *sampling investigation* only part of the population is examined. Such investigations have wide application, for example, in connection with opinion polls, experiments and statistical quality control. All experimental activity can be said to consist of sampling investigations. In fact, the research worker can perform only limited sequences of measurements, which may be regarded as samples from a conceptually infinite sequence of observations. The idea that very large sets of data must be available for drawing reliable conclusions is totally incorrect; sometimes even a very small sample can produce sufficient information.

A sampling investigation is generally more imprecise than a complete investigation. However, this is not always so. When mass produced items are checked by humans, periods of fatigue can make 100% control illusory; a well-planned sampling inspection scheme can give a better result.

Let us add that, in the above, we have used the term sampling investigation, which has the advantage that it is very general and covers both experimental and other activities. Another term, which does not include experiments, is *sampling survey*. It is often used when sampling from human populations.

(f) Comparative and Noncomparative Investigations

The meaning of these terms is obvious. A comparative investigation consists of a comparison of two or more populations. Examples: Do cars of make A consume more petrol than cars of make B? Is drug A better than drug B? In a noncomparative investigation only one population is studied. Examples: How much petrol do cars of make A need? Which is the effect of drug A? The planning of the two types of investigation is quite different, as will be discussed in Chapter 16.

10.3. Examples of Sampling Investigations

We shall now give four examples of sampling investigations, two noncomparative and two comparative ones. They are intended to show how statistical problems arise in practice. The first two examples are closely related to Example 1 and Example 2 of Chapter 1.

Example 1. Batch Containing Defective Units

Consider a batch of N units. The fraction of defectives p, that is, the ratio of the number of defective units in the batch to the total number N, is unknown. (When p is expressed as a percentage it is usually called the percent defective.) In statistical language p is said to be an unknown parameter. To get information about p we take n units from the batch and examine them; assume that x units turn out to be defective. What can be said about p?

This is an example of a noncomparative sampling investigation. The population consists of N elements, which can be conveniently represented by N ones and zeros, where "one" denotes a unit is defective, "zero" a unit is satisfactory. The sample consists of n binary numbers. Example:

Population:	1 0 0 0 1 0 0 0 1 1 0 1 0 0 0 0	$N = 16$; $p = 5/16$
	↓ ↓ ↓ ↓	
Sample:	0 1 0 0	$n = 4$; $x = 1$

Clearly, we have here a sample taken without replacement from a finite population.

The question that we formulated about p is vague and will now be made more precise in three different ways.

(a) Point estimation

Give a point estimate, that is, a number that estimates the actual value of p. One possibility is to use the observed ratio x/n of the number of defectives to the number in the sample. Evidently, x/n may lie far from p, but if the sample is random and n is large enough, it may be expected that this point estimate will be close to p.

(b) Interval estimation

Give an interval that covers the unknown value p with given probability, for example, an interval of the form $(x/n - d_1, x/n + d_2)$. Such an interval is called a confidence interval.

(c) Testing of hypotheses

Consider the hypothesis that $p \leq p_0$, where p_0 is a given number $(0 \leq p_0 \leq 1)$, for example, 1/10. We want to test this hypothesis by means of the sample, that is, we want to see if the sample is compatible with this hypothesis or if the hypothesis should be rejected in favour of $p > p_0$. It seems reasonable to formulate such a procedure as follows: Reject the hypothesis that $p \leq p_0$ if x is large. The idea behind this formulation is that, if the hypothesis is true, it is improbable that the number of defectives is large.

With $p_0 = 1/10$ and $n = 3$, a value of $x = 3$ is rather improbable. Hence a large value of x indicates that the hypothesis is false. The procedure just described is called a test of significance. We shall discuss such tests in more detail later.

In order to study the questions arising in the above three situations it is necessary, as in every sampling investigation, to translate the practical problems involved into theoretical ones; that is, we must construct a model. This is easy in this example, at least in principle. Assume that the sample is taken at random so that all units have an equal chance of being included in the sample. This is more difficult in practice than it may seem at first sight, but we can try to do the best we can to achieve this. Theoretically, the meaning of the assumption is clear, and its consequences are well known from probability theory. We know from Chapter 9 that, when n units are taken at random without replacement from N units, then the number, X, of units with the property we are studying has a hypergeometric distribution

$$p_X(k) = \binom{Np}{k}\binom{Nq}{n-k}\bigg/\binom{N}{n}.$$

If the model is correct, the value x that we have obtained may be regarded as an observation on X. It is very important to distinguish between x and X in this situation. Once the sample has been selected, the value x is a given number, whereas X is a rv that can assume various values with given probabilities. That is why we say that x is an observation on X. Sometimes we say, equivalently, that x is a random sample of size 1 from the distribution of X.

The model described above can be used for solving the different problems in (a)–(c). We shall return to this later. □

Example 2. Measuring a Physical Constant

At a laboratory, five measurements were made of a physical constant. The following values were obtained:

2.13 2.10 2.05 2.11 2.14.

This is also a noncomparative sampling investigation. The sample consisting of five measurements is considered to have been taken from a hypothetical infinite population of all values that would have been obtained if measurement had been repeated indefinitely.

Call the unknown physical constant m. Generally speaking, the question is: What can be said about m? We shall make the question more precise in three different ways.

(a) Point estimation

Find a point estimate of m. An estimate that seems natural is the arithmetic mean of the values, which we denote by $\bar{x}$; in this case, we have $\bar{x} = 2.11$.

(b) Interval estimation

Find a confidence interval that covers m with a given large probability. For example, it seems possible to take the interval $(\bar{x} - d, \bar{x} + d)$, where d is suitably chosen.

(c) Testing of hypotheses

Let us assume that it is generally considered by the research workers in the field that the physical constant m has a certain value m_0. It then seems natural to formulate the hypothesis that $m = m_0$, and to find out if the observed measurements are compatible or not with this hypothesis. To test the hypothesis, we could use the following test of significance: Reject the hypothesis if $|\bar{x} - m_0| > \Delta$, but not otherwise. The determination of Δ presents a problem of its own, to which we will return later. In fact, it is not clear that a test of significance should have exactly this form, and there are other possibilities.

In this example we again need a model. The choice of model is not as obvious as in the previous example. As we know, it is rather common that measurements of this type are approximately normally distributed. If that is the case, it seems natural to choose the following model: Let $N(m, \sigma^2)$ be a normal distribution with unknown parameters m and σ^2. The measurements $x_1, \ldots, x_n$ are assumed to be observations on the independent rv's $X_1, \ldots, X_n$, which are all $N(m, \sigma^2)$. We call these measurements a random sample of n values from $N(m, \sigma^2)$.

A consequence of the model is that $E(X) = m$; that is, the expectation of the rv X is the unknown constant m. It is not self-evident that one should choose a model with this type of agreement between expectation and constant m, since the model implies that the observations have no systematic error (see the end of §6.3). If it were known that each measurement has a known constant systematic error δ, it would be more realistic to assume that $X \sim N(m + \delta, \sigma^2)$. It would then be possible to subtract the error δ from each measurement and assume that the values then obtained are observations from $N(m, \sigma^2)$.

By this choice of model, we have replaced the practical problems by theoretical ones, which can be treated by means of statistical theory. As will be seen later, such a model also provides an approach to the important question of how many measurements are required. □

This example illustrates clearly the correspondence between model and reality:

Empirical world	Model world
Measurements $x_1, \ldots, x_n$	Independent rv's $X_1, \ldots, X_n$ with observations $x_1, \ldots, x_n$
Unknown physical constant m	The constant m is the expectation of $X_1, \ldots, X_n$ (if systematic errors are absent)

Example 3. Comparison of Two Methods of Measurement

We shall modify the previous example in order to obtain a comparative investigation. Let us assume that interest lies not in the constant m itself, but rather in two methods A and B used for its determination. We want to know if there is a systematic difference between the methods. We then supplement the measurements already made (concerning method A, say) with additional measurements made according to method B and obtain

Method A: 2.13 2.10 2.05 2.11 2.14

Method B: 2.30 2.22 2.25 2.31 2.29.

The B-series has arithmetic mean 2.27, which is rather different from that of the A-series, which was 2.11. The difference $2.27 - 2.11 = 0.16$ is a point estimate of the difference $m_2 - m_1$ between the constants m_2 and m_1 obtained according to method B and method A, respectively. (Both may be different from the true value m, and therefore new symbols are introduced.) We may also want an interval estimate of this difference or a test of hypothesis, for example, that $m_2 - m_1 = 0$, which means there is no systematic difference between the methods.

The planning of the investigation now becomes more complicated. It is important to avoid disturbing systematic errors which have nothing to do with the comparison itself. For example, if the methods demand long training, the same person should perhaps perform all measurements. Otherwise, we may arrive at the wrong conclusions and assert that there is a real difference between the two methods when, in fact, the difference is due to personal errors. Later we shall discuss planning in more detail.

Also the choice of model is more complicated than in the previous example, since there are now two series of measurements that may come from different distributions, for example, from two approximately normal distributions. □

Example 4. Fluoride and Caries

We shall give one more example of a comparative investigation. A dentist wants to find out the effect of fluoride on the occurrence of caries in children. To this end, he divides 1,000 children, say, into two equally large groups A and B. In group A a solution of fluoride is applied regularly to the teeth of the children, and in B a solution is applied that does not contain fluoride, for example, distilled water. After some time has elapsed, the dentist examines the teeth of the children by noting the number of cavities each child has developed since the beginning of the study.

Here is a portion of his results (fictitious data):

Group A

Child number:	1	2	3	4	5	6	7	8
New cavities:	0	6	0	2	3	0	2	1

Group B

Child number:	501	502	503	504	505	506	507	508
New cavities:	4	11	6	3	0	0	0	1

Even from this small part of the investigation it is seen that there is a considerable variation between the children. Biological data very often behave like this. Nevertheless, when all the data have been suitably collected and analysed, it should be possible to determine the effect, if any, of the fluoride on the incidence of caries.

A condition for properly determining this effect is that the planning of the investigation is well executed. All disturbing factors must be avoided. For example, it would be quite incorrect to choose the entire group A from one school and group B from another school, for there may be general differences between the children from different schools. For example, the food may tend to be different. It would also be inappropriate to let one dentist examine the children in A and another dentist those in B, for there is always a subjective element present when cavities are recorded. For these and other reasons it is safest to divide the children at random into two groups and perform the study in such a way that it is not known either to the children, nurses, dentists, and so on, which children have obtained the fluoride. This can be achieved by writing code numbers on bottles that are deciphered when all the data have been collected and the statistical analysis begins. (For this reason, the numbering we used above of the children in groups A and B is not practically useful!)

A possible general model for a correctly planned and executed investigation of this type is the following: The new cavities of the children in A and B are regarded as observations on independent discrete rv's $X_1, \ldots, X_{500}$ and $Y_1, \ldots, Y_{500}$ with distribution functions $F_X(x)$ and $F_Y(y)$. The expectations are assumed to be m_1 and m_2, respectively. The interesting question is whether m_1 is smaller than m_2, for in that case there is a real effect in the treatment with fluoride. As in the previous examples, various problems in point estimation, interval estimation and the testing of hypotheses can be formulated and studied by statistical theory. □

The examples given above have something important in common. They illustrate the following general scheme for sampling investigations:

Empirical world	Model world
1. Formulate practical problem	
	2. Construct a probability model
3. Collect data	
	4. Perform a statistical analysis
5. Draw practical conclusions	

In the following we mostly discuss the theory on the right (numbers 2 and 4). There is perhaps a risk that the practical problems (numbers 1, 3 and 5) are kept unduly in the background. *Remember that all five parts are important*!

10.4. Main Problems in Statistical Theory

We shall now look somewhat more closely at the probability model that we study in a sampling investigation. The model may contain one or several distributions. For simplicity, we assume that there is just one single distribution.

The distribution is partly unknown, for it depends on an unknown *parameter*, which we call θ. (Sometimes there are several such parameters.) The different values that θ can attain form a space, which we call the *parameter space A*.

The important concept of a parameter is illustrated in the examples in the previous section. The parameter belongs to the model and hence to the model world, but often has a concrete interpretation in the empirical world. In each problem it is of crucial importance to find out what is model and what is reality. This is not some kind of philosophizing for its own sake; we have in §1.2 already warned the reader not to confuse model and reality.

In order to get information about the parameter one plans some kind of sampling investigation and collects data $x_1, \ldots, x_n$ by measurements or counts during stable conditions. (In Example 2 in §10.3 we have described such a situation.) In the model world, it is then natural to assume that the values are observations on independent rv's $X_1, \ldots, X_n$ with the same distribution F. We then say that we have a random sample from F. Hence we introduce the following:

Definition. A *random sample* $x_1, \ldots, x_n$ from F consists of observations on independent rv's $X_1, \ldots, X_n$, each with distribution F. □

Sometimes a single letter $\mathbf{x}$ is used for the sample, which saves space. It is possible to conceive $\mathbf{x} = (x_1, \ldots, x_n)$ as an n-dimensional vector. In the same way sometimes it is suitable to introduce an n-dimensional rv $\mathbf{X} = (X_1, \ldots, X_n)$ that $\mathbf{x}$ is an observation of.

(A conscientious reader possibly dislikes our writing "with distribution F" in the above definition instead of the more complete "with distribution function $F(x)$". We do this intentionally in the sequel in order to simplify the writing.)

The definition of a random sample is of basic importance in statistical theory, and not very easy to digest, even if we have done our best by the preparations in §10.3. Read this section again if you have time. We shall assist the reader by considering some very simple examples.

A random sample of $n = 5$ values from a normal distribution $N(0, 2^2)$ may take the form $-1.31, 0.64, 0.90, -0.24, 1.02$. A random sample of $n = 10$ values from a Poisson distribution Po(3) may be 5, 5, 0, 1, 3, 1, 4, 3, 2, 6.

Remark 1. Extended Definition of Random Sample

Later we will need the following extended definition. A random sample $x_1, \ldots, x_n$ from $F_1, \ldots, F_n$ consists of observations on independent rv's $X_1, \ldots, X_n$ with distributions $F_1, \ldots, F_n$. Note that we allow the values in the sample to come from different distributions. □

Remark 2. Random Sample from Multidimensional Distribution

The first definition of a random sample can be extended immediately to multi-dimensional distributions. For example, consider a two-dimensional rv (X, Y) with distribution function $F_{X,Y}(x, y)$. A random sample then consists of pairs of numbers $(x_1, y_1), \ldots, (x_n, y_n)$ which are observations on the independent two-dimensional rv's $(X_1, Y_1), \ldots, (X_n, Y_n)$ each with distribution F. (The n rv's are independent, but note that within each pair there is a dependence described by $F_{X,Y}(x, y)$.) As an example of a random sample of this type we may mention height and weight of 100 randomly chosen seven-year-old children. In this book we are not much concerned with samples from multidimensional distributions. □

When the model is chosen, various problems can be stated, depending on the practical questions of interest:

Point estimation	How to estimate θ
Interval estimation	How to construct an interval that covers θ with prescribed probability
Testing a hypothesis	How to test a given hypothesis concerning θ

Problems in these areas are usually called *inference problems* and the theory developed for handling them we call *theory of statistical inference*. Chapters 12–16 will be devoted to such problems.

10.5. Some Historical Notes

While probability theory has a long history (see §1.3 for some glimpses), statistical theory is much younger. However, descriptive statistics is old. Collection of data and construction of tables have been carried out in many countries for several hundred years; the original meaning of "statistics" is the "science of state", in German "Staatenkunde".

Two lines of development can be discerned in the history of statistical theory, one in physics and the related sciences, the other in the biological sciences.

At the beginning of the nineteenth century, Gauss (1777–1855) developed a theory of errors of observation, partly founded on the normal distribution, and advocated the so-called method of least squares for point estimation of unknown parameters. Gauss's methods became a standard tool for physicists, geodesists, and others, and have remained so; the connection to the other line has not always been noted.

The other—and more important—line of development also has its origin in the nineteenth century when people began to collect biological data of various kinds and wanted to analyse them mathematically and statistically. The Englishmen F. Galton (1822–1911) and K. Pearson (1857–1936) are excellent representatives of this new interest in the quantitative study of biological phenomena. The latter developed, at the turn of the century, a general approximate theory for large samples and derived, among many other things, the so-called χ^2 test. The modern statistical theory and methodology are to a very large extent due to the English geneticist and statistician R.A. Fisher (1890–1962). He is one of the originators of the theory of statistical inference and has also developed the analysis of variance and the analysis of regression; furthermore, he has formulated important principles of the design of experiments. Important contributions to statistical inference have also been made by the Polish-born statistician J. Neyman (1894–1981), who later lived in England and the USA, and E.S. Pearson (1895–1980), son of K. Pearson.

By the activities and contributions of these and many other research workers, statistical theory has been developed and expanded to a science of its own, based on probability theory and having wide application in different parts of society.

CHAPTER 11

Descriptive Statistics

11.1. Introduction

In almost every statistical investigation descriptive statistics is used in some form. This important field of statistics concerns tabulation, graphical representation and numerical treatment of data. The idea is to describe the data in a convenient way, to condense them in a manner that makes life easier for the "data consumer". Also collection of data can, if we like, be regarded as a part of descriptive statistics.

It is an art to present data, graphically or numerically, in a way that is easy to understand. The task is to help the consumer grasp the essentials of the message that the data are meant to convey. Everyday readers of journals and watchers of TV are fed such messages—more or less successfully.

Computers have contributed to the development of descriptive statistics, and much of what is said in this chapter is well known to anybody working professionally with computers.

The role of descriptive statistics varies from investigation to investigation. In an investigation involving complete enumeration descriptive statistics plays an important part. On the other hand, in a sampling investigation the use of descriptive statistics constitutes only one stage of the investigation, often mathematical statistics being needed as well.

As to the actual collection of data, we only point out that there are many aids available nowadays such as cards, forms, tapes, and other devices that can be read by computers. The collection of data should always be planned with great care, especially when a large investigation is to be performed. Much time should be devoted to the choice between different alternatives, to the construction of forms, the formulation of questions, and so on. Many investigations have failed because these time-consuming preparations have been neglected.

In §11.2 we discuss tabulation and graphical presentation, and in §11.3 computation of measures of location and dispersion. The last two sections, §11.4 and §11.5, contain some remarks about terminology and numerical computations.

We assume that the reader possesses a pocket calculator giving sums of squares, and perhaps arithmetic mean, variance and standard deviation.

11.2. Tabulation and Graphical Presentation

(a) Raw Data

The number of matches in each of 35 match boxes was determined, and the result was

51	52	49	51	52	51	53
52	48	52	50	53	49	50
51	53	51	52	50	51	53
53	55	50	49	53	50	51
51	52	48	53	50	49	51

This is an example of *raw data.*

(b) Frequency Tabulated Data

A large raw data set is difficult to absorb. From the previous example we may produce *frequency tabulated data*:

48	‖
49	‖‖
50	卌 |
51	卌 ‖‖
52	卌 |
53	卌 ‖
54	
55	|

The result can be summarized in a *frequency table* (see Table 11.1). Besides *frequencies* f_i of the different values y_i one usually computes *relative frequencies* $p_i = f_i/n$. The relative frequencies are sometimes expressed as per-

Table 11.1. Frequency table for the number of matches in match boxes.

Class y_i	Frequency f_i	Rel. fr. $100\ p_i$	Cumul. rel. fr. $100\ P_i$
48	2	5.7	5.7
49	4	11.4	17.1
50	6	17.1	34.3
51	9	25.7	60.0
52	6	17.1	77.1
53	7	20.0	97.1
54	0	0.0	97.1
55	1	2.9	100.0
Sum	35	100.0	

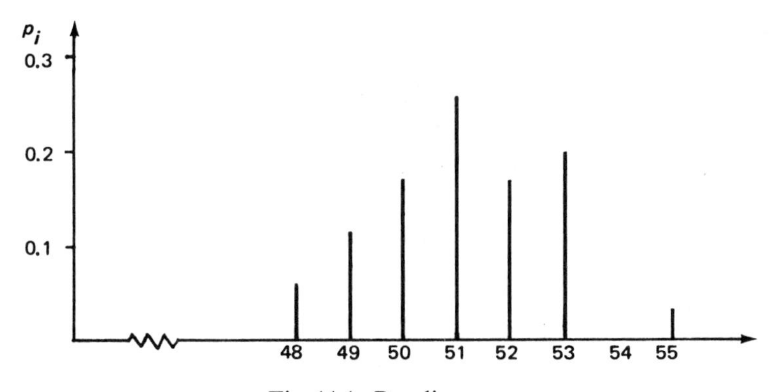

Fig. 11.1. Bar diagram.

centages. Sometimes one also computes *cumulative relative frequencies* P_i, which give the proportion of boxes with at most 48 matches, at most 49 matches, and so on. The values P_i are computed by successive addition of the relative frequencies p_i in the table.

Graphical methods are very helpful in the presentation of data. Fig. 11.1 contains a *bar diagram* showing the relative frequencies p_i. Of course, it is possible to plot the numbers f_i instead; only the vertical scale is then changed. Figure 11.2 contains a *cumulative bar diagram* illustrating the quantities P_i.

(c) Grouped Data

The capacitance of 630 condensers was measured. A small part of this large set of data is reproduced in Table 11.2.

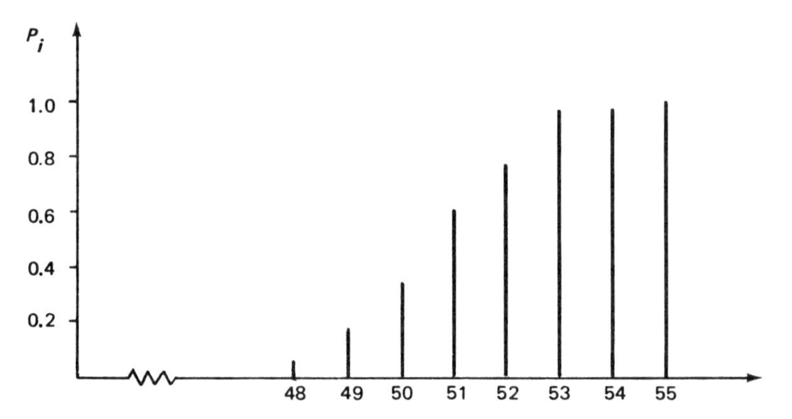

Fig. 11.2. Cumulative bar diagram.

Table 11.2. Capacitance of condensers.

2.504	2.616	2.627	2.541	2.618	2.476	2.328	2.404	2.506	2.413
2.578	2.638	2.596	2.362	2.519	2.520	2.372	2.483	2.501	2.395
2.572	2.407	2.370	2.517	2.530	2.465	2.535	2.540	2.609	2.499
2.547	2.574	2.528	2.460	2.467	2.418	2.427	2.451	2.602	2.546
2.424	2.446	2.491	2.475	2.601	2.352	2.621	2.558	2.322	2.459
2.305	2.518	2.464	2.446	2.506	2.495	2.455	2.496	2.473	2.633
2.458	2.476	2.504	2.391	2.525	2.562	2.639	2.673	2.524	2.435
2.484	2.585	2.479	2.510	2.448	2.495	2.446	2.504	2.557	2.521
2.547	2.540	2.512	2.589	2.563	2.546	2.518	2.533	2.513	2.576
2.619	2.539	2.560	2.485	2.495	2.445	2.518	2.561	2.495	2.491

In order to get a good overview of the data it is usual to *group* the data, which means that data of about the same size are entered into the same class. Note the terminology: Frequency tabulation means that equal values are entered into the same class. Grouping signifies that data of about the same size are entered into the same class.

The result is then presented in a table with a suitable number of columns. Data for the condensers have been grouped in Table 11.3. Observe the notation used in that table; for example, y_i for the ith midpoint.

When choosing class limits, the following rules should be followed:

(a) The class widths $h_i = g_{i+1} - g_i$ should be constant: $h_i = h$ for all i.
(b) So-called open classes, that is, classes of the type "observation less than 2.3195", should be avoided. Hence always give the class limits even for the lowest and the highest classes.

One should always choose limits that make the grouping easy. Never group age data into 13-year intervals!

Table 11.3. Capacitance of 630 condensers.

Class limits g_i g_{i+1}	Midpoint y_i	Frequency f_i	Rel. fr. p_i	Cumul. rel. fr. (%) $100\,P_i$
2.2995–2.3195	2.3095	2	0.0032	0.3
31 – 33	32	7	0.0111	1.4
33 – 35	34	9	0.0143	2.9
35 – 37	36	22	0.0349	6.4
37 – 39	38	14	0.0222	8.6
39 – 41	40	30	0.0476	13.3
41 – 43	42	28	0.0444	17.8
43 – 45	44	42	0.0667	24.4
45 – 47	46	45	0.0714	31.6
47 – 49	48	81	0.1286	44.4
49 – 51	50	90	0.1429	58.7
51 – 53	52	59	0.0937	68.1
53 – 55	54	63	0.1000	78.1
55 – 57	56	50	0.0794	86.0
57 – 59	58	31	0.0492	91.0
59 – 61	60	25	0.0397	94.9
61 – 63	62	10	0.0159	96.5
63 – 65	64	10	0.0159	98.1
65 – 67	66	7	0.0111	99.2
2.6795–2.6995	2.6895	5	0.0080	100.0

When stating exact positions of the class limits one has to consider what a certain value, say 2.300, really means. Assume that all values have been rounded off in such a way that each measurement in the interval (2.2995, 2.3005) is given as 2.300. When all values for which $2.300 \le x_j \le 2.319$ are entered in the lowest class, this means that the actual class limits are (2.2995, 2.3195), as indicated in the table. Hence the midpoint of this interval is 2.3095.

The exact positions of the class limits must be determined in each particular case, and the correction caused by rounding used in the above example is not always present. For example, if age data are to be grouped into five-year intervals, suitable class limits are [0, 5), [5, 10), ..., meaning that a 4-year-old belongs to the lowest class and a 5-year-old to the next class.

All columns in Table 11.3 may not be needed; only columns that are really informative should be included. The total number of observations should always be given. Tables showing only class limits and relative frequencies are an abomination of the same kind as the advertisers' slogan: nine film stars out of ten use the soap SOANDSO.

Grouped data can be illustrated graphically in about the same way as frequency tabulated data, but there are some important modifications.

A *histogram* (Fig. 11.3) can be constructed with rectangles whose areas are proportional to the relative frequencies in each interval. If the class width is

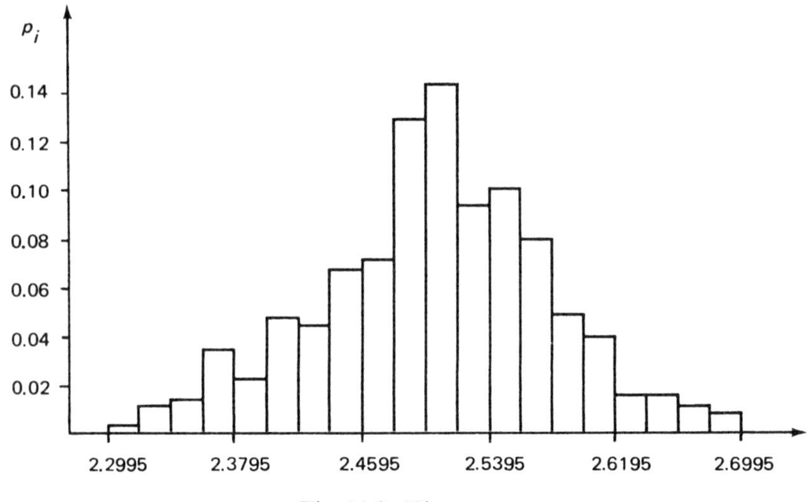

Fig. 11.3. Histogram.

constant, say equal to h, this means that the height of the rectangle is proportional to p_i. (The histogram corresponds to the bar diagram used for frequency tabulated data.)

It is also possible to construct a *cumulative histogram* (Fig. 11.4).

The procedures described here can be varied in many ways depending on practical circumstances and needs. For example, data are often recorded directly in frequency tabulated or grouped form. Careful planning is then very important. Information lost by grouping cannot be recovered without collect-

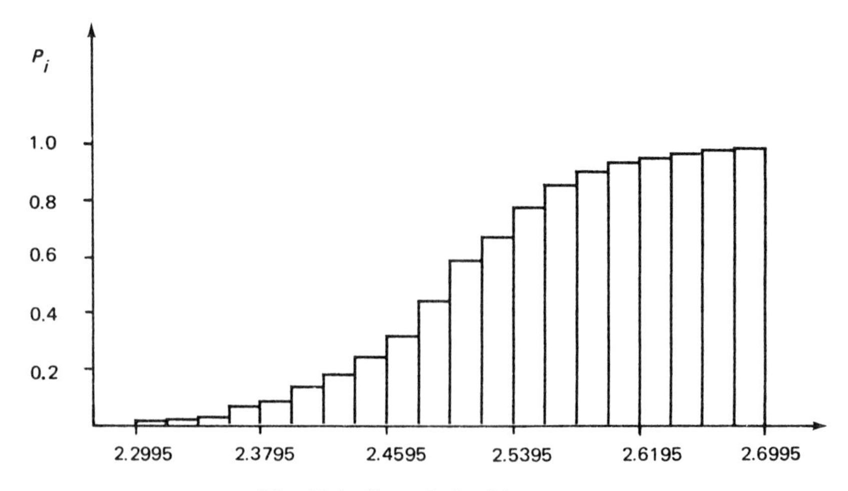

Fig. 11.4. Cumulative histogram.

ing fresh data. Example: recording ages in 10-year intervals when 5-year intervals may be needed later.

11.3. Measures of Location and Dispersion

(a) Raw Data

The values

2.504 2.616 2.627 2.541 2.618 2.476 2.328 2.404 2.506

were obtained by measuring the capacitance of nine condensers marked "2.5 μF". For such a sample we often want to find a *measure of location* and a *measure of dispersion.* The measure of location should indicate what the capacitance is "on the average", while the measure of dispersion should show how the capacitances differ from one another.

Let $x_1, \ldots, x_n$ be the data. As a measure of location we often use the *arithmetic mean* (often simply called the mean)

$$\bar{x} = \frac{1}{n}\sum_{1}^{n} x_j. \tag{1}$$

In the example we have $\sum x_j = 22.620$ and $\bar{x} = 22.620/9 = 2.5133$.

Another measure of location is the *median* $\tilde{x}$, which is the middlemost value in the data, that is, the observation having equally many values above it as below it. (We could also call it the central value in the data.) If n is odd, the median is uniquely determined by this description; if n is even, the median can be defined as the arithmetic mean of the two middle values.

The median is often used when one wants to prevent very low or very high values from affecting the measure of location too much. (In the condenser example the median is 2.506. This figure does not change if, for example, the largest value, 2.627, is increased by an arbitrary amount.) For this reason the median is often used, for example, in income statistics.

As a measure of dispersion we often use the *variance*

$$s^2 = \frac{1}{n-1}\sum_{1}^{n} (x_j - \bar{x})^2, \tag{2}$$

or the *standard deviation*

$$s = \sqrt{\frac{1}{n-1}\sum_{1}^{n} (x_j - \bar{x})^2}, \tag{3}$$

which is the square root of the variance.

By taking the square root, the same dimension is obtained as for the original values. For example, if the data are expressed in cm, then s^2 is expressed in cm^2 and hence s in cm.

It is easy to see that the standard deviation is an appropriate measure of dispersion. If the sample values are close together, the differences $x_j - \bar{x}$, and hence s, will be small; if the values are more spread out, s will be larger.

That the sum of squares is divided by $n - 1$ instead of n may seem strange, and we admit that from the point of view of descriptive statistics, there is no reason for doing so. The division by $n - 1$ has, however, good reason from a theoretical point of view discussed later (see §12.3).

We often set

$$Q = \sum_1^n (x_j - \bar{x})^2 \tag{4}$$

and call this the *sum of squares about the arithmetic mean.* We then have

$$s^2 = Q/(n - 1). \tag{2'}$$

We find

$$\sum_1^n (x_j - \bar{x})^2 = \sum_1^n x_j^2 - 2\bar{x} \cdot \sum_1^n x_j + n\bar{x}^2 = \sum_1^n x_j^2 - 2n\bar{x}^2 + n\bar{x}^2$$
$$= \sum_1^n x_j^2 - n\bar{x}^2.$$

Since $n\bar{x}^2 = (\sum x_j)^2/n$ we obtain

$$Q = \sum_1^n x_j^2 - \frac{1}{n}\left(\sum_1^n x_j\right)^2. \tag{4'}$$

This relation is often used for numerical computations.

In the example of the nine capacitances we have $\sum x_j^2 = 56.934618$ and $Q = 56.934618 - (22.620)^2/9 = 0.083018$. Hence by (2) and (3), $s^2 = 0.083018/8 = 0.01038$ and $s = \sqrt{0.01038} = 0.102$.

Certain pocket calculators automatically produce $\bar{x}$, s^2 and s and hence are very useful for statistical calculations.

The standard deviation is sometimes expressed as a percentage of the mean. We then obtain the *coefficient of variation*

$$100 \cdot s/\bar{x}, \tag{5}$$

which is, evidently, a dimensionless quantity, that is, it does not depend on the unit of measurement. This coefficient is only used for nonnegative data.

The *range* is defined by

$$R = x_{\max} - x_{\min},$$

where $x_{\min}$ is the smallest and $x_{\max}$ the largest value. When n is small (not larger than about 10), R can be used as a measure of dispersion instead of the standard deviation s. The range has, of course, the advantage over s that it is very quick to compute. If n is large, R is of more limited interest.

The *interval of variation* is defined as the interval

$$(x_{\min}, x_{\max}).$$

It is often informative to give the interval of variation as a complement to the above-mentioned measures of location and dispersion.

(b) Frequency Tabulated Data

When computing measures of location and dispersion for frequency tabulated data it is, of course, possible to disregard the frequency tabulation and treat the raw data in the way shown in (a). However, at least for large sets, it is more convenient to use a frequency table for the calculations. (On the other hand, it is seldom worth the trouble to use this approach unless the table is already available.)

If we define $\bar{y}$ as

$$\bar{y} = \frac{1}{n}\sum_{1}^{k} f_i y_i, \tag{6}$$

where k is the number of different values, it is realized that $\bar{y} = \bar{x}$, where $\bar{x}$ is the arithmetic mean of the original values. (In fact, since there are f_1 observations of size y_1, f_2 of size y_2, and so on, we get the term y_1 with frequency f_1, the term y_2 with frequency f_2, and so on.)

We also set

$$Q = \sum_{1}^{k} f_i (y_i - \bar{y})^2$$

and find that this is the same quantity as $\sum (x_j - \bar{x})^2$, since in the latter sum the term $(y_i - \bar{y})^2$ appears exactly f_i times. The simplest way to perform the calculations is to use the relation

$$Q = \sum_{1}^{k} f_i y_i^2 - \frac{1}{n}\left(\sum_{1}^{k} f_i y_i\right)^2. \tag{7}$$

Taking the matches data set of Table 11.1, we find

$$\sum f_i y_i^2 = 91533; \qquad \sum f_i y_i = 1789,$$

and hence

$$\bar{x} = 1789/35 = 51.1; \qquad Q = 91533 - 1789^2/35 = 89.543;$$

$$s^2 = 89.543/(35 - 1) = 2.63; \qquad s = \sqrt{2.63} = 1.62.$$

(c) Grouped Data

We want to determine the arithmetic mean, variance and standard deviation for a grouped set of data. We are actually interested in $\bar{x}$, s^2 and s for the original set of data, but, of course, want to work with the grouped set instead. We then approximate the grouped set by a frequency tabulated set with f_i values y_i $(i = 1, 2, \ldots)$. As a consequence, we replace each value x_j by the midvalue y_i of the class to which x_j belongs. We then obtain

$$\bar{x} = \frac{1}{n} \sum f_i y_i,$$

which we call the arithmetic mean of the grouped data (which is an approximation to the arithmetic mean of the raw data). In the same way we find

$$Q = \sum f_i y_i^2 - \frac{1}{n}\left(\sum f_i y_i\right)^2,$$

and then as usual $s^2 = Q/(n-1)$ and $s = \sqrt{Q/(n-1)}$.

For the grouped data of Table 11.3 we obtain

$$\sum f_i y_i^2 = 3957.068598; \qquad \sum f_i y_i = 1578.2450,$$

and hence

$$\bar{x} = 1578.2450/630 = 2.505,$$

$$Q = 3957.068598 - (1578.2450)^2/630 = 3.32688,$$

$$s^2 = 3.32688/629 = 0.005289; \qquad s = \sqrt{0.005289} = 0.0727.$$

11.4. Terminology

Several of the terms used in this chapter were already introduced in Chapter 6 to denote various properties of random variables. There we used the terms median, variance, standard deviation and coefficient of variation. As the reader has surely discovered by now, there are close analogies between the two uses: For example, in both cases, variance and standard deviation are measures of dispersion. When confusion may arise, it is safest to mention explicitly that, in the cases discussed in this chapter, computations concern data sets. The variance based on data may be called the sample variance, the standard deviation may be called the sample standard deviation, and so on.

Later in this book we will show that the related terminology has a deeper significance: When estimating, say, the unknown standard deviation of a rv, it is often advantageous to use the sample standard deviation, and analogously for other unknown parameters.

11.5. Numerical Computation

Numerical computation is a vast field, and we have only one remark to make here.

It is often difficult to know how many digits should be kept when doing numerical calculations.

No rules can be given for the appropriate number of digits to retain in one's *data*, for this depends on the precision of the measurements, and on many other factors. Counts should never be rounded off.

In the course of the *numerical work*, it is advisable to keep several extra digits. A small difference between two large numbers requires special care.

The *final result* should not be given to more digits than is reasonable. Definite rules cannot be formulated, but the following may serve as guidelines: Give frequencies exactly, relative frequencies as percentages to one decimal, occasionally to two decimals. Record arithmetic means to one more significant figure than the original values, occasionally to two more. The standard deviation usually should be given to two or three significant figures.

Exercises

1101. For 22 persons the caries index was determined, that is, the number of decayed surfaces among the 100 areas obtained if the wisdom teeth and all lingual surfaces are disregarded. Result:

41	47	66	73	48	52	49	54	61	62	47
52	65	61	69	31	54	53	50	47	36	69

Compute the arithmetic mean, median, variance, standard deviation and coefficient of variation for these data. (§11.3)

1102. The following experiment was performed with a weak radioactive substance. The number of emitted α-particles was determined for each of 100 time periods of length 30 seconds. Result:

17	10	8	20	13	9	14	23	11	13	16	16	13	15	16
9	14	10	15	17	18	17	18	13	21	17	18	14	20	16
11	11	16	16	13	4	11	19	22	14	19	13	12	27	14
17	19	14	11	16	14	12	12	15	15	18	8	18	15	10
17	17	13	17	12	11	13	19	13	13	12	12	17	16	15
7	14	25	16	11	12	12	14	22	16	21	10	15	15	21
18	6	13	21	18	15	19	10	20	15					

(a) Tabulate the frequencies and make a bar diagram.
(b) Use the result in (a) for computing arithmetic mean and standard deviation. (§11.3)

1103. A zoologist studied a certain type of parasite in sand lizards and measured the length (unit: 10^{-3} mm) of 176 such parasites. Result:

220	205	175	215	190	175	190	200	155	230	185	165
115	190	205	160	190	160	250	100	215	170	195	155
200	195	190	190	200	180	105	200	190	195	155	150
195	110	135	190	195	195	155	180	115	200	180	190
195	140	190	130	185	195	160	185	195	240	200	160
195	195	170	200	190	180	190	150	165	120	160	170
195	155	200	85	175	135	185	190	200	180	200	160
140	170	150	190	185	165	185	150	200	120	140	95
185	180	140	120	185	160	220	200	185	145	215	140
180	155	190	190	165	125	160	180	150	170	235	150
175	185	195	135	175	205	230	140	175	200	95	145
215	125	180	215	155	175	150	130	155	180	145	155
170	155	175	180	145	155	210	235	175	180	105	145
165	170	190	135	160	205	225	200	135	215	175	145
240	140	220	235	230	140	260	180				

(a) Group the material (choose 82.5–97.5 as the lowest interval) and construct a histogram.

(b) Compute arithmetic mean and standard deviation for the grouped set. (§11.3)

*1104. For two data sets $x_1, \ldots, x_4$ and $y_1, \ldots, y_5$ the arithmetic means and variances are

$$\bar{x} = 1.38; \qquad \bar{y} = 2.10,$$

$$s_x^2 = 0.0552; \qquad s_y^2 = 0.0792.$$

Compute the arithmetic mean and variance, assuming that all nine values are regarded as one data set.

*1105. The arithmetic mean and standard deviation in a data set were found to be $\bar{x} = 9.496$, $s = 0.345$. There were 800 values. A check showed that a value which ought to have been 9.56 had been entered as 1.56. Correct the arithmetic mean and standard deviation.

CHAPTER 12

Point Estimation

12.1. Introduction

Point estimation is a common activity in everyday life; only the name is new. For example, one sees a man and estimates his age from how he looks, or one measures the length of some object and obtains an estimate of its real length. We shall show in this chapter how data are used to construct point estimates of unknown parameters.

In §12.2 we introduce the basic definitions, in §12.3 we discuss estimation of the mean and variance of a rv, in §12.4 and §12.5 we introduce the method of maximum likelihood and the method of least squares, respectively. §12.6 and §12.7 contain applications to the normal distribution and the binomial and related distributions, respectively. §12.8 treats the important concept of standard errors, and §12.9 the use of normal probability paper. In §12.10 we briefly discuss estimation of probability function, density function and distribution function. Finally, in §12.11 we solve some estimation problems when the parameter has a so-called prior distribution.

12.2. General Ideas

In order to understand this important section the reader must be well acquainted with the ideas behind the terms parameter and random sample (see §10.4).

Let F be the distribution of a rv X. The distribution depends on an unknown parameter θ with parameter space A. We have a random sample $\mathbf{x} = (x_1, \ldots, x_n)$ from F and want to use it to estimate θ.

Remember that the parameter θ is an abstract quantity which often has a concrete interpretation in the real world. For example, θ may correspond to a physical or chemical constant, and the estimation procedure is performed in order to gain knowledge about its value.

For this purpose, we form a suitably chosen function of the sample values. We call this function a *point estimate* of θ. As a general symbol for this function we introduce θ^* (theta-star). We sometimes write $\theta^*(\mathbf{x})$ in order to emphasize that the estimate is a function of the sample values.

It is of basic importance to understand that a point estimate changes if a new sample is collected. For example, assume that θ is the expectation of a continuous rv X, that we have collected 5 values and that we take as our point estimate the arithmetic mean $\theta^* = \sum_1^5 x_i/5$. (Generally we call this $\bar{x}$ but prefer just now the more formal general notation.) The result might be the following:

Sample	x_1	x_2	x_3	x_4	x_5	θ^*
1	2.13	2.10	2.05	2.11	2.14	2.11

Repeated collection of observations would perhaps result in

Sample	x_1	x_2	x_3	x_4	x_5	θ^*
2	2.20	2.11	2.22	2.16	2.10	2.16
3	2.31	2.21	2.24	2.17	2.15	2.22
⋮	⋮	⋮	⋮	⋮	⋮	⋮

In fact, we have only sample 1 at our disposal, but the statistical treatment is based on the idea that the observed values are considered as a sample selected from a hypothetical set of all samples that would result upon repeating the collection of observations indefinitely. A consequence of this way of looking at the sampling procedure is that the estimate θ^* (in this case, 2.11) is itself regarded as an observation on a rv which assumes different values (2.11, 2.16, 2.22, ...).

In the general case this means the following. The estimate $\theta^*(\mathbf{x})$ is an observation on a rv $\theta^*(\mathbf{X})$ which we call a *sample variable*. (Another term that is frequently encountered is *statistic*.) This rv has a certain distribution, which, in the continuous case, is illustrated in Fig. 12.1.

To sum up: In statistical theory, a point estimate $\theta^*(\mathbf{x})$ is regarded as an observation on a sample variable $\theta^*(\mathbf{X})$.

Let us stress the following: It is important to distinguish between $\theta^*(\mathbf{x})$, which is a *numerical value* computed from the sample, and $\theta^*(\mathbf{X})$ which is a *random variable*. However, it is not very convenient to attach to each θ^* a parenthesis with a symbol in it; when there is no risk of confusion we therefore write just θ^*.

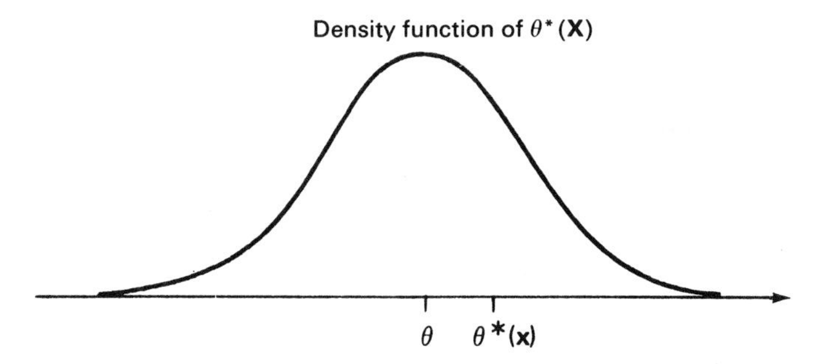

Fig. 12.1. An estimate is an observation on a sample variable.

To find the distribution of a sample variable is an important and sometimes difficult task. It can be solved in two different ways: by analytic method or by simulation. In the former case, the distribution of θ^* is derived by means of probability theory, exactly or approximately. In the latter case, the sampling procedure is repeated a large number of times (by analogy with the example), and an approximation to the distribution is then obtained by tabulating all values of θ^* in the spirit of descriptive statistics. In the following we use only the analytic method.

Example 1. Radioactive Decay

A radioactive substance is assumed to decay with an intensity such that a Geiger–Müller counter registers X particles in an interval of length 1, where $X \sim \text{Po}(\theta)$ (see §9.4). Two time intervals, each of length 1, resulted in x_1 and x_2 particles. We regard $\mathbf{x} = (x_1, x_2)$ as a random sample of two values from $\text{Po}(\theta)$.

In order to estimate θ we take as our point estimate $\theta^*(\mathbf{x}) = (x_1 + x_2)/2$, that is, the arithmetic mean. It is an observation on the sample variable $\theta^*(\mathbf{X}) = (X_1 + X_2)/2$. By Theorem 8 in §9.4, we have $X_1 + X_2 \sim \text{Po}(2\theta)$, that is,

$$P(X_1 + X_2 = i) = e^{-2\theta}(2\theta)^i/i! \qquad (i = 0, 1, \ldots).$$

This gives

$$P[\tfrac{1}{2}(X_1 + X_2) = i] = e^{-2\theta}(2\theta)^{2i}/(2i)! \qquad (i = 0, 1/2, 1, 3/2, \ldots),$$

which is the probability function of the sample variable $\theta^*(\mathbf{X})$. □

It is generally possible to find many different estimates θ^* for the parameter θ. We shall now discuss principles for the choice of a good estimate. Let us begin with an illustration:

Example 2. Measurements

When measuring a physical constant θ, the values 113.1, 112.9 and 119.2 were obtained. We assume that they constitute a random sample from a distribution F with mean θ. Three possible estimates of θ are: the arithmetic mean, the median and the arithmetic

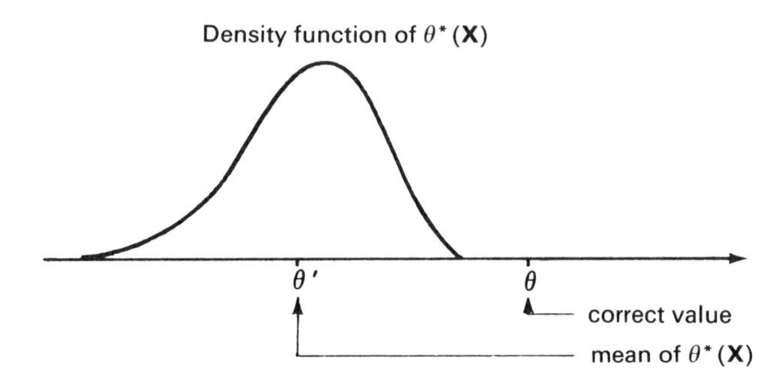

Fig. 12.2. Biased estimate.

mean of the first two values (in the last case we have omitted the third value which differs considerably from the other two); then θ^* would be 115.1, 113.1 and 113.0, respectively. Which estimate is best? □

It seems reasonable to choose an estimate $\theta^*(\mathbf{x})$ such that the distribution of $\theta^*(\mathbf{X})$ is concentrated around θ. Then the estimate will be near this unknown value with large probability, which is desirable. This is a vague condition, which must be made more precise.

We shall introduce three definitions:

Definition. A point estimate $\theta^*(\mathbf{x})$ is said to be *unbiased* if the corresponding sample variable has expectation θ, that is, if for each $\theta \in A$

$$E[\theta^*(\mathbf{X})] = \theta.$$ □

If the expectation is different from θ the estimate is said to be *biased*. It is quite reasonable to demand that an estimate should be unbiased. It seems unsound to select an estimate $\theta^*(\mathbf{x})$ with large bias, that is, an estimate with an expectation $\theta' = E[\theta^*(\mathbf{X})]$ very different from θ (see Fig. 12.2). On the other hand, a small bias is generally not very serious. For example, if the height of a person is overestimated by a few millimeters, the consequence cannot be serious.

Definition. If, for any fixed $\theta \in A$ and for any given $\varepsilon > 0$,

$$P(|\theta^*(\mathbf{X}) - \theta| > \varepsilon) \to 0$$

as the sample size n goes to infinity, then the point estimate $\theta^*(\mathbf{x})$ is said to be *consistent*. □

Vaguely speaking, the definition implies that the distribution of the sample variable becomes more and more concentrated about the true value θ as n increases. If the sample is large, it is then likely that the point estimate is near

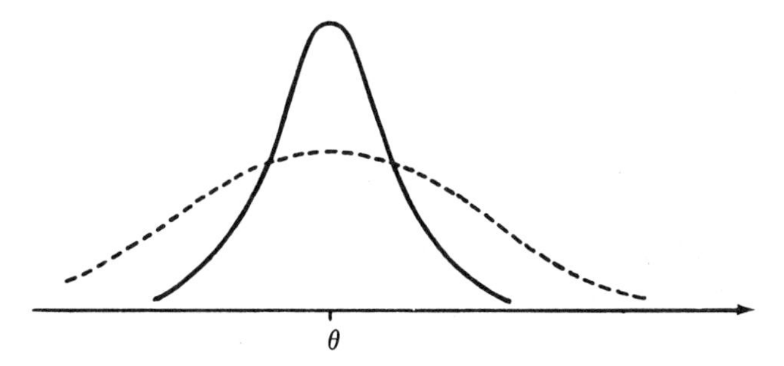

Fig. 12.3. An estimate with more concentrated distribution (heavy line) is preferable.

this true value. Consistency is a rather weak property; most of the estimates we consider in the sequel are consistent.

As a measure of concentration we often use the variance $V[\theta^*(\mathbf{X})]$, or, equivalently, the standard deviation $D[\theta^*(\mathbf{X})]$ of the sample variable. If this variance (standard deviation) is small, the estimate is considered to be good. Hence the following definition seems natural:

Definition. If two estimates θ_1^* and θ_2^* are unbiased and

$$V[\theta_1^*(\mathbf{X})] \leq V[\theta_2^*(\mathbf{X})]$$

for all $\theta \in A$ (with strict inequality for some $\theta \in A$), then θ_1^* is said to be *more efficient* than θ_2^*. □

Or, succinctly: An estimate with smaller variance is better (see Fig. 12.3).

Sometimes we form the ratio of the variance of the better estimate to that of the poorer estimate and call this ratio the *efficiency* of the poorer estimate relative to the better one. Clearly, the ratio is always between 0 and 1.

Example 1. Radioactive Decay (continued)

Two examples of unbiased estimates of θ are

$$\theta_1^* = (x_1 + x_2)/2; \qquad \theta_2^* = x_1.$$

That, say, the first estimate is unbiased follows from

$$E[\theta_1^*(\mathbf{X})] = E[\tfrac{1}{2}(X_1 + X_2)] = \tfrac{1}{2}E(X_1) + \tfrac{1}{2}E(X_2) = \tfrac{1}{2}\theta + \tfrac{1}{2}\theta = \theta.$$

The second estimate is rather foolish, since x_2 is not used at all. However, we shall determine its efficiency relative to θ_1^*. We find

$$V_1 = V[\tfrac{1}{2}(X_1 + X_2)] = \tfrac{1}{4}V(X_1) + \tfrac{1}{4}V(X_2) = \tfrac{1}{4}\theta + \tfrac{1}{4}\theta = \tfrac{1}{2}\theta;$$

$$V_2 = V(X_1) = \theta.$$

Hence the efficiency of θ_2^* relative to θ_1^* is $V_1/V_2 = 1/2$. It is possible to show that θ_1^*

in this example has a much stronger property: it is more efficient than all other unbiased estimates. The proof is given in more advanced textbooks on mathematical statistics. □

We sometimes have observations $x_1, \ldots, x_n$ from different distributions $F_1, \ldots, F_n$ all depending on the same parameter. Even in this more general case it is possible to construct point estimates by choosing suitable functions of the observations, and all that has been said in this section can be applied. (Example: One of the acute angles in a right-angled triangle is θ. In order to estimate θ, the acute angles are both measured. The values obtained, x_1 and x_2, are estimates of θ and $\pi/2 - \theta$, respectively.)

12.3. Estimation of Mean and Variance

As shown in the previous section, many different estimates of an unknown parameter can be found. The question then arises whether there exists some general method for deriving a good estimate. In the next section, we shall describe a general method which assumes that the mathematical form of the distribution is known. In the present section, our level of ambition is lower. We shall discuss estimates which are not "tailored" with respect to the distribution but can be used regardless of its form. We cannot then expect to get the best possible estimate in all cases.

The discussion will be limited to two special but important situations: The unknown parameter is assumed to be either the mean m or the variance σ^2 of the distribution.

(a) Estimation of m

As a point estimate of m we take the sample mean, that is,

$$m^* = \bar{x}.$$

We shall determine the mean and variance of this estimate. We regard $\bar{x} = \sum x_i/n$ as an observation on a rv

$$m^*(\mathbf{X}) = \sum_1^n X_i/n,$$

where the rv's X_i are independent and have the same distribution, each with mean m and standard deviation σ. According to Theorem 4 in §7.3

$$E[m^*(\mathbf{X})] = m; \qquad V[m^*(\mathbf{X})] = \sigma^2/n.$$

The first relation states that the estimate is unbiased, and the second shows that the variance is small if n is large enough. Hence, at least for large n, the estimate will presumably be near the true value m.

This can be expressed more precisely. Applying the law of large numbers to the rv's X_i we find (see Theorem 5 in §7.3)

$$P\left(\left|\frac{1}{n}\sum_1^n X_i - m\right| > \varepsilon\right) \to 0 \quad \text{as} \quad n \to \infty.$$

This means that the estimate is consistent. Hence we have proved:

Theorem 1. *The sample mean $\bar{x}$ is an unbiased and consistent estimate of m.* □

(b) Estimation of σ^2

Now assume that σ^2 is the unknown parameter and that m is also unknown. We take as an estimate the sample variance

$$s^2 = \frac{1}{n-1}\sum_1^n (x_i - \bar{x})^2.$$

This expression is known from Chapter 11, where it was used as a measure of dispersion of a set of data. We shall now use it for a new purpose.

Theorem 2. *The sample variance s^2 is an unbiased estimate of σ^2.* □

PROOF. The theorem states that if $s^2(\mathbf{X})$ is the sample variable corresponding to s^2, then

$$E[s^2(\mathbf{X})] = E\left[\frac{1}{n-1}\sum_1^n (X_i - \bar{X})^2\right] = \sigma^2,$$

where $\bar{X} = \sum X_i/n$. One obtains successively

$$\begin{aligned}
\sum (X_i - \bar{X})^2 &= \sum [(X_i - m) - (\bar{X} - m)]^2 \\
&= \sum (X_i - m)^2 - 2(\bar{X} - m)\sum (X_i - m) + n(\bar{X} - m)^2 \\
&= \sum (X_i - m)^2 - 2n(\bar{X} - m)^2 + n(\bar{X} - m)^2 \\
&= \sum (X_i - m)^2 - n(\bar{X} - m)^2.
\end{aligned}$$

Using, among other results, Theorem 4 in §7.3, we further find

$$\begin{aligned}
E[\sum (X_i - \bar{X})^2] &= E[\sum (X_i - m)^2] - nE[(\bar{X} - m)^2] \\
&= nE[(X - m)^2] - nE[(\bar{X} - m)^2] \\
&= nV(X) - nV(\bar{X}) = n\sigma^2 - n\frac{\sigma^2}{n} = (n-1)\sigma^2.
\end{aligned}$$

Dividing by $n - 1$ we get the theorem. □

The theorem explains why we divide by $n - 1$ when calculating s^2; in doing so, we obtain an unbiased estimate of σ^2. If n is reasonably large, division by n or $n - 1$ makes no great difference, but it is better to be consistent and always use the latter factor.

In practice one often wants to estimate the standard deviation σ (not the variance σ^2). It is then natural to use the sample standard deviation

$$s = \sqrt{\frac{1}{n-1}\sum_{1}^{n}(x_i - \bar{x})^2}.$$

The estimate s is not unbiased, but this is of no great importance in practice, because the systematic error $E[s(\mathbf{X})] - \sigma$ is small, in general, if n is reasonably large.

Remark.

Consider the sum of squares about the arithmetic mean, that is, $Q = \sum(x_i - \bar{x})^2$. It follows from Theorem 2 that the corresponding sample variable satisfies $E[Q(\mathbf{X})] = (n-1)\sigma^2$. □

We have now shown that $\bar{x}$ and s can be used for two purposes. First, as in Chapter 11, they can be used for descriptive purposes. Second, they can be used for getting information about an unknown mean or an unknown standard deviation in a distribution. In the latter case, it should be remembered that $\bar{x}$ and s are not always the best possible estimates. For certain distributions $\bar{x}$ and s are good; for other distributions there may exist better estimates.

12.4. The Method of Maximum Likelihood

Up to this point we have used our intuition in finding estimates. Fortunately, there exist objective methods, the most important being the *method of maximum likelihood* (abbreviated: *ML method*). The idea can be traced to Legendre in the nineteenth century, but the first scientist who used it systematically was R.A. Fisher.

As before, let $x_1, \ldots, x_n$ be a random sample from a distribution $F(x; \theta)$ that depends upon an unknown parameter θ with parameter space A. If X is a continuous rv it has a density function $f(x; \theta)$, and if it is discrete a probability function $p(x; \theta)$. We have added the symbol θ in order to emphasize that θ is an unknown parameter in these functions. We now assume that the density function, or the probability function, is known, apart from the actual value of θ.

We introduce the following:

Definition. The function

$$L(\theta) = \begin{cases} f(x_1; \theta) \cdot f(x_2; \theta) \cdot \ldots \cdot f(x_n; \theta) & \text{(continuous case)} \\ p(x_1; \theta) \cdot p(x_2; \theta) \cdot \ldots \cdot p(x_n; \theta) & \text{(discrete case)} \end{cases} \quad (1)$$

is called the *likelihood function* (abbreviated: *L function*). □

In the discrete case, $L(\theta)$ is just the probability of obtaining the particular sample $x_1, \ldots, x_n$. In the continuous case, $L(\theta)$ is the value assumed by the n-dimensional density function of $\mathbf{X} = (X_1, \ldots, X_n)$ at $x_1, \ldots, x_n$ (see §10.4).

The idea behind the ML method is the following: In the L function we let the argument θ assume all possible values in A and determine which value of θ makes the function as large as possible. This value is called θ^* and is taken as our estimate. The motivation is perhaps clearest in the discrete case: $L(\theta)$ is, as we have just said, the probability of obtaining the sample $x_1, \ldots, x_n$ that has arisen. Hence we choose as an estimate a value such that this probability is maximized (for given $x_1, \ldots, x_n$). Thus we have:

Definition. The value θ^* for which $L(\theta)$ assumes its largest value within A is called the *ML estimate* of θ. □

(We apologize for the misuse of notation which can be traced here and in the sequel: The symbol θ is sometimes the unknown value of the parameter and sometimes the argument in the L function.)

Example 3. Tossing a Coin

It is known that when tossing a certain coin a head appears either with probability $p = 1/2$ or with probability $p = 1/4$. In order to estimate p the coin is tossed twice. Head is obtained at both tosses. Let us apply the ML method.

If the probability of a head is p, the probability that both tosses result in heads is

$$L(p) = p^2.$$

We have $A = \{1/2, 1/4\}$ and so the L function is

$$L(1/2) = (1/2)^2 = 1/4; \qquad L(1/4) = (1/4)^2 = 1/16.$$

The first value is the larger; hence the ML estimate is $p^* = 1/2$. □

The ML estimate possesses several good properties: It is in general consistent. It is not always unbiased but can often be adjusted so that it becomes unbiased; we then have an *adjusted ML estimate.* The ML method often produces estimates that are more efficient than all other unbiased estimates (see Definition, p. 195). Unfortunately, we must be vague at this point, for the theory is complicated and cannot be given here. In the sequel, we therefore regard the ML method as an intuitively attractive method for constructing estimates.

Example 3. Tossing a Coin (continued)

We now consider a coin which is completely unknown to us, that is, A is $0 \le p \le 1$. In order to estimate p we toss the coin n times and get head x times. The model is well known: x is considered as an observation on X, where $X \sim \text{Bin}(n, p)$.

We employ the ML method and obtain (see (1) in §9.2)

$$L(p) = p_X(x) = \binom{n}{x} p^x(1-p)^{n-x}.$$

(a) $x \neq 0$ or n

Taking logarithms we find

$$\ln L(p) = \ln\binom{n}{x} + x \ln p + (n-x)\ln(1-p).$$

A differentiation with respect to p results in

$$\frac{d \ln L(p)}{dp} = \frac{x}{p} - \frac{n-x}{1-p}.$$

Setting the derivative equal to zero and solving the equation, we obtain the ML estimate

$$p^* = x/n.$$

It is not difficult to check that this value maximizes $L(p)$. In this case we get a familiar estimate, namely the relative frequency. In the introduction to probability theory (see the beginning of §2.3) we suggested that it is natural to use the relative frequency as an approximation to p when the number of tosses is large. We now get this estimate, for *any* n, as a consequence of the ML principle. □

(b) $x = 0$ or n

When $x = 0$ the L function reduces to

$$L(p) = (1-p)^n.$$

This function has its maximum when p is zero. Hence we have $p^* = 0$.

Similarly, it is shown that when $x = n$, the ML estimate is given by $p^* = 1$.

Hence in both these cases the ML estimate is, again, the relative frequency; that is, in all cases treated in (a) and (b) we have $p^* = x/n$. □

Example 4. Lifetimes of Bulbs

We switch on n new electric bulbs and record their lifetimes as $x_1, \ldots, x_n$. These values are considered as a random sample from an exponential distribution with density function

$$f(x) = \frac{1}{\theta} e^{-x/\theta} \qquad (x \ge 0),$$

where θ is unknown. We know that $E(X) = \theta$ (see Example 4, p. 98). We shall now estimate this parameter using the ML method. The parameter space is $0 < \theta < \infty$. We find

$$L(\theta) = \prod_{i=1}^{n} \frac{1}{\theta} e^{-x_i/\theta} = \frac{1}{\theta^n} e^{-\sum x_i/\theta}.$$

Taking logarithms and calculating the first derivative, we find that maximum is attained for

$$\theta^* = \sum x_i/n.$$

Hence we estimate the expectation of the lifetime by taking the arithmetic mean of the lifetimes in the sample. It can be shown that this estimate is unbiased (this is an easy exercise!) and also that it is more efficient than all other unbiased estimates. □

The *ML* method can be generalized in two directions (see examples in §12.6):

(a) The distribution contains several unknown parameters.

If these parameters are, say, θ_1 and θ_2, the L function will be a function of both. The values θ_1^* and θ_2^* which simultaneously maximize L are called the *ML* estimates of θ_1 and θ_2, respectively.

(b) The values $x_1, \ldots, x_n$ are observations on rv's $X_1, \ldots, X_n$ with different distributions all depending on the same parameter.

Even in this general case the procedure is the same as before.

12.5. The Method of Least Squares

The *method of least squares* (abbreviated: *LS method*) was developed by Gauss at the beginning of the nineteenth century, long before any coherent statistical theory existed.

Let $x_1, \ldots, x_n$ be a random sample from a distribution with mean $E(X) = m(\theta)$, where $m(\theta)$ is a known function and θ an unknown parameter with parameter space A. Let

$$Q(\theta) = \sum_{i=1}^{n} [x_i - m(\theta)]^2 \tag{2}$$

be the sum of the squares of the deviations of the observations from $m(\theta)$.

Definition. The value θ^*, for which $Q(\theta)$ assumes its least possible value within A, is called the *LS estimate* of θ. □

Example 5. Accelerated Motion

An experimenter drops an object n times independently from a given point and determines after time t the distance it has fallen. The distances $x_1, \ldots, x_n$ vary because of various errors connected with the experiment. We regard the x_i's as a random sample from a distribution with mean $\theta t^2/2$ and standard deviation σ, where σ measures the experimental errors.

We shall estimate the acceleration θ by means of the *LS* method. We find

$$Q(\theta) = \sum_{1}^{n} (x_i - \theta t^2/2)^2.$$

The parameter space is $0 < \theta < \infty$. Differentiating $Q(\theta)$ we obtain

$$\frac{dQ(\theta)}{d\theta} = -t^2 \sum (x_i - \theta t^2/2).$$

Setting this derivative equal to zero and solving for θ we obtain the LS estimate

$$\theta^* = \frac{\bar{x}}{t^2/2}.$$

It is easy to show that Q attains its minimum for this value.

Note that we could not have used the ML method in this example, for the distribution of X is not specified. □

The LS method can be generalized in at least two directions:

(a) The distribution contains several unknown parameters.

The expression Q is then a function of these parameters, and an estimate of each parameter is found such that Q is minimized.

(b) The values $x_1, \ldots, x_n$ are observations on rv's $X_1, \ldots, X_n$ with different distributions depending on the same unknown parameter.

We assume that $E(X_i) = m_i(\theta)$, where the functional forms of $m_i(\theta)$ are known. In this case, we set

$$Q(\theta) = \sum_{i=1}^{n} \lambda_i [x_i - m_i(\theta)]^2, \tag{2'}$$

where $\lambda_1, \ldots, \lambda_n$ are weights. Minimizing this function we obtain the LS estimate.

The weights can be chosen in various ways. Sometimes $V(X_i) = kw_i$, where the quantities w_i are known and k is a constant of proportionality which may be unknown. Then we take $\lambda_i = 1/w_i$. An observation on a rv with large variance will then have less influence on the estimate than an observation on one with small variance. If the variances are equal we set the weights equal and hence the expression to be minimized becomes

$$Q(\theta) = \sum_{i=1}^{n} [x_i - m_i(\theta)]^2.$$

If there are several unknown parameters the procedure is similar.

Example 5. Accelerated Motion (continued)

The experimenter now chooses different times $t_1, \ldots, t_n$ for each observation. We suppose that the precision is the same throughout the experiment so that we can take equal weights in (2′). Then we have to minimize

$$Q(\theta) = \sum_{i=1}^{n} (x_i - \theta t_i^2/2)^2.$$

For brevity we set $u_i = t_i^2/2$. Differentiating with respect to θ we obtain

$$\frac{dQ}{d\theta} = -2 \sum_{i=1}^{n} u_i(x_i - \theta u_i).$$

Setting the derivative equal to zero and solving the resulting equation for θ, we get the *LS* estimate

$$\theta^* = \sum_{i=1}^{n} u_i x_i \Big/ \sum_{i=1}^{n} u_i^2. \qquad \square$$

The *LS* method often gives different results from the *ML* method. Since the *ML* method often produces good estimates, we generally prefer it, except possibly when the *ML* estimate is difficult to calculate. However, the *LS* method has the great advantage that the distribution need not be completely known.

12.6. Application to the Normal Distribution

Since the normal distribution often appears in models used in practice, we shall discuss some of the most common cases of point estimates connected with this distribution.

(a) A Single Sample

Let $x_1, \ldots, x_n$ be a random sample from $N(m, \sigma^2)$. We first assume that *m is unknown and σ^2 known.* The parameter space A is $-\infty < m < \infty$. We then have (see Chapter 8)

$$f(x; m) = \frac{1}{\sigma\sqrt{2\pi}} e^{-(x-m)^2/2\sigma^2},$$

and hence

$$L(m) = \prod_{1}^{n} \frac{1}{\sigma\sqrt{2\pi}} e^{-(x_i-m)^2/2\sigma^2} = ce^{-\sum (x_i-m)^2/2\sigma^2},$$

where c does not depend on m. The function $L(m)$ attains its maximum when the function $\sum (x_i - m)^2$ attains its minimum, which happens for $m = \bar{x}$. Hence the *ML* estimate is given by

$$m^* = \bar{x}.$$

It is easily seen that the *LS* method provides the same estimate.

It follows from Theorem 1 in §12.3 that m^* is unbiased and consistent. It can also be shown that m^* is more efficient than all other unbiased estimates.

We now turn to the case when σ^2 *is unknown and m known*. The parameter space A is $0 < \sigma^2 < \infty$. We then find

$$L(\sigma^2) = \frac{1}{(2\pi\sigma^2)^{n/2}} e^{-\sum (x_i - m)^2/2\sigma^2}.$$

By taking logarithms and differentiating with respect to σ^2 we find that this function of σ^2 attains its maximum in A for

$$(\sigma^2)^* = \sum_1^n (x_i - m)^2/n.$$

We leave it to the reader to show that the estimate is unbiased. It can also be proved that it is consistent and more effective than all other unbiased estimates.

We shall now investigate a third alternative. Here it is assumed that *both m and* σ^2 *are unknown*, which in practice is the most common situation. We then have

$$L(m, \sigma^2) = \frac{1}{(2\pi\sigma^2)^{n/2}} e^{-\sum (x_i - m)^2/2\sigma^2}.$$

By taking logarithms and differentiating with respect to m and σ^2 we find

$$\frac{\partial \ln L}{\partial m} = \frac{1}{\sigma^2} \sum_1^n (x_i - m),$$

$$\frac{\partial \ln L}{\partial \sigma^2} = -\frac{n}{2\sigma^2} + \frac{1}{2\sigma^4} \sum_1^n (x_i - m)^2.$$

Setting the derivatives equal to zero and solving the resulting pair of equations, we obtain the ML estimates

$$m^* = \bar{x}; \qquad (\sigma^2)^* = \frac{1}{n} \sum_1^n (x_i - \bar{x})^2.$$

The first estimate is unbiased, but the second has to be adjusted for bias. The adjusted ML estimate is (see Theorem 2 in §12.3)

$$s^2 = \frac{1}{n-1} \sum_1^n (x_i - \bar{x})^2.$$

Let us pause and note how remarkable the results of these derivations are. In §12.3 the estimates $\bar{x}$ and s^2 were chosen more or less by intuition, and were found to be unbiased and consistent for any distribution. We have now shown that, when the distribution is normal, they can be derived objectively by means of the ML method (and, in the case of $\bar{x}$, alternatively by means of the LS method).

When estimating σ instead of σ^2 in the case of a normal distribution, one generally uses the sample standard deviation s (in spite of the fact that s is not quite unbiased).

(b) Two Samples

It is common in practice to have two independent samples from normal distributions, one from $N(m_1, \sigma_1^2)$, the other from $N(m_2, \sigma_2^2)$. If all four parameters are unknown, each sample can be treated according to the method in (a). However, it is often reasonable to assume in advance that $\sigma_1^2 = \sigma_2^2$, and then a new problem arises: How should the common variance $\sigma_1^2 = \sigma_2^2$ (which we may call σ^2) be estimated using both samples? We shall solve this problem by the *ML* method.

Thus let $x_1, \ldots, x_{n_1}$ and $y_1, \ldots, y_{n_2}$ be random samples from $N(m_1, \sigma^2)$ and $N(m_2, \sigma^2)$, respectively, where m_1, m_2 and σ^2 are unknown parameters. We have

$$L(m_1, m_2, \sigma^2) = L_1 \cdot L_2,$$

where

$$L_1 = \frac{1}{(2\pi\sigma^2)^{n_1/2}} e^{-\sum (x_i - m_1)^2/2\sigma^2},$$

and analogously for L_2.

Proceeding in the usual way, we obtain the *ML* estimates

$$m_1^* = \bar{x}; \qquad m_2^* = \bar{y},$$

$$(\sigma^2)^* = \left[\sum_1^{n_1} (x_i - \bar{x})^2 + \sum_1^{n_2} (y_i - \bar{y})^2\right] \Big/ (n_1 + n_2).$$

The first two estimates are unbiased, but the third one has to be adjusted. If we call the two sums of squares Q_1 and Q_2, the adjusted *ML* estimate is found to be

$$s^2 = \frac{Q_1 + Q_2}{(n_1 - 1) + (n_2 - 1)}. \tag{3}$$

Its unbiasedness follows from the Remark, p. 198. As an estimate of σ we take the square root of this expression.

(c) Several Samples

We shall generalize the result in (b) to k samples which, using new notation, we write as follows:

Sample	Observations	Distribution
1	$x_{11} x_{12} \ldots x_{1n_1}$	$N(m_1, \sigma^2)$
2	$x_{21} x_{22} \ldots x_{2n_2}$	$N(m_2, \sigma^2)$
$\vdots$	$\vdots \quad \vdots \quad \vdots$	$\vdots$
k	$x_{k1} x_{k2} \ldots x_{kn_k}$	$N(m_k, \sigma^2)$

It is seen from the table that the variance is assumed to be the same in all samples. The adjusted ML estimate of σ^2 is given by

$$s^2 = \frac{Q_1 + \cdots + Q_k}{(n_1 - 1) + \cdots + (n_k - 1)}. \tag{4}$$

Here Q_i is the sum of squares about the sample mean for the ith sample, that is,

$$Q_i = \sum_{j=1}^{n_i} (x_{ij} - \bar{x}_i)^2,$$

where

$$\bar{x}_i = \sum_{j=1}^{n_i} x_{ij}/n_i.$$

Alternatively, s^2 in (4) may be written in the form

$$s^2 = \sum_{i=1}^{k} (n_i - 1)s_i^2 / \sum_{i=1}^{k} (n_i - 1), \tag{4'}$$

where s_i^2 is the sample variance for the ith sample, that is,

$$s_i^2 = Q_i/(n_i - 1).$$

It is seen from (4′) that s^2 is a weighted mean of the sample variances.

The model we have described here and the resulting point estimate are often used in practice.

Example 6. Triple Measurements

Each of ten objects was measured three times, and the results were:

Object	Measurements			T_i
1	11.9	10.8	15.2	37.9
2	9.9	11.1	11.0	32.0
3	11.0	11.4	12.5	34.9
4	12.0	8.5	12.0	32.5
5	10.3	12.6	12.4	35.3
6	12.1	12.4	16.4	40.9
7	9.6	11.9	15.0	36.5
8	11.7	16.9	16.4	45.0
9	8.0	10.6	10.8	29.4
10	12.3	10.6	9.5	32.4

Let us assume that the measurements of object i can be regarded as observations from $N(m_i, \sigma^2)$, where σ^2 is a measure of the precision of the procedure (that is, of the variation from measurement to measurement of the same object). Let us estimate σ.

It is possible to follow formula (4) literally and calculate each Q_i separately. However, when the number of observations in each sample is the same, say n, it is preferable to proceed as follows. Calculate

$$\sum_{i=1}^{k} Q_i = \sum_{i,j} (x_{ij} - \bar{x}_i)^2 = \sum_{i,j} x_{ij}^2 - \frac{1}{n} \sum_i T_i^2,$$

where $T_i = \sum_{j=1}^{n} x_{ij}$ is the sum of the values in sample i. (It is easy to verify that the sum of the Q_i's can be rewritten in this way.)

Using this procedure we find

$$\sum_{i,j} x_{ij}^2 = 4{,}383.84; \qquad \sum_i T_i^2 = 12{,}924.94,$$

$$\sum_i Q_i = 4{,}383.84 - \frac{1}{3} 12{,}924.94 = 75.53,$$

$$s^2 = \frac{75.53}{10(3-1)} = 3.776; \qquad s = \sqrt{3.776} = 1.94. \qquad \square$$

12.7. Application to the Binomial and Related Distributions

(a) Binomial Distribution

In Example 3, p. 200, we encountered a classical situation. We had an observation x on X, where $X \sim \text{Bin}(n, p)$, p unknown. We derived the *ML* estimate, which is the relative frequency

$$p^* = x/n.$$

Since

$$Q(p) = (x - np)^2$$

attains its minimum for $p = p^*$, the *ML* estimate and the *LS* estimate coincide.

We shall have a closer look at p^*. As usual, we regard p^* as an observation on a rv $p^*(X) = X/n$. We have studied this rv already in subsection (d) of §9.2. If we apply what we found there to the present situation, we see that the estimate is consistent. In fact, according to Bernoulli's theorem we have

$$P\left(\left|\frac{X}{n} - p\right| > \varepsilon\right) \to 0 \quad \text{as} \quad n \to \infty.$$

Furthermore, formulae (6) of §9.2 show that

$$E(X/n) = p; \qquad V(X/n) = pq/n.$$

Hence the estimate p^* is unbiased. It can also be proved that p^* is more efficient than all other unbiased estimates of p.

(b) Hypergeometric Distribution

We now consider the same problem as in (a) but for a hypergeometric distribution. The probability function is then

$$p_X(k) = \binom{Np}{k}\binom{Nq}{n-k}\bigg/\binom{N}{n},$$

where $q = 1 - p$ and p is unknown. According to Theorem 6 in §9.3 we have $E(X) = np$. Hence, when applying the LS method we have to minimize

$$Q = (x - np)^2,$$

that is, the same expression as in (a). This implies that the LS estimate is also the same as before, namely

$$p^* = x/n.$$

The discussion of the properties of the estimate is similar to the binomial case. For example, by (11) in §9.3 we have

$$E(X/n) = p; \qquad V(X/n) = d_n^2 pq/n.$$

Hence p^* is unbiased; it can easily be shown that it is also consistent.

(c) Poisson Distribution

Let $x_1, \ldots, x_n$ be a random sample from Po(m). The probability function is then (see §9.4)

$$p_X(k) = e^{-m}m^k/k! \qquad (k = 0, 1, \ldots).$$

For estimating m we use the ML method and obtain

$$L(m) = \prod_{i=1}^{n} e^{-m}m^{x_i}/x_i! = ce^{-nm}m^{\sum x_i},$$

where c does not depend on m. Taking logarithms, differentiating with respect to m, and setting the derivative equal to zero we find

$$-n + \frac{\sum x_i}{m} = 0,$$

and hence

$$m^* = \bar{x}.$$

Thus the ML estimate is the sample mean. It is a good exercise to prove that $\bar{x}$ is unbiased and consistent; moreover, the estimate is more efficient than any other unbiased estimate.

The situation can be generalized to the following: Assume that $x_1, \ldots, x_n$ are observations on different rv's $X_1, \ldots, X_n$, where $X_i \sim$ Po(mt_i) and the quantities $t_1, \ldots, t_n$ are known beforehand. We leave it to the reader to show that the ML estimate is given by

$$m^* = \sum_1^n x_i / \sum_1^n t_i.$$

Example 7. Radioactive Decay

Let us use the same model as in Example 1, p. 193. Assume that x_1 particles have been registered in a time interval t_1 and x_2 particles in another interval t_2. If the decay intensity is λ, we realize that, according to the model used, x_1 and x_2 are observations on rv's $X_1 \sim \text{Po}(\lambda t_1)$ and $X_2 \sim \text{Po}(\lambda t_2)$, respectively. Using the formula given above, we obtain the ML estimate

$$\lambda^* = (x_1 + x_2)/(t_1 + t_2).$$

This is a reasonable result: The total number of particles registered is divided by the total time of observations. □

12.8. Standard Error of an Estimate

Throughout this chapter we have used the variance $V(\theta^*)$ or, equivalently, the standard deviation $D(\theta^*)$, as a measure of precision of an estimate θ^*. The smaller the variance, the more satisfactory the estimate. These two quantities will play a central role also in the following chapters.

The following complication sometimes arises: The variance and the standard deviation may themselves be unknown because they depend on the parameter to be estimated (and perhaps on other unknown parameters). This may seem an insuperable difficulty, but there is a way out of the dilemma. If, say, $D(\theta^*)$ is unknown, we estimate this quantity also, although it is of secondary interest to us. Consequently, we do not attempt to obtain an exact measure of the precision, only an approximate one.

Definition. An estimate of $D(\theta^*)$ is called the *standard error* of θ^* and is written $d(\theta^*)$. □

The problem of choosing a standard error satisfactorily has no unique solution and has to be attacked case by case. Of course, a consistent estimate of $D(\theta^*)$ should be chosen. Sometimes we simply write d instead of $d(\theta^*)$.

It may certainly look puzzling that the standard deviation of an estimate itself needs to be estimated, and it is important for the reader to keep the following concepts apart: θ^* is an estimate of θ and d is an estimate of D.

Example 8. Tossing a Coin

Let us again consider a coin that shows head with the unknown probability p. The coin is tossed n times, resulting in x heads. The estimate $p^* = x/n$ has standard deviation $D(p^*) = \sqrt{pq/n}$, which depends on p. (For example, if $n = 100$ the standard deviation is 0.05 for $p = 0.5$ but 0.03 for $p = 0.1$ or 0.9; quite generally, the standard deviation is smaller when p is near zero or one than when p is near 0.5.) It is easy to construct an estimate of $D(p^*)$: We may take $d(p^*) = \sqrt{p^*q^*/n}$; that is, we use p^* instead of p and $q^* = 1 - p^*$ instead of q. If n is reasonably large, we expect that $d(p^*)$ will not differ very much from $D(p^*)$. □

Example 9. Normal Distribution with Unknown Mean and Unknown Variance

The estimate $m^* = \bar{x}$ of m has standard deviation $D(m^*) = \sigma/\sqrt{n}$, which depends on the unknown σ. It is natural to replace σ by the sample standard deviation s, that is, to choose the standard error $d(m^*) = s/\sqrt{n}$. If n is at least moderately large, this standard error will presumably not differ much from $D(m^*)$. □

The concept of standard error is very important, as Chapter 13, in particular, will show.

Remark. Confusion of Terms

The definition of standard error given in this book is not used everywhere. Certain authors call the standard deviation itself, that is $D(\theta^*)$, the standard error. The reader is therefore advised to check an author's use of the term. □

12.9. Graphical Method for Estimating Parameters

So far in this chapter we have used only numerical methods for estimating parameters. Sometimes a graphical method is useful.

We limit the discussion to the general normal distribution with

$$F_X(x) = \Phi\left(\frac{x - m}{\sigma}\right). \tag{5}$$

So-called *normal probability paper* is commercially available (see Fig. 12.4). Such paper has a horizontal scale of the usual type, but the vertical scale is constructed in such a way that the graph of the function $y = \Phi(x)$ is a straight line. If the vertical scale had been the usual linear one, the graph of $y = \Phi(x)$ would have been a curve, as we know from Chapter 8. Hence this linear scale has been modified so that the curve is replaced by a straight line. That such linearization is possible is not at all remarkable, and there are many similar situations. (For example, if we want the graph of $y = e^{ax+b}$ to be a straight line, we use paper with vertical log-scale.)

Let us now assume temporarily that m and σ in (5) are *known*. It is then clear that the function

$$y = \Phi\left(\frac{x - m}{\sigma}\right)$$

corresponding to a general normal distribution will also be a straight line on

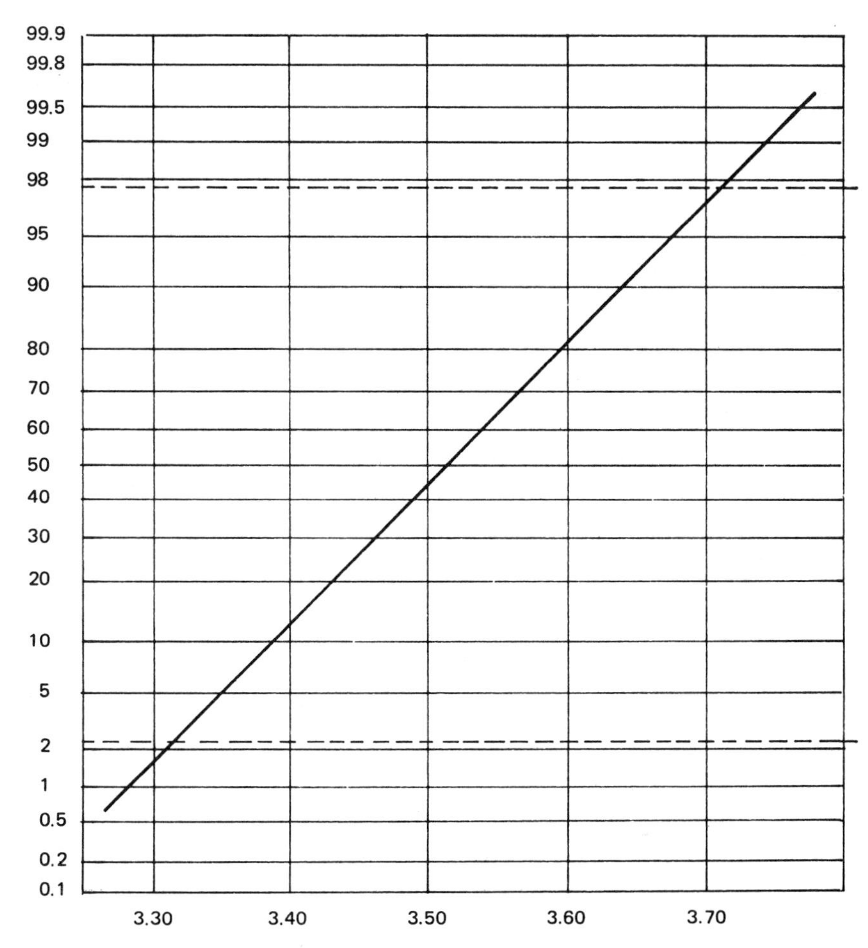

Fig. 12.4. Normal probability paper showing the distribution function of $N(m, \sigma^2)$, where $m = 3.52$ and $\sigma = 0.10$.

normal probability paper. In Fig. 12.4 this line is drawn for the special values $m = 3.52$, $\sigma = 0.10$.

In order to understand how the paper works it is instructive to consider how to draw this line. The horizontal scale can be selected in many ways, but it should be chosen in such a manner that there is ample space for all x-values in the interval $(m - 3\sigma, m + 3\sigma)$. We then plot two values at some suitable distance, for example, $x = m - 2\sigma$ and $x = m + 2\sigma$. The corresponding y-values are (see Chapter 8) $y = \Phi(-2) \approx 0.023$ and $y = \Phi(2) \approx 0.977$. (It is convenient to choose these values, since the lines $y = 0.023$ and $y = 0.977$ are represented as dashed lines on normal probability paper.) As a check of the correctness of the drawing we may note that for $x = m$ we have $y = 0.50$ (see expression (5)).

After these preparations it is easy to describe how to estimate m and σ in (5) when these parameters are *unknown*.

The values in the random sample $x_1, \ldots, x_n$ are first ordered from the smallest to the largest. These ordered values are written $x_{(1)}, \ldots, x_{(n)}$. Thereafter a suitable horizontal x-scale is chosen. The plotting is now performed as follows: The ith sample value $x_{(i)}$ is plotted along the horizontal axis, and the number $(i - 1/2)/n$ along the vertical axis according to the scale provided. (We cannot discuss here why the plotting is done in this way; the idea is to get the best possible estimates.) In this way n points are obtained which lie on a straight line, apart from chance fluctuations. A straight line is drawn through the points as accurately as possible. Let us call this line L^*.

In order to estimate m we mark the point of intersection of L^* and the horizontal line marked 50 (the "50 line"). As an estimate we take m^* = the x-coordinate of this point of intersection.

In order to estimate σ there are several possibilities, for example, the following: Mark the point of intersection of L^* and the "upper two-sigma line" (the dashed line close to the 98 line) and the point of intersection of L^* and the "lower two-sigma line" (the dashed line close to the 2 line). Call the x-coordinates of these points x_{+2} and x_{-2}, respectively. As an estimate of σ we take

$$\sigma^* = (x_{+2} - x_{-2})/4.$$

The precision of the estimates depends to a large extent on the sample size (and on the person who draws the line L^*!). The larger the sample, the more nearly the points will lie on a straight line and the easier it is to draw the line L^*.

Example 10

Assume that we have collected the following data:

i	1	2	3	4	5	6	7	8	9	10
$x_{(i)}$	0.55	0.95	1.22	1.50	1.81	2.12	2.17	2.31	2.43	2.77
$(i - 1/2)/n$	0.025	0.075	0.125	0.175	0.225	0.275	0.325	0.375	0.425	0.475

i	11	12	13	14	15	16	17	18	19	20
$x_{(i)}$	2.94	3.00	3.18	3.59	3.66	3.82	3.88	3.95	4.16	4.72
$(i - 1/2)/n$	0.525	0.575	0.625	0.675	0.725	0.775	0.825	0.875	0.925	0.975

The values have been plotted on normal probability paper in Fig. 12.5, and the reader is advised to study the details. As an estimate of m one reads $m^* = 2.7$. Moreover, one obtains $x_{-2} = 0.35$ and $x_{+2} = 5.05$, which gives $\sigma^* = (5.05 - 0.35)/4 = 1.2$. □

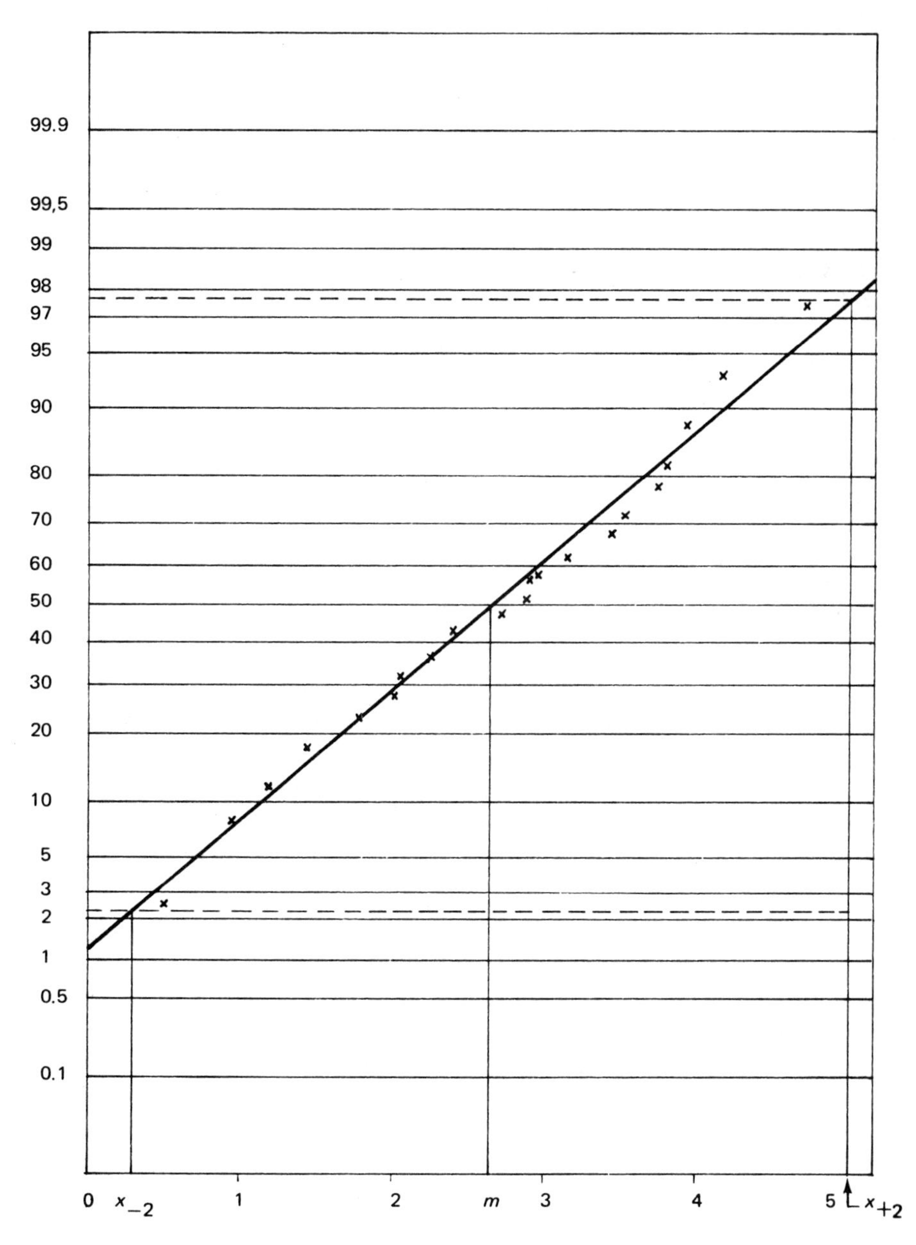

Fig. 12.5. Normal probability paper with a straight line obtained from a random sample.

Remark. Grouped Sample

If n is large it is impractical to plot each individual value. Instead, the sample is grouped (see Chapter 11). Then one point is plotted for each class, such that the upper class limit is plotted along the horizontal axis and the corresponding cumulative relative frequency along the vertical axis. (These terms are explained in Chapter 11.) Thereafter the procedure is the same as before. □

Normal probability paper can also be used for determining whether a large sample has come from a normal distribution. To do this, one inspects the graph of the sample and tries to decide whether the data appear to agree with a linear function or if the data would be better represented by a curve. In the latter case, the sample cannot come from a normal distribution. Hence the graphical method may give dual information, namely, about the form of the distribution and about the values of the parameters. A weakness of the method, as described here, is that the information may be rather vague; for example, we have not stated how large the deviations from a straight line are allowed to be before we reject the normal distribution. (For more information on this point, see larger textbooks.)

12.10. Estimation of Probability Function, Density Function and Distribution Function[1]

As usual, let $x_1, \ldots, x_n$ be a random sample from the distribution F. We want sometimes to estimate the whole probability function, density function or distribution function. We shall describe briefly how this is done, using known results.

(a) The Probability Function

Let $p(k)$, $k = 0, 1, \ldots$, be the unknown probability function. Tabulate the frequencies of the sample according to the description in subsection (b) of §11.2. We then obtain relative frequencies f_k/n which are estimates of the probabilities $p(k)$; see Fig. 12.6.

Note that each f_k is an observation from $\text{Bin}(n, p(k))$. Thus according to the beginning of §12.7 we have for all $k = 0, 1, \ldots$,

$$E(f_k/n) = p(k); \qquad V(f_k/n) = p(k)[1 - p(k)]/n.$$

Hence the relative frequencies are unbiased estimates of the unknown probabilities in the probability function. The larger n is, the smaller the variances

[1] Special section which may be omitted.

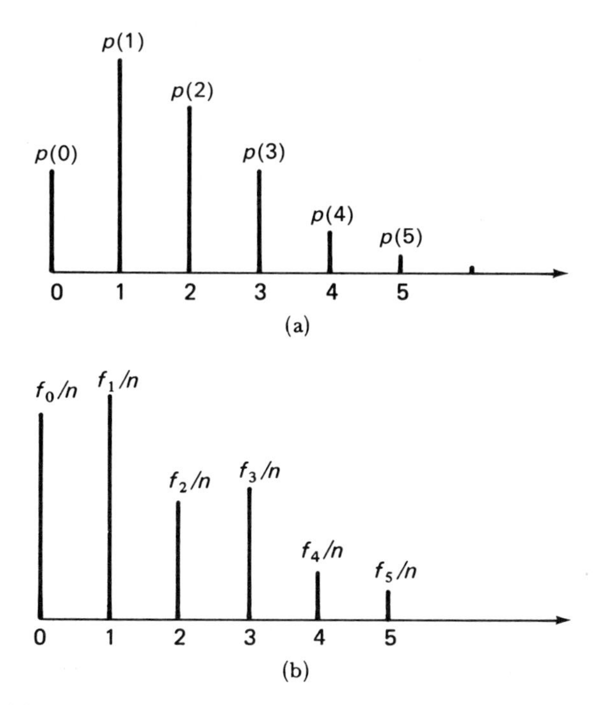

Fig. 12.6. (a) Unknown probability function. (b) Estimate of probability function.

and the better the estimates. This valuable result is evidently true for the probabilities in any probability function, independently of its form.

The standard errors of the estimates f_k/n are obtained by replacing the parameter $p(k)$ in $V(f_k/n)$ by f_k/n and taking the square root (see §12.8). Hence we obtain

$$d(f_k/n) = \sqrt{f_k(n - f_k)/n^3}.$$

(b) The Density Function

We imitate the procedure in (a) as far as possible.

Group the observations into r classes according to the description in subsection (c) of §11.2. Let $p_1, \ldots, p_r$ be the unknown areas under the density function $f(x)$; see Fig. 12.7. Draw a histogram, using the grouped sample, and denote the relative frequencies then obtained by $f_1/n, \ldots, f_r/n$. These frequencies are unbiased estimates of $p_1, \ldots, p_r$ with variances analogous to those in (a). Hence the conclusions are similar, but note that we do not obtain an estimate of the density function $f(x)$ itself, only estimates of certain areas under this function. By choosing r large (which requires a sufficiently large sample), we nevertheless obtain a good picture of the form of the density function.

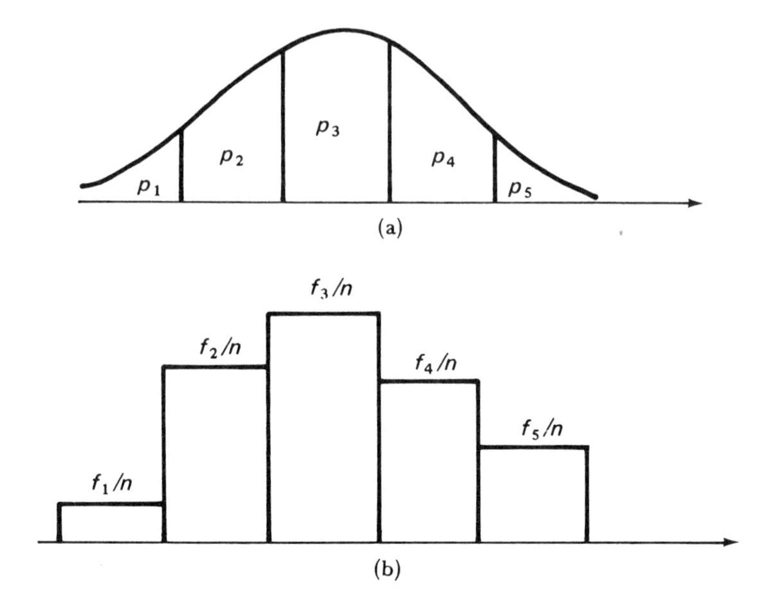

Fig. 12.7. (a) Unknown density function. (b) Estimate of density function.

(c) The Distribution Function

When the calculations in (a) or (b) have been performed, it is possible, by summation from the left in the diagrams in Fig. 12.6(b) or Fig. 12.7(b), to obtain information about the distribution function $F(x)$.

Direct calculations are also possible. Assume that we want to estimate $F(x)$ for a given value x. We then count the number g_x of values in the sample that are $\leq x$ and form the estimate

$$F^*(x) = g_x/n.$$

We then have (see the estimate of $p(k)$ in (a))

$$E[F^*(x)] = F(x); \qquad V[F^*(x)] = F(x)[1 - F(x)]/n.$$

Hence the estimate is unbiased and has a small variance when n is large enough. By performing this calculation for several x-values a picture of the distribution function is obtained.

If the sample is small, or only moderately large, it is sometimes useful to perform the above calculation for all the n sample points $x_1, \ldots, x_n$. This implies that, proceeding from small to large x, we estimate $F(x)$ for $x = x_{(1)}$, $x = x_{(2)}$ and so on, where

$$x_{(1)} \leq x_{(2)} \leq \cdots \leq x_{(n)}$$

is the ordered sample.

It is easy to see that the estimates become

$$F^*(x_{(i)}) = i/n \qquad (i = 1, 2, \ldots, n).$$

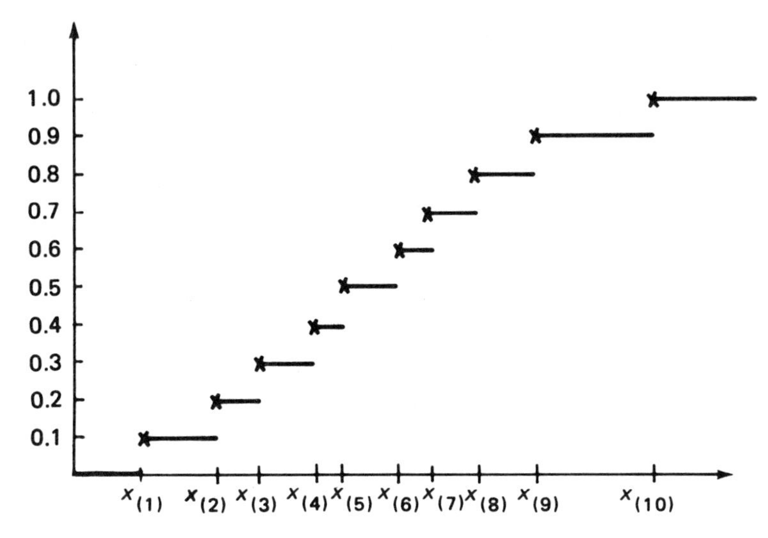

Fig. 12.8. Empirical distribution function ($n = 10$).

The same estimate, i/n, applies to $F^*(x)$ for x between $x_{(i)}$ and $x_{(i+1)}$. In this way we obtain an estimate of the whole function $F(x)$ called the *empirical distribution function* (see Fig. 12.8).

The methods described in this section are intended to be used when the functional form of $F(x)$ is not known. If this form is known, other methods are generally used. For example, we have seen earlier in this chapter how to proceed if it is known beforehand that the distribution is normal with unknown m and unknown σ^2.

12.11. Parameter with Prior Distribution[1]

In this chapter we have assumed up to now that no information is available beforehand about θ; all the information has come from the observations. If prior information is available, it should, of course, be used when making inferences. The information may take the form that the parameter space is restricted in some way or that θ has been selected in some special manner. We explain this best by an example.

Example 11. Tossing a Randomly Chosen Coin

Two coins M_1 and M_2 are available that come up heads with known probabilities p_1 and p_2, respectively. With known probability w_i, person A_1 chooses the coin M_i

[1] Special section, which may be omitted.

($i = 1, 2$; $w_1 + w_2 = 1$). He gives the coin to person A_2 who performs n tosses in order to estimate the probability θ that the coin will come up heads. (Hence A_2 knows that θ must be p_1 or p_2 and that, if w_1 should be near 1, it is very probable that $\theta = p_1$.) Hence, in this problem, two types of probabilities occur that must be distinguished.

In this case the ML method is inappropriate; instead Bayes' theorem should be used (Theorem 6 in §2.5). Let A be the event "x heads are obtained" and H_i the event "the tossed coin is M_i". Since

$$P(H_i) = w_i; \qquad P(A|H_i) = \binom{n}{x} p_i^x q_i^{n-x},$$

it is found after reduction that

$$P(\theta = p_i|A) = w_i p_i^x q_i^{n-x}/(w_1 p_1^x q_1^{n-x} + w_2 p_2^x q_2^{n-x}) \qquad (i = 1, 2).$$

This is the conditional probability that the coin is M_i (that is, that $\theta = p_i$) given x heads. As an estimate of θ it seems reasonable to take the value p_i that corresponds to the largest of these probabilities.

Suppose, for example, that $p_1 = 1/4$, $p_2 = 1/2$, $w_1 = 2/3$, $w_2 = 1/3$ and that four tosses with the coin all give heads. We obtain $P(\theta = 1/2|A) = 1/3 \cdot (1/2)^4/(2/3 \cdot (1/4)^4 + 1/3 \cdot (1/2)^4) = 8/9$; similarly, we find $P(\theta = 1/4|A) = 1/9$. Hence with a rather large probability it can be stated that coin M_2 was selected, that is, that $\theta = 1/2$ (in spite of the fact that, at the beginning, M_1 was the more probable!).

Let us compare with the ML method. We get $L(p_1) = (1/4)^4$, $L(p_2) = (1/2)^4$, that is, $p_2 = 1/2$ maximizes the L function. Hence the ML method gives the same answer (which is not always the case).

The method can, of course, be used for more than two coins. □

In the example we have just discussed, there is what is generally called a known *prior distribution* of the unknown parameter. This distribution is discrete here (a two-point distribution with probabilities w_1, w_2), but it can also be continuous.

The distribution which is determined by means of the joint information given by the prior distribution and the sample is called the *posterior distribution*. In the above example, this distribution consists of the two probabilities $P(\theta = p_i)$ and hence is discrete, but this distribution also can be continuous.

We shall now give an example involving a continuous prior distribution.

Example 11. Tossing a Randomly Chosen Coin (continued)

Let us now assume that θ has a prior density function $g(p)$ ($0 \le p \le 1$), where we let p be the argument. Suppose again that n tosses with the selected coin result in x heads. The posterior distribution is then found to be (by an extension of Bayes' theorem to a "dense" set of alternatives)

$$f(p) = c \cdot g(p) p^x q^{n-x} \qquad (0 \le p \le 1),$$

where c is a suitably chosen constant. If p is determined so that $f(p)$ is maximized, it is realized that the result will generally be different from that obtained by the ML method. (However, the same result arises in the interesting special case $g(p) = 1$; the prior distribution is then uniform over the interval (0, 1).) □

Remark. Bayesian Statistics

In the Remark, p. 14, we mentioned subjective probabilities. The so-called Bayesians, who are named after the Englishman Bayes (the author of Bayes' theorem), favour this concept. They assert that *every* model with an unknown parameter should contain a prior distribution for the parameter. According to their view, this distribution should be chosen subjectively, that is, by the individual user.

To a Bayesian, calculations of the type we described in Example 11 are of central importance in all statistical problems. First, a prior distribution for the parameter is chosen, then observations are performed, and finally the posterior distribution is determined by Bayes' theorem. The latter distribution amalgamates the prior knowledge with the information furnished by the observations.

The claim of the Bayesians that this is a universal statistical method has given rise to much discussion and criticism. The critics retort that it is impossible to state a prior distribution for an unknown constant and even absurd, especially since it is—a constant! □

Exercises

1201. Two independent measurements x_1 and x_2 are performed on the gravitational constant g. They are assumed to constitute a random sample from $N(g, 15^2)$. As an estimate of g we take the arithmetic mean $\bar{x} = (x_1 + x_2)/2$. Find the distribution of the sample variable $\bar{X} = (X_1 + X_2)/2$. (§12.2)

1202. Observations x_1 and x_2 are available on the independent rv's X_1 and X_2, which are both Bin(1, p). Consider the following two estimates of p:

$$p_1^* = x_1; \qquad p_2^* = (x_1 + x_2)/2.$$

(a) Which values can p_1^* and p_2^* assume?
(b) Are the estimates unbiased?
(c) Find the variances of the estimates and the efficiency of p_1^* relative to p_2^*. (§12.2)

1203. Consider the random sample $x_1, \ldots, x_n$ from $N(m, \sigma^2)$, where m and σ^2 are unknown. We form the estimates

$$m_1^* = \bar{x} = (x_1 + x_2 + \cdots + x_n)/n,$$

$$m_2^* = (x_1 + x_n)/2.$$

(a) Show that both estimates are unbiased.
(b) Determine the efficiency of m_2^* relative to m_1^*. (§12.2)

1204. The distance between two points is measured on four occasions. Result (in meters):

1,456.3 1,458.5 1,457.7 1,457.2.

This is regarded as a random sample from $N(m, \sigma^2)$, where m is the true distance and σ^2 measures the precision of the method. Find an unbiased estimate of σ^2 if:

(a) it is known that $m = 1457.0$;
(b) m is unknown. (§12.3)

1205. The discrete rv X has probability function $p_X(k) = \theta(1-\theta)^{k-1}$ for $k = 1$, 2, ..., where $0 < \theta < 1$. Consider a random sample 4, 5, 4, 6, 4, 1 from this distribution.
   (a) Find the likelihood function $L(\theta)$.
   (b) Find the value θ which maximizes $L(\theta)$. Consequently, what is the *ML* estimate of θ? (§12.4)

1206. The rv X has density function $f_X(x) = \theta(1+x)^{-\theta-1}$ for $x \geq 0$. It is known beforehand that θ is either 2, 3 or 4. Let 0.2, 0.8 be a random sample of two values from this distribution.
   (a) Find the L function for the three possible values of θ.
   (b) Determine the *ML* estimate of θ. (§12.4)

1207. Let $x_1, \ldots, x_n$ be a random sample from a distribution with density function $f_X(x) = \theta x^{\theta-1}$ for $0 < x < 1$. Find the *ML* estimate of θ. (§12.4)

1208. The rv X has density function $f_X(x) = (x/a)e^{-x^2/2a}$ for $x \geq 0$, where a is an unknown positive parameter. (This is a Rayleigh distribution; see §3.7.) Let $x_1, \ldots, x_n$ be a random sample from this distribution. Find the *ML* estimate of a. (§12.4)

1209. The rv X has density function

$$f_X(x) = \theta(1+x)^{-\theta-1} \qquad \text{if} \quad x \geq 0,$$

where θ is either 2, 3 or 4.
   (a) Find $E(X)$ as a function of θ.
   (b) Find the *LS* estimate of θ based on the random sample 0.2, 0.8. Compare the result with that of Exercise 1206. (§12.5)

1210. A physicist wants to determine the wavelength θ of a certain line in a metal spectrum by performing measurements on photographic plates. The resulting measurements $x_1, x_2, \ldots$ are regarded as observations on independent rv's $X_1, X_2, \ldots$ with the same mean $E(X_i) = \theta$ but standard deviations $D(X_i) = \sigma_i$ that vary from plate to plate. The physicist has collected the following data; also shown are the corresponding values of σ_i:

x_i	79.1	80.0	81.3	81.9	81.7
$100\sigma_i$	2	1	2	3	1

Find the *LS* estimate of θ.
Hint: Take $\lambda_i = 1/\sigma_i^2$ in formula (2′). (§12.5)

1211. Let $x_1, \ldots, x_7$ and $y_1, \ldots, y_4$ be random samples from $N(m_1, \sigma^2)$ and $N(m_2, \sigma^2)$, respectively. It is found that

$$\sum_1^7 x_i = 42; \qquad \sum_1^7 x_i^2 = 276,$$

$$\sum_1^4 y_i = 8; \qquad \sum_1^4 y_i^2 = 19.$$

Use this information for computing an estimate of σ. (§12.6)

1212. In order to test a measuring instrument the following three series of measurements were performed on different pharmaceutical preparations:

Series 1	0.16	0.18	0.19	0.18	0.21
Series 2	0.22	0.21	0.24	0.24	0.25
Series 3	0.17	0.15	0.15	0.18	0.14

The result of a series is regarded as a random sample from a normal distribution with unknown mean, and a variance that is unknown but the same for the three distributions.
(a) Estimate the means of the distributions.
(b) Estimate the common standard deviation. (§12.6)

1213. Two research workers, A and B, have estimated the fraction p of colour-blindness in a large population of males: A has chosen 1,000 persons and found 77 colour-blind, while B has examined 2,000 and found 124 colour-blind. Determine the ML estimate of p. (§12.7).

1214. The number of ships that, during a time interval of length t (unit: minute), pass Elsinore on their way to the south through the sound between Sweden and Denmark is considered to be $\text{Po}(\lambda t)$. The number of ships passing in non-overlapping intervals are assumed to be independent. A person counts the number of ships in three different time intervals. Result:

Period of observation	30	30	40
Number of ships	10	12	18

(a) Find the ML estimate λ^* of λ.
(b) Determine the standard deviation of λ^*. (§12.7)

1215. Let $x = 16$ be an observation on $X \sim \text{Bin}(25, p)$.
(a) Estimate p.
(b) Find the standard deviation of the estimate.
(c) Find a standard error of the estimate. (§12.8)

1216. In an urn there are N balls, Np of which are white, where p is unknown. Fifty balls were drawn at random without replacement from the urn, 32 of which were found to be white.
(a) Find an estimate of p.
(b) Find a standard error of the estimate if $N = 100$.
(c) Answer the same question if $N = 1{,}000$ (§12.8)

1217. A measurement of the melting-point of naphthalene may be regarded as an observation on a rv X with mean m (= the unknown melting-point) and unknown standard deviation σ. A random sample of nine values gave $\bar{x} = 80.9$ and $s = 0.3$. Estimate m and find a standard error of the estimate. (§12.8)

1218. Consider the following ordered sample of 20 values from $N(m, \sigma^2)$.

3.33	4.06	4.69	5.13	5.93	6.34	6.66	6.71	7.00	7.79
8.07	8.20	8.60	9.34	9.57	9.78	9.96	9.99	10.74	11.44

Estimate m and σ using normal probability paper. (§12.9)

*1219. In order to determine the area θ of a square, independent measurements $x_1, \ldots, x_n$ on the side of the square were performed. They are assumed to be observations from $N(m, \sigma^2)$, where m is unknown and σ^2 is known. (Evidently, we have $\theta = m^2$.)
(a) Find the ML estimate of θ.
(b) Find the ML estimate of θ adjusted for bias.

*1220. Two methods are available for determining a certain chemical quantity θ. Using these methods, we have obtained two independent estimates θ_1^* and θ_2^*, which are both unbiased. The standard deviations of the estimates are 0.4 and 0.6, respectively. Combine the estimates into a single unbiased estimate by forming the expression

$$\theta^* = p\theta_1^* + (1 - p)\theta_2^*,$$

where p is between 0 and 1.
(a) Prove that θ^* is unbiased for any p.
(b) Find the value of p for which the variance of θ^* is as small as possible.

*1221. An urn contains N balls, all of different colours, where N is unknown. Balls are drawn one by one with replacement until for the first time a ball is obtained that has been drawn before. The result of one such trial is that the first three balls drawn are of different colours, while the fourth ball drawn has been obtained before. Find the ML estimate N^* of N. (N^* should be an integer.)

*1222. Let $X \sim \text{Bin}(n, p)$, where p is unknown. We have an observation x on X. Find an unbiased estimate of the variance $V(X)$.

*1223. Let $x_1, \ldots, x_n$ be independent observations on a rv with density function $e^{-(x-\theta)}$ $(x \geq \theta)$. The parameter θ is unknown. Consider the estimates $\theta_1^* = \bar{x} - 1$ and $\theta_2^* = \min(x_i) - (1/n)$.
(a) Show that both estimates are unbiased.
(b) Find the efficiency of θ_1^* relative to θ_2^*.

CHAPTER 13

Interval Estimation

13.1. Introduction

A point estimate does not always give sufficient information about an unknown parameter. It is often more appropriate to use an *interval estimate*, or, as we often say, a *confidence interval*. The idea has already been presented in the examples in §10.3.

In §13.2 a definition and an example are given. §13.3 describes a general method for constructing confidence intervals. In §13.4 the method is applied to the normal distribution, and in §13.5 we show how the normal approximation can be used for constructing interval estimates. In §13.6 applications are made to the binomial distribution and related distributions.

13.2. Some Ideas About Interval Estimates

Let $\mathbf{x} = (x_1, \ldots, x_n)$ be a random sample from a distribution that depends on the unknown parameter θ (or perhaps several samples from distributions depending upon θ).

We begin with a vague definition that will later be made more precise: An interval I_θ that covers θ with probability $1 - \alpha$ is called a *confidence interval* for θ with *confidence level* $1 - \alpha$.

The left and right endpoints of the interval, the so-called *confidence limits*, are denoted by $a_1(\mathbf{x})$ and $a_2(\mathbf{x})$, respectively. They are functions of the values in the sample and hence observations on the sample variables $a_1(\mathbf{X})$ and $a_2(\mathbf{X})$. Using these rv's the definition given above implies that

$$P[a_1(\mathbf{X}) < \theta < a_2(\mathbf{X})] = 1 - \alpha.$$

Hence a confidence interval $(a_1(\mathbf{x}), a_2(\mathbf{x}))$ can be said to be an "observation" on an interval with random endpoints. The idea is illustrated in Fig. 13.2.

The main problem is to select the two functions. We seek functions that make the interval short and hence make the interval estimate precise.

We have here assumed that both endpoints are finite; the interval is then called *two-sided*. We sometimes use a *one-sided* interval $(-\infty, a_2(\mathbf{x}))$ or $(a_1(\mathbf{x}), \infty)$. It may seem curious that such long (!) intervals can be of any use; we shall show that later by means of examples.

In principle, the probability $1 - \alpha$ can be fixed at any value. However, the confidence interval is of practical interest only if this probability is large. Usually we select one of the values 0.95, 0.99 and 0.999, depending on the desired level of certainty. If we then assert that the confidence interval covers θ, we run a risk of 0.05, 0.01 and 0.001, respectively, of making a wrong assertion.

We illustrate the idea with an example.

Example 1. Measurement

A research worker measures a quantity whose unknown value is θ and gets the value x. The error of measurement is assumed to be $N(0, \sigma^2)$, where σ is known. Hence x is an observation on X, where $X \sim N(\theta, \sigma^2)$. We want to construct a two-sided confidence interval for θ with confidence level 0.95 (more briefly: a 95% confidence interval).

As $X \sim N(\theta, \sigma^2)$, the probability is 0.95 that (see §8.4)

$$\theta - 1.96\sigma < X < \theta + 1.96\sigma,$$

or, equivalently, that

$$X - 1.96\sigma < \theta < X + 1.96\sigma.$$

Hence the interval

$$I_\theta = (x - 1.96\sigma, x + 1.96\sigma)$$

is a 95% confidence interval (see Fig. 13.1).

It is seen that the length of the interval depends on σ. The larger the precision, the smaller is σ and the shorter is the interval. The length also depends on the confidence

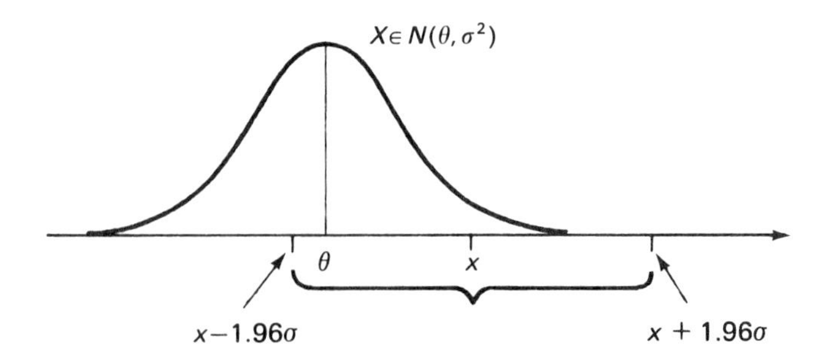

Fig. 13.1. 95% confidence interval for θ.

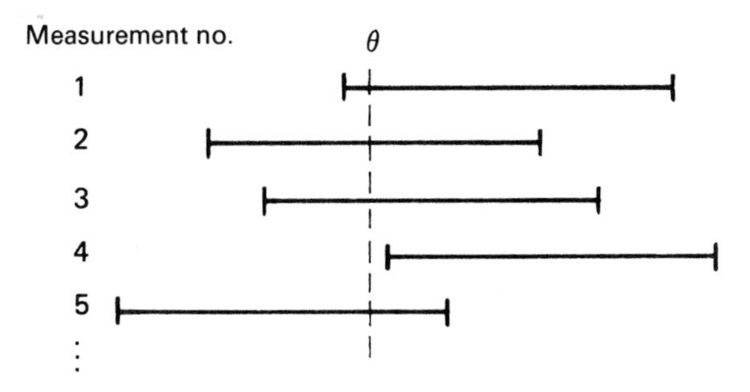

Fig. 13.2. The confidence interval covers the unknown parameter θ with probability $1 - \alpha$.

level. If, instead, we had chosen a 99% confidence level, the constant 1.96 would have been replaced by 2.58 and the interval would have become approximately 30% longer. □

It is instructive to consider what happens if we repeat the collection of samples several times and each time determine, say, a two-sided 95% interval. (The reader may think of Example 1, for concreteness.) In the long run, 95% of the intervals cover the unknown θ and the rest, 5%, miss it.

When we say in the sequel that an interval covers a parameter with confidence level $1 - \alpha$, we mean just what Fig. 13.2 is intended to show: We use a method that leads to a correct statement with probability $1 - \alpha$.

Example 1. Measurement (continued)

Let us now instead construct a one-sided 95% confidence interval; for example, one with a lower confidence limit.

The probability is 0.95 that (see (12) in §8.4)

$$X < \theta + 1.64\sigma,$$

or, equivalently, that

$$X - 1.64\sigma < \theta.$$

Hence the interval $(X - 1.64\sigma, \infty)$ covers θ with probability 0.95; that is, $I_\theta = (x - 1.64\sigma, \infty)$ is a one-sided 95% confidence interval for θ.

As was said earlier, it may seem strange that an interval with infinite length can be of any use. The following situation shows that this can be the case. The measurement may concern a product for which it is desirable that θ is large. The interval then provides valuable information: With 95% certainty we may assert that the correct value is at least $x - 1.64\sigma$. □

Note that, in this example and all similar cases, we do *not* say that "the probability that θ falls in the confidence interval is $1 - \alpha$". Such a formulation is misleading, since it gives the impression that θ is a rv instead of an unknown constant. In fact, it is the endpoints of the confidence interval that are rv's or, more correctly: observations on two sample variables (as illustrated by Fig. 13.2).

13.3. General Method

First, let θ^* be a point estimate with a distribution of *continuous* type. (If the distribution contains other unknown parameters, we suppose that the estimate has been selected in such a manner that its distribution does not depend on these parameters.) Assume that we have been able to determine the density function of the estimate. We may then find two quantities $b_1(\theta)$ and $b_2(\theta)$ such that (see Fig. 13.3)

$$P[b_1(\theta) < \theta^* < b_2(\theta)] = 1 - \alpha. \tag{1}$$

This can be done in many ways, for we need only draw any two vertical lines such that the area between them, below the density function, is $1 - \alpha$. To get a unique result, it is customary to take the same probability mass to the left of b_1 and to the right of b_2 (see again Fig. 13.3). We now assume that the values b_1 and b_2 are monotone functions of θ (both strictly increasing or both strictly decreasing). Then inverse functions $b_1^{-1}(\cdot)$ and $b_2^{-1}(\cdot)$ to $b_1(\theta)$ and $b_2(\theta)$ exist, and the inequality in (1) can be rewritten

$$P[a_1(\theta^*) < \theta < a_2(\theta^*)] = 1 - \alpha. \tag{2}$$

For example, if the limits in (1) are increasing, we have $a_1(\cdot) = b_2^{-1}(\cdot)$ and $a_2(\cdot) = b_1^{-1}(\cdot)$, and similarly if they are decreasing. The problem is then solved, for relation (2) implies that $I_\theta = (a_1, a_2)$ is a two-sided confidence interval for θ with confidence level $1 - \alpha$. One-sided intervals can be obtained similarly.

In the *discrete* case a complication arises: It is not always possible to find limits b_1 and b_2 that satisfy relation (1) exactly, and the confidence level will only be approximately $1 - \alpha$. From a practical point of view a small change of the level is unimportant.

Remark. Modified Method

The above method is sometimes modified as follows: A function $g(\mathbf{x}; \theta)$ is found that contains θ but whose distribution does not depend on θ or other unknown parameters.

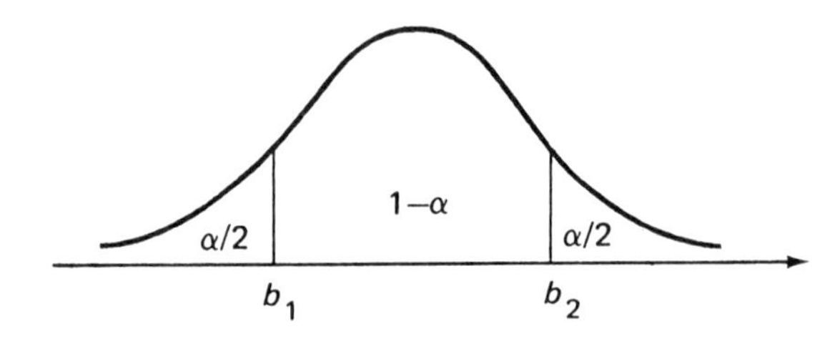

Fig. 13.3. Density function of θ^*.

A relation of type (1) is then derived, that is $P(b_1 < g < b_2) = 1 - \alpha$. If it can be rewritten as a relation of type (2), that is, $P(a_1 < \theta < a_2) = 1 - \alpha$, the problem is again solved. An illustration is given on p. 232. □

13.4. Application to the Normal Distribution

This section is important, for anyone handling statistical analyses will sooner or later encounter questions of the type described here. In fact, it is very common to construct an interval estimate based on data from a normal distribution.

We restrict the discussion to the most important problems concerning means and dispersions connected with this distribution. As often happens in this book, we cannot develop the whole mathematical apparatus required for complete proofs.

(a) Auxiliary Tools

We begin with two "auxiliary distributions" needed later. (The reader may begin with subsection (b) and consult (a) when necessary.)

The first distribution is the χ^2 distribution (chi-square):

Definition. A rv X with a density function of the form

$$kx^{f/2-1}e^{-x/2} \qquad (x \geq 0), \tag{3}$$

is said to have a χ^2 *distribution* with f degrees of freedom. □

We write $X \sim \chi^2(f)$. By setting $f = 1, 2, \ldots$ we obtain a whole family of distributions. For $f = 2$ we have an exponential distribution. All distributions in the family are skew, but the larger is f, the more symmetric is the distribution (see Fig. 13.4). By analogy to the quantile λ_α of the normal distribution we

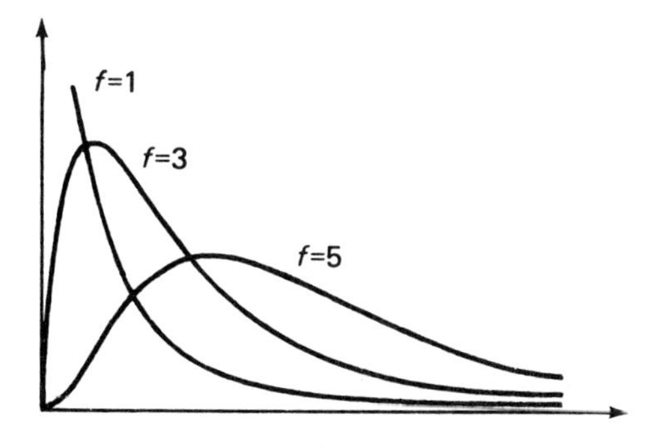

Fig. 13.4. χ^2 distribution.

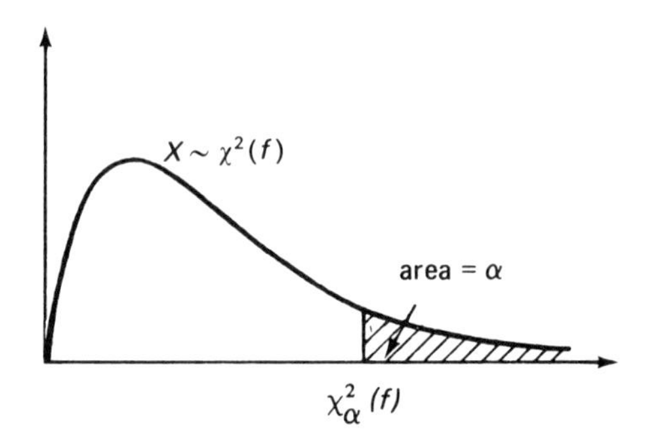

Fig. 13.5. Definition of $\chi^2_\alpha(f)$.

introduce the notation $\chi^2_\alpha(f)$ for the value with area α to the right (see Fig. 13.5). Table 4 at the end of the book gives quantiles for different α and f.

Remark. Relationship to the Gamma Distribution

The χ^2 distribution is a special case of the gamma distribution $\Gamma(p, a)$ for $p = f/2$, $a = 2$, that is, $\chi^2(f) \equiv \Gamma(f/2, 2)$. From the density function of the gamma distribution in §3.7 it follows that the constant in the definition of the χ^2 distribution is given by $k = 2^{-f/2}/\Gamma(f/2)$. □

The χ^2 distribution has an interesting property: If X and Y are independent and $\chi^2(f_1)$ and $\chi^2(f_2)$, respectively, then $X + Y \sim \chi^2(f_1 + f_2)$. Hence convolution of two χ^2 distributions results in a new χ^2 distribution. This can be proved by means of the convolution formula (11) in §5.3.

The reason why the χ^2 distribution is important, in the statistical problems to be solved in this section, is apparent from

Theorem 1. *If* X_i, $i = 1, \ldots, n$, *are independent* rv's *and* $X_i \sim N(0, 1)$, *then*

$$\sum_1^n X_i^2 \sim \chi^2(n), \tag{4}$$

and

$$\sum_1^n (X_i - \bar{X})^2 \sim \chi^2(n-1) \qquad \left(\bar{X} = \sum_1^n X_i/n\right). \tag{4'}$$

□

Hence the theorem states that both the sum of squares of independent standard normal variables and the sum of squares about their arithmetic mean have a χ^2 distribution, but with different degrees of freedom. This elegant result was first given by W. Helmert at the end of the nineteenth century. No proof will be given.

Theorem 1 can be generalized to

Theorem 1′. *If the* rv's X_i, $i = 1, \ldots, n$, *are independent and* $N(m, \sigma^2)$, *then*

$$\frac{1}{\sigma^2}\sum_1^n (X_i - m)^2 \sim \chi^2(n), \tag{5}$$

and

$$\frac{1}{\sigma^2}\sum_1^n (X_i - \bar{X})^2 \sim \chi^2(n-1). \qquad \square$$

PROOF. Use Theorem 1 in §8.4 and the previous theorem. □

The other auxiliary distribution is the *t distribution*, which was first arrived at by the English brewery chemist W.S. Gosset at the beginning of the century. (He used the pen-name "Student", whence the alternative name Student distribution.)

Definition. A rv X with density function

$$k\left(1 + \frac{x^2}{f}\right)^{-(f+1)/2} \qquad (-\infty < x < \infty), \tag{6}$$

is said to have a *t distribution* with f degrees of freedom. □

We write $X \sim t(f)$. Here f is a positive integer. The distributions for $f = 1, 2, \ldots$ are all symmetric about $x = 0$. When f increases to infinity, the t distribution tends to a standard normal distribution. Quantiles $t_\alpha(f)$ are defined in the usual way (see Fig. 13.6) and are tabulated in Table 3 at the end of this book.

It may be said that with his t distribution Gosset opened the door to the world of modern statistics. He did this by deriving a result which, in modern terminology, can be written as:

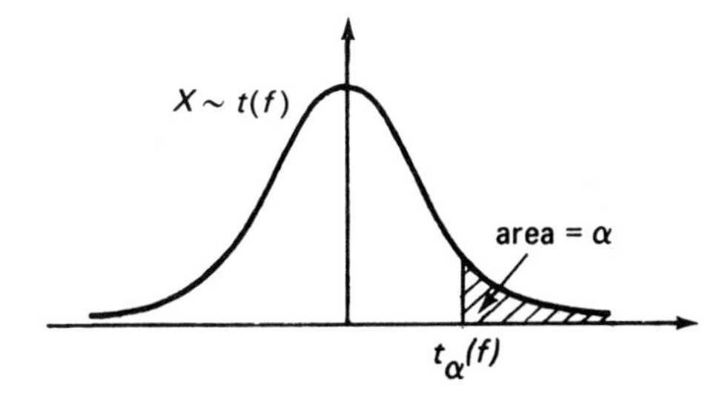

Fig. 13.6. Definition of $t_\alpha(f)$.

Theorem 2. *If $X \sim N(0, 1)$ and $Y \sim \chi^2(f)$, where X and Y are independent, then the* rv

$$Z = \frac{X}{\sqrt{Y/f}}$$

has a t distribution with f degrees of freedom. □

The reason why it is important to deal with such an artificial looking rv will emerge later.

*PROOF. We combine several results from Chapter 5:

(1) If in Example 2, p. 81, we take $a = 1/f$, $b = 0$, we find, using the density function of Y in (6), that Y/f has density function

$$f_{Y/f}(x) = kx^{f/2-1}e^{-fx/2},$$

where k is a constant.

(2) If in Example 4, p. 82, we take $X = Y/f$, we find that $\sqrt{Y/f}$ has density function

$$f_{\sqrt{Y/f}}(x) = kx(x^2)^{f/2-1}e^{-fx^2/2} = kx^{f-1}e^{-fx^2/2}.$$

(3) It follows from (15) in §5.5 that the rv given in the theorem has density function

$$f_Z(x) = \int_0^\infty y\varphi(yx)f_{\sqrt{Y/f}}(y)\,dy = \int_0^\infty y\cdot\frac{1}{\sqrt{2\pi}}e^{-y^2x^2/2}ky^{f-1}e^{-fy^2/2}\,dy$$

$$= \frac{k}{\sqrt{2\pi}}\int_0^\infty y^f e^{-(x^2+f)y^2/2}\,dy.$$

The substitution of variables $t = y\sqrt{x^2 + f}$ gives

$$f_Z(x) = \frac{k}{\sqrt{2\pi}(x^2 + f)^{(f+1)/2}}\cdot\int_0^\infty t^f e^{-t^2/2}\,dt$$

$$= k'\cdot\left(1 + \frac{x^2}{f}\right)^{-(f+1)/2},$$

and the theorem is proved. □

We are now prepared to attack various statistical problems connected with samples from normal distributions.

(b) A Single Random Sample. Confidence Interval for the Mean

Let $x_1, \ldots, x_n$ be a random sample from $N(m, \sigma^2)$; that is, x_i, $i = 1, \ldots, n$, are observations on the independent rv's X_i, where $X_i \sim N(m, \sigma^2)$. We want to construct a confidence interval for the mean m.

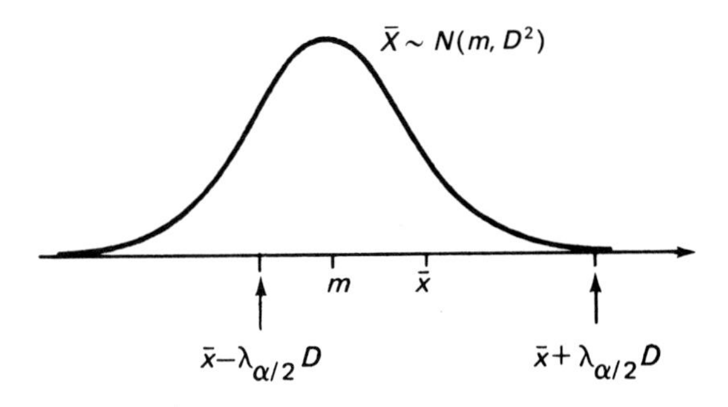

Fig. 13.7. Confidence interval for m.

First assume that the *standard deviation σ is known*. In Example 1 we have already solved the problem for the special case $n = 1$, $1 - \alpha = 0.95$. In the general case we argue as follows:

By Corollary 1 of Theorem 3′ in §8.5, the arithmetic mean $\bar{X}$ is normally distributed: $\bar{X} \sim N(m, D^2)$, where $D^2 = \sigma^2/n$. Consequently, the inequality

$$m - \lambda_{\alpha/2}D < \bar{X} < m + \lambda_{\alpha/2}D$$

is satisfied with probability $1 - \alpha$. The limits are strictly increasing functions of m. The inequality can be rewritten as

$$\bar{X} - \lambda_{\alpha/2}D < m < \bar{X} + \lambda_{\alpha/2}D.$$

Hence this inequality is also satisfied with probability $1 - \alpha$. This means that

$$I_m = (\bar{x} - \lambda_{\alpha/2}D, \bar{x} + \lambda_{\alpha/2}D)$$

is a two-sided confidence interval for m with confidence level $1 - \alpha$ (see Fig. 13.7).

By means of a similar method it may be shown (exercise!) that both $(\bar{x} - \lambda_\alpha D, \infty)$ and $(-\infty, \bar{x} + \lambda_\alpha D)$ are one-sided confidence intervals for m with confidence level $1 - \alpha$.

Remark. Why Does the Method Work?

As a preparation for subsequent problems we shall modify the above derivation slightly. The difference may appear subtle, but is important in principle. From Theorem 1 in §8.4, it follows that

$$\frac{\bar{X} - m}{D} \sim N(0, 1).$$

Hence with probability $1 - \alpha$

$$-\lambda_{\alpha/2} < \frac{\bar{X} - m}{D} < \lambda_{\alpha/2}.$$

Rewriting this, we find as before

$$\bar{X} - \lambda_{\alpha/2}D < m < \bar{X} + \lambda_{\alpha/2}D$$

and the conclusion is the same.

We see that the solution works because (1) the distribution of the ratio $(\bar{X} - m)/D$ is known, and (2) the ratio is a strictly monotone function of m. □

Now assume instead that the *standard deviation σ is unknown.* (This was the situation studied by Gosset.) We then use

Lemma. *The ratio $(\bar{x} - m)/d$, where $d = s/\sqrt{n}$, is an observation on a* rv *that has a t distribution with $n - 1$ degrees of freedom.* □

*PROOF. The ratio can be rewritten

$$\frac{\bar{x} - m}{d} = \frac{\bar{x} - m}{s/\sqrt{n}} = \frac{\sqrt{n}(\bar{x} - m)/\sigma}{\sqrt{\dfrac{\sum (x_i - \bar{x})^2/\sigma^2}{n - 1}}}.$$

This is an observation on the rv $Y\Big/\sqrt{\dfrac{Z}{n-1}}$, where $Y = \sqrt{n}(\bar{X} - m)/\sigma$ and $Z = \Sigma (X_i - \bar{X})^2/\sigma^2$.

It now follows from Theorem 1 in §8.4 and Theorem 1′ of the present chapter that $Y \sim N(0, 1)$ and $Z \sim \chi^2(n - 1)$, respectively. It can be shown that Y and Z are independent. Thus we are entitled to apply Theorem 2 and the lemma is proved. □

Now argue exactly as in the Remark, p. 231. By the lemma, the probability is $1 - \alpha$ that

$$-t_{\alpha/2}(f) < \frac{X - m}{d} < t_{\alpha/2}(f) \qquad (f = n - 1),$$

or

$$\bar{X} - t_{\alpha/2}(f)d < m < \bar{X} + t_{\alpha/2}(f)d.$$

Hence the confidence interval is given by

$$I_m = (\bar{x} - t_{\alpha/2}(f)d, \bar{x} + t_{\alpha/2}(f)d) \qquad (d = s/\sqrt{n}, \quad f = n - 1).$$

Comparison with the interval for known σ shows that we have replaced the standard deviation D by the standard error d and $\lambda_{\alpha/2}$ by $t_{\alpha/2}(f)$.

One-sided confidence intervals can be found in a similar way, but we give no details.

We sum up what we have found in

Theorem 3. *Let $x_1, \ldots, x_n$ be a random sample from $N(m, \sigma^2)$, where m is unknown. Then*

$$I_m = (\bar{x} - \lambda_{\alpha/2}D, \bar{x} + \lambda_{\alpha/2}D) \qquad \textit{if } \sigma \textit{ is known} \quad (D = \sigma/\sqrt{n}), \tag{7'}$$

and

$$I_m = (\bar{x} - t_{\alpha/2}(f)d, \bar{x} + t_{\alpha/2}(f)d) \qquad \textit{if } \sigma \textit{ is unknown} \quad (d = s/\sqrt{n}, f = n - 1) \tag{7''}$$

are two-sided confidence intervals for m with confidence level $1 - \alpha$. □

This theorem is typical of what happens in statistical theory. Using some fairly complicated mathematics we prove theorems which, apart from the theoretical background, look very simple and are also numerically easy to apply to given data. Another, and more difficult, question is whether the data have been collected in such a way that the assumptions of the theorems are fulfilled. We will discuss this question to some extent in Chapter 16.

Remark 1. Numerical Comparison of "Known σ" and "Unknown σ"

It is instructive to study the result in Theorem 3 numerically. Assume, for example, that $1 - \alpha = 0.95$ and $n = 10$.

When σ is known, the interval is $\bar{x} \pm 1.96D$ and hence has constant length $2 \cdot 1.96D$.

When σ is unknown, we first find that $t_{0.025}(9) = 2.26$ (see Table 3 at the end of the book); thus the interval now becomes $\bar{x} \pm 2.26d$. The length of the interval is no longer constant, for $d = s/\sqrt{n}$ depends on the sample. If we collect samples repeatedly and each time compute a confidence interval, the intervals have varying lengths. The long run average length of the intervals is, approximately, $2 \cdot 2.26D$. (That we add "approximately" is due to the fact that $E(s)$ is not exactly equal to σ; see p. 198.) It is found that the average length of the interval is greater than when σ is known. Hence the lack of knowledge about σ causes a loss of information about m, which seems reasonable.

The loss of information is smaller for large n. For example, when $n = 30$ we have $t_{0.025}(29) = 2.05$, which is not much larger than 1.96. This also seems reasonable; for larger n, s is a better estimate of σ, and d will generally deviate less from D. □

Remark 2. A Word of Warning

In certain fields of application, particularly in biology and medicine, it is common to report the result of an experiment as the sample mean followed by D or d:

$$\bar{x} \pm D \qquad \text{or} \qquad \bar{x} \pm d.$$

This practice cannot be criticized in itself provided no misunderstanding arises. However, it may be tempting to believe that $(\bar{x} - D, \bar{x} + D)$ and $(\bar{x} - d, \bar{x} + d)$ are confidence intervals that cover the unknown parameter with high probability.

As Theorem 3 shows, that is a terrible mistake, for one then "forgets" to multiply D by $\lambda_{\alpha/2}$ and d by $t_{\alpha/2}(f)$, both of which are ≥ 1.96 if the confidence level is $\geq 95\%$. By neglecting these factors the level becomes far too low. For example, the probability that $(\bar{X} - D, \bar{X} + D)$ covers m is equal only to $\Phi(1) - \Phi(-1) = 0.68$, that is, somewhat more than 2/3, and still smaller for the other interval, especially when n is small. □

Remark 3. How Large a Sample?

It generally costs money to perform a statistical investigation. Therefore, it is important when determining n in advance that the right amount of information is obtained, neither too much nor too little. If σ is known we may select n such that the confidence interval has a prescribed length. As shown by (7) we then obtain an equation of the form

$$2\lambda_{\alpha/2}\sigma/\sqrt{n} = \text{prescribed length.}$$

From this equation n can be found.

Note that, if we find that a confidence interval is too long and want an interval half as long, we must make the sample *four times* larger. Hence we must collect three times as many observations as we already have, and then compute a new confidence interval based on all the data. This costly truth should be familiar to any research worker!

The corresponding steps when σ is unknown will not be discussed here. A preliminary investigation has to be made before the final sample is collected. □

Example 2. Sugar-Content

From a large batch of sugar-beets 60 beets were chosen at random and the percentage sugar content was determined for each of them. Result:

15.70	14.30	12.65	13.20	16.85	16.05	16.55	16.05
16.60	17.05	17.70	16.40	17.65	17.20	17.90	17.20
17.45	17.65	17.15	17.85	17.35	17.30	16.30	15.75
16.95	17.45	16.30	16.30	17.65	15.20	16.65	15.55
14.75	16.95	17.20	14.65	18.20	15.75	17.45	16.10
17.90	17.35	15.35	15.05	17.15	16.10	18.30	13.45
17.45	17.25	17.10	17.20	16.85	17.50	16.85	16.35
17.40	16.40	14.25	16.50				

We denote the unknown average sugar content in the batch by m and assume that the data can be regarded, approximately, as a random sample from $N(m, \sigma^2)$. We use Theorem 3 for constructing a confidence interval for m under two different conditions:

(1) Known σ

It is assumed known from earlier investigations that $\sigma = 1.2$. We find by computation $\bar{x} = 16.51$, and, further $D = 1.2/\sqrt{60} = 0.155$. Hence a 99% confidence interval for m is given by

$$I_m = (16.51 \pm 2.58 \cdot 0.155) = (16.11, 16.91).$$

(2) Unknown σ

We find $Q = \sum (x_i - \bar{x})^2 = 89.69$, $s = \sqrt{89.69/59} = 1.23$. We have, further, $t_{0.005}(59) = 2.66$ and $d = 1.23/\sqrt{60} = 0.159$. A 99% confidence interval for m is given by

$$I_m = (16.51 \pm 2.66 \cdot 0.159) = (16.09, 16.93).$$

Let us add a remark. Assume that, in (1), we are not satisfied with the length 0.8 of the confidence interval, but want the length to be 0.5. The number n of sugar-beets then required is obtained from the equation

$$2 \cdot 2.58 \cdot 1.2/\sqrt{n} = 0.5$$

giving n as approximately 150. Hence about 90 further beets should be analysed. □

(c) A Single Sample. Confidence Interval for the Standard Deviation

We shall now construct a confidence interval for σ. Again there are two alternatives, m known and m unknown. Since the former is rather uncommon in practice, we confine the discussion to the latter.

Let us first consider σ^2. The sum of squares about the sample mean, $Q = \sum (x_i - \bar{x})^2$, is an observation on the rv $\sum (X_i - \bar{X})^2$, whose distribution we know from Theorem 1′:

$$\frac{1}{\sigma^2} \sum (X_i - \bar{X})^2 \sim \chi^2(n-1).$$

It follows from the definition of $\chi^2(f)$ (see Fig. 13.5) that the probability mass between $\chi^2_{1-\alpha/2}(f)$ and $\chi^2_{\alpha/2}(f)$ is $1 - \alpha$. Hence the probability is $1 - \alpha$ that

$$\chi^2_{1-\alpha/2}(f) < \frac{1}{\sigma^2} \sum (X_i - \bar{X})^2 < \chi^2_{\alpha/2}(f) \qquad (f = n - 1),$$

or

$$\frac{1}{\chi^2_{\alpha/2}(f)} \sum (X_i - \bar{X})^2 < \sigma^2 < \frac{1}{\chi^2_{1-\alpha/2}(f)} \sum (X_i - \bar{X})^2.$$

It is apparent from this inequality that

$$I_{\sigma^2} = (Q/\chi^2_{\alpha/2}(f), Q/\chi^2_{1-\alpha/2}(f))$$

is a confidence interval for σ^2 with confidence level $1 - \alpha$. In order to get the desired interval for σ, we need only take the square roots of the limits (please check that this is allowed). After a simple reordering we finally obtain:

Theorem 4. *Let $x_1, \ldots, x_n$ be a random sample from $N(m, \sigma^2)$. Then*

$$I_\sigma = (k_1 s, k_2 s), \tag{8}$$

where

$$k_1 = \sqrt{f/\chi^2_{\alpha/2}(f)}; \qquad k_2 = \sqrt{f/\chi^2_{1-\alpha/2}(f)} \qquad (f = n - 1),$$

is a confidence interval for σ with confidence level $1 - \alpha$. □

Remark 1.

For large n we have approximately

$$k_1 = 1 - \lambda_{\alpha/2} \frac{1}{\sqrt{2f}}; \qquad k_2 = 1 + \lambda_{\alpha/2} \frac{1}{\sqrt{2f}}. \tag{9}$$

The approximation is generally good enough if n exceeds 30. It has the advantage that no special table is needed, only the usual table of quantiles of the normal distribution. □

Remark 2.

Starting from s, it is often practical to give the endpoints of the confidence interval as a percentage increase and decrease. For large n the interval $(k_1 s, k_2 s)$ then simply becomes $(-P, P)$, where

$$P = 100\lambda_{\alpha/2}/\sqrt{2f}.$$ □

Remark 3.

If n is small, the confidence interval for σ is unfortunately rather long, which is inevitable. Hence the information obtained about σ is not very accurate. It is generally of no use calculating the interval unless n is at least 20 (except for showing friends of small samples what happens!). □

Remark 4.

Assume n is so large that the approximate procedure can be used. It appears from the expressions for k_1 and k_2 that the length $k_2 s - k_1 s$ is then approximately inversely proportional to $\sqrt{n}$. Hence, by making the size of the sample four times as large, we achieve a halving of the confidence interval. (See Remark 3, p. 233, where a similar phenomenon was observed.) □

Example 2. Sugar Content (continued)

We shall use Theorem 4 to construct a 95% confidence interval for σ (which measures the variation of sugar content of the sugar-beets in the given batch). We obtained earlier that $s = 1.23$. Linear interpolation in Table 4 at the end of the book yields for $f = n - 1 = 60 - 1 = 59$

$$\chi^2_{0.975}(59) = 39.7; \qquad \chi^2_{0.025}(59) = 82.1,$$

so that

$$k_1 = \sqrt{59/82.1} = 0.85; \qquad k_2 = \sqrt{59/39.7} = 1.22.$$

Hence the confidence interval becomes

$$I_\sigma = (0.85 \cdot 1.23,\ 1.22 \cdot 1.23) = (1.0,\ 1.5).$$

It is a good exercise to check that the approximate formulae (9) give roughly the same result. □

(d) Two Samples. Confidence Interval for Difference Between Means

The following model is often employed in practice: Two independent samples have been collected,

$$x_1, \ldots, x_{n_1} \qquad \text{from} \quad N(m_1, \sigma_1^2),$$

$$y_1, \ldots, y_{n_2} \qquad \text{from} \quad N(m_2, \sigma_2^2).$$

A confidence interval for the difference $m_1 - m_2$ is required (see Fig. 13.8).

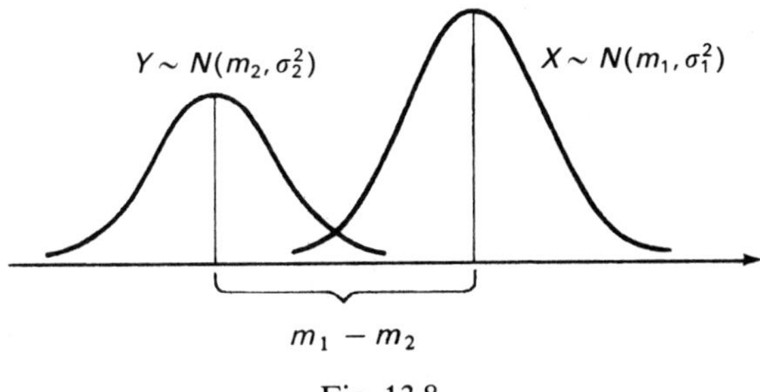

Fig. 13.8.

We now have four parameters, which seems troublesome. Indeed, the general problem with four unknown parameters is difficult to solve, and we discuss only two special cases, the first of which is a preparation for the second.

First, assume that *the standard deviations are known.* As a point estimate of $m_1 - m_2$ we take the difference $\bar{x} - \bar{y}$ of the sample means. We then repeat the single-sample argument. The difference $\bar{x} - \bar{y}$ is an observation on the rv $\bar{X} - \bar{Y}$, which by Corollary 2 of Theorem 3′ in §8.5 is normally distributed with mean $m_1 - m_2$ and standard deviation

$$D = \sqrt{\sigma_1^2/n_1 + \sigma_2^2/n_2}.$$

Arguing as in (b) we obtain the interval

$$I_{m_1 - m_2} = (\bar{x} - \bar{y} - \lambda_{\alpha/2} D, \bar{x} - \bar{y} + \lambda_{\alpha/2} D),$$

where D is given above. The confidence level is $1 - \alpha$.

Second, we turn to the following more important situation. Assume that *the standard deviations are unknown but equal.* Let us set $\sigma_1 = \sigma_2 = \sigma$. We must now replace

$$D = \sqrt{\sigma_1^2/n_1 + \sigma_2^2/n_2} = \sigma\sqrt{1/n_1 + 1/n_2}$$

by a suitable standard error. As an estimate of σ we take (see (4) in §12.6)

$$s = \sqrt{\frac{Q_1 + Q_2}{(n_1 - 1) + (n_2 - 1)}}, \tag{10}$$

where Q_1 and Q_2 are the sums of squares about the sample means, and hence obtain

$$d = s\sqrt{1/n_1 + 1/n_2}.$$

Similarly, as in the one-sample case, we form the ratio $[(\bar{x} - \bar{y}) - (m_1 - m_2)]/d$. Using the same ideas as in the lemma, p. 232, it may be proved that this ratio is an observation on a rv which has a t distribution with $f = (n_1 - 1) + (n_2 - 1)$ degrees of freedom (see the denominator in (10) where this expression appears). It is now clear how to construct the confidence interval. We state it, together with the interval previously derived, in:

Theorem 5. *Let $x_1, \ldots, x_{n_1}$ and $y_1, \ldots, y_{n_2}$ be independent random samples from $N(m_1, \sigma_1^2)$ and $N(m_2, \sigma_2^2)$. If σ_1 and σ_2 are known, then*

$$I_{m_1-m_2} = (\bar{x} - \bar{y} - \lambda_{\alpha/2} D, \bar{x} - \bar{y} + \lambda_{\alpha/2} D) \tag{11'}$$

is a two-sided confidence interval for $m_1 - m_2$ with confidence level $1 - \alpha$; here $D = \sqrt{\sigma_1^2/n_1 + \sigma_2^2/n_2}$. If $\sigma_1 = \sigma_2 = \sigma$, where σ is unknown, then

$$I_{m_1-m_2} = (\bar{x} - \bar{y} - t_{\alpha/2}(f)d, \bar{x} - \bar{y} + t_{\alpha/2}(f)d) \tag{11''}$$

is a two-sided confidence interval for $m_1 - m_2$ with confidence level $1 - \alpha$. Here $d = s\sqrt{1/n_1 + 1/n_2}$, where s is given by (10), while $f = (n_1 - 1) + (n_2 - 1)$.

□

Example 3. Comparison of Sugar Content in Two Batches

For comparing the mean sugar content in two large batches A and B, 60 sugar-beets have been taken from A and 30 from B. The amount of sugar content in the beets are assumed to be observations from $N(m_1, \sigma^2)$ and $N(m_2, \sigma^2)$, where σ is unknown. Let us assume that the sample from A is that given in Example 2, p. 234, and that the sample from B has $\bar{y} = 16.04$, $Q_2 = 46.90$. We find

$$s = \sqrt{\frac{89.69 + 46.90}{(60 - 1) + (30 - 1)}} = 1.246,$$

$$d = 1.246\sqrt{\frac{1}{60} + \frac{1}{30}} = 0.28.$$

We now construct a 99.9% two-sided confidence interval for $m_1 - m_2$. For $\alpha = 0.001$ and $f = (60 - 1) + (30 - 1) = 88$ we find $t_{\alpha/2}(f) = 3.4$. Thus the interval becomes

$$I_{m_1-m_2} = (16.51 - 16.04 \pm 3.4 \cdot 0.28) = (0.47 \pm 0.95) = (-0.5, 1.4).$$

According to this result, the mean percentage of sugar content of A cannot be more than 0.5 smaller and not more than 1.4 larger than that of B.

Let us also solve a planning problem. We want to take n sugar-beets from each batch where n is such that the resulting confidence interval for $m_1 - m_2$ has length 1.0. The confidence level should be 99%. We find the relation (please show this!)

$$2.58 \cdot 1.2\sqrt{\frac{1}{n} + \frac{1}{n}} = 0.5$$

so that $n = 77$. Hence we ought to take approximately 80 sugar-beets from each batch.

□

(e) Paired Samples

We shall now treat a situation which is easily mishandled unless one is careful. In fact, it resembles the situation discussed in (d), but the analysis is quite different.

It is instructive to start with a practical problem. Assume that two technicians A and B both perform repeated measurements on *the same object.* If each performs the same number of measurements, they obtain values $x_1, \ldots, x_n$ and $y_1, \ldots, y_n$, which may be regarded as two independent samples. If the distribution is normal and the technicians work with the same unknown precision, we obtain the model discussed in (d), and we can, for example, construct a confidence interval for the systematic difference $m_1 - m_2$ between the two means.

Let us now assume, instead, that each technician performs one measurement on each of n *different objects.* The resulting data can be written:

	Object			
	1	2	...	n
A	x_1	x_2	...	x_n
B	y_1	y_2	...	y_n

As before, we have two series of measurements, but the previous model is now worthless since differences may exist between the objects, regardless of whether there are differences between the technicians or not. Evidently, the observations on the same object should be paired in some way.

The following model is then often appropriate: The value x_j from the jth object comes from a normal distribution $N(m_j, \sigma_1^2)$ and y_j from another normal distribution $N(m_j + \Delta, \sigma_2^2)$ (see Fig. 13.9). There are then $n + 3$ unknown parameters $m_1, \ldots, m_n, \sigma_1, \sigma_2$ and Δ. Differences between $m_1, \ldots, m_n$ measure

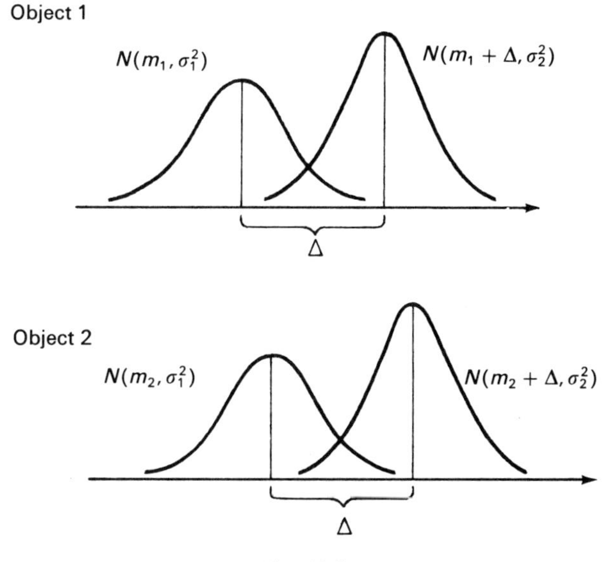

Fig. 13.9.

the differences between the objects, while Δ indicates the systematic difference between measurements performed by B and A (in this order.). If $\Delta > 0$, B's values are, on the average, larger than A's. Here we assume that the systematic error does not depend on the individual object, which may be realistic if the differences between the objects are not very large. Also note that we allow A and B to work with different precision (σ_1 and σ_2 may be different).

Usually Δ is the most interesting parameter. To estimate Δ we use a trick. Form the differences "within objects", that is, $z_1 = y_1 - x_1, \ldots, z_n = y_n - x_n$. It is not difficult to see that these differences are observations from a normal distribution $N(\Delta, \sigma^2)$, where $\sigma^2 = \sigma_1^2 + \sigma_2^2$. Hence in this way we have got rid of most of the parameters, and there are, in fact, only two left, namely Δ and σ.

Clearly, we have replaced our problem by that discussed in (b) and can use the formulae given there. (Remember that x should everywhere be replaced by z.) For example, we obtain a two-sided confidence interval for Δ by computing

$$I_\Delta = (\bar{z} \pm t_{\alpha/2}(f)d) \qquad (d = s/\sqrt{n}, \quad f = n - 1). \tag{12}$$

Here s is computed from $z_1, \ldots, z_n$.

Example 4. Measuring Different Objects

A and B performed measurements on 11 different objects (values in mm):

Person	Object 1	2	3	4	5	6	7	8	9	10	11
A	20.2	22.0	19.7	21.4	16.3	17.0	24.5	15.6	16.0	13.2	19.0
B	21.6	22.9	20.0	23.6	15.7	17.5	27.3	19.8	16.4	14.7	20.1
Difference	1.4	0.9	0.3	2.2	−0.6	0.5	2.8	4.2	0.4	1.5	1.1

The measurements of object j are assumed to be observations from $N(m_j, \sigma_1^2)$ and $N(m_j + \Delta, \sigma_2^2)$, where Δ is the systematic difference between B and A. One finds by usual computations

$$\bar{z} = 1.34; \qquad s = 1.33; \qquad d = s/\sqrt{11} = 0.401;$$

$$f = 11 - 1 = 10; \qquad t_{0.025}(10) = 2.23.$$

Hence a 95% confidence interval for Δ becomes

$$I_\Delta = (1.34 \pm 2.23 \cdot 0.401) = (1.34 \pm 0.89) = (0.45, 2.23).$$

According to this result B's measurements exceed those of A, on the average, by an amount lying approximately between 0.5 and 2.2 mm. □

We shall conclude this section with a remark showing the common thread of all arguments used here. In principle, all problems concerning means have been solved in the same manner. The confidence interval for θ may in all cases be written

$$(\theta^* \pm A_1 D) \qquad \text{or} \qquad (\theta^* \pm A_2 d),$$

where A_1 and A_2 are suitably chosen factors. The former version is used when D is completely known, the latter version otherwise.

It follows from the Remark, p. 231, that a great many other problems concerning means or other unknown parameters θ can be solved in the same way, namely all problems for which $(\theta^* - \theta)/D$ or $(\theta^* - \theta)/d$ has a known distribution.

13.5. Using the Normal Approximation

In practice, distributions often occur that are not at all like the normal. Examples: distribution of capital assets of different people, age distributions, distribution of times spent in hospital.

Nonnormal distributions can often be handled by means of the same technique as normal ones. Remarkably enough, it can be shown that several of the methods described in the previous section are approximately true even if the distribution is nonnormal. A prerequisite is, however, that the samples are reasonably large.

The idea is quite simple: the normal approximation is applied to the estimate one is interested in. Even the explanation why this works is uncomplicated.

Let us consider a random sample from an unspecified distribution (or, more generally, several samples from different distributions). The distribution depends on an unknown parameter θ. Assume that we have found a point estimate θ^* which is approximately normally distributed with mean θ and standard deviation $D = D(\theta^*)$. By Theorem 1 in §8.4 we then have

$$\frac{\theta^* - \theta}{D} \sim N(0, 1).$$

If D does not depend on θ, we can repeat the argument used in the Remark, p. 231: We obtain the confidence interval $(\theta^* \pm \lambda_{\alpha/2} D)$, but now with *approximate* confidence level $1 - \alpha$.

If D should depend on θ (and perhaps on additional parameters), D is as usual replaced by some suitable standard error d. By means of special methods it can be shown that the confidence level is still approximately equal to $1 - \alpha$. However, the approximation is often less accurate than before.

Hence we have:

Theorem 6. *Assume that a point estimate θ^* of θ is based on one or several random samples. Further, assume that this estimate is approximately normally* *distributed with mean θ and standard deviation D. Then*

$$\begin{aligned} I_\theta &= (\theta^* - \lambda_{\alpha/2} D, \theta^* + \lambda_{\alpha/2} D) && \text{if } D \text{ does not depend on } \theta, \\ I_\theta &= (\theta^* - \lambda_{\alpha/2} d, \theta^* + \lambda_{\alpha/2} d) && \text{if } D \text{ depends on } \theta, \end{aligned} \tag{13}$$

(and d is chosen appropriately) are confidence intervals for θ with approximate confidence level $1 - \alpha$. □

The approximation in this theorem is generally better the larger the samples. It is very difficult to formulate a brief statement about the accuracy, since the required number n depends on the functional form of the given distribution. Generally, samples of at least 20 observations can usually be analysed by the technique given here.

In applications, Theorem 6 is one of the most important and useful theorems of statistical theory. We shall give two special cases in the form of corollaries.

Corollary 1. *Let $x_1, \ldots, x_n$ be a large random sample from a distribution with mean m and standard deviation σ. Then*

$$\begin{aligned} I_m &= (\bar{x} - \lambda_{\alpha/2} D, \bar{x} + \lambda_{\alpha/2} D) && \text{if } \sigma \text{ is known} \quad (D = \sigma/\sqrt{n}), \\ I_m &= (\bar{x} - \lambda_{\alpha/2} d, \bar{x} + \lambda_{\alpha/2} d) && \text{if } \sigma \text{ is unknown} \quad (d = s/\sqrt{n}), \end{aligned} \tag{14}$$

are confidence intervals for m with approximate confidence level $1 - \alpha$. □

The first interval is obtained using the central limit theorem. By the corollary to Theorem 4 in §8.6, $\bar{x}$ is an observation on a rv X which is, approximately, $N(m, \sigma^2/n)$. The second interval requires a special proof.

Compare this corollary with Theorem 3 in §13.4, which is similar. There is, however, a difference. In the latter theorem we assumed that X is exactly normally distributed and could then derive an exact result; we now employ an approximation. Also note that in Theorem 3 we used the quantiles of the normal distribution in the first formula and those of the t distribution in the second; we now use the quantiles of the normal distribution in both cases.

Example 5. Work Study

A work study in a textile factory was performed as follows. The time required for mending weft breakages when weaving cotton cloth was measured. Four hundred observations were collected and grouped; see Table 13.1. Clearly, the distribution is very skew.

We regard this as a random sample of 400 values from a distribution with mean m and standard deviation σ. A computation shows that $\bar{x} = 47.01$, $s = 22.84$. An approximate 99% confidence interval for m is given by $I_m = (\bar{x} - 2.58d, \bar{x} + 2.58d)$, where $d = s/\sqrt{n}$. In this case $d = 22.84/\sqrt{400} = 1.14$, and hence the interval is found to be $I_m = (44.1, 49.9)$. □

Table 13.1. Work study at textile factory.

Time (in 10^{-2} min)	Number
15–19	3
20–24	17
25–29	49
30–34	57
35–39	50
40–44	46
45–49	39
50–54	35
55–59	27
60–64	17
65–69	19
70–74	9
75–79	2
80–84	3
85–89	12
90–94	2
95–99	4
100–104	2
105–109	
110–114	2
115–119	1
...	
135–139	1
...	
165–169	1
...	
175–179	1
...	
240–244	1
Sum	400

The other corollary deals with two large independent samples:

Corollary 2. *Let $x_1, \ldots, x_{n_1}$ and $y_1, \ldots, y_{n_2}$ be large random samples, independent of one another, from two distributions with means and standard deviations m_1, σ_1 and m_2, σ_2, respectively. If σ_1 and σ_2 are known, then*

$$I_{m_1 - m_2} = (\bar{x} - \bar{y} - \lambda_{\alpha/2} D, \bar{x} - \bar{y} + \lambda_{\alpha/2} D) \tag{15'}$$

is a confidence interval for $m_1 - m_2$ with approximate confidence level $1 - \alpha$; here $D = \sqrt{\sigma_1^2/n_1 + \sigma_2^2/n_2}$. If σ_1 and σ_2 are unknown, we take instead

$$I_{m_1 - m_2} = (\bar{x} - \bar{y} - \lambda_{\alpha/2} d, \bar{x} - \bar{y} + \lambda_{\alpha/2} d), \tag{15''}$$

where

$$d = \sqrt{s_1^2/n_1 + s_2^2/n_2}. \qquad \square$$

Here s_1 and s_2 are the sample standard deviations.

The theorem corresponds to Theorem 5 in §13.4 but is less restrictive concerning the standard deviations: in the second formula above it is not necessary that $\sigma_1 = \sigma_2$.

Example 6. Comparison of Anaesthetics

A dentist tested two local anaesthetics A and B by giving certain doses to 22 and 30 patients, respectively. The duration of the anaesthesia was observed for each patient. Result (unit: minute):

A	195	240	154	95	65	82	132	155	125	119	155	345
	145	200	130	223	145	207	183	190	137	210		
B	88	73	165	188	145	158	195	165	140	145	203	196
	230	225	128	190	170	158	72	135	105	155	165	120
	138	125	188	145	208	75						

Let us regard the observations from A and B as random samples from distributions with means and standard deviations m_1, σ_1 and m_2, σ_2, respectively. It is found that $\bar{x} = 165.09$, $s_1 = 60.96$, $\bar{y} = 153.10$, $s_2 = 43.08$, $\bar{x} - \bar{y} = 11.99$,

$$d = \sqrt{\frac{60.96^2}{22} + \frac{43.08^2}{30}} = 15.19.$$

Hence according to (15″), taking $1 - \alpha = 0.99$,

$$I_{m_1 - m_2} = (11.99 - 2.58 \cdot 15.19, 11.99 + 2.58 \cdot 15.19) = (-27, 51).$$

Because of the large variation between patients within groups, the interval is very long, and hence the information about the effect of the treatments limited. For a more informative result, more patients must be examined, or another experimental plan used (see Chapter 16). $\square$

13.6. Application to the Binomial and Related Distributions

(a) The Binomial Distribution

Let x be an observation on X, where $X \sim \text{Bin}(n, p)$ and p is unknown. In §12.7 we took as a point estimate the relative frequency

$$p^* = x/n.$$

We shall now take a further step and construct an interval estimate. Unfortunately, due to the discreteness of the distribution it is impossible to find an interval which has, exactly, the desired confidence level. However, it is possible to choose the interval such that it covers p with probability at least $1 - \alpha$; we are then on the safe side. Such confidence limits are tabulated in A. Hald, *Statistical Tables and Formulas* (Wiley, New York, 1952).

For large n we can use the approximate procedure of §13.5, which finds an important application here. By the normal approximation (see §9.2) we have for large n, approximately, that $X/n \sim N(p, D^2)$, where $D = \sqrt{pq/n}$. Evidently, the estimate p^* is an observation on this rv. As a standard error we take $d = \sqrt{p^*q^*/n}$ (see Example 8 in §12.8) and obtain the interval

$$I_p = (p^* - \lambda_{\alpha/2}d, p^* + \lambda_{\alpha/2}d) \qquad (d = \sqrt{p^*q^*/n}). \tag{16}$$

Example 7. Interviews

When interviewing $n = 250$ persons, taken at random from a very large population, 42 had a certain opinion H. As a point estimate of the relative number p in the population with this opinion we take $p^* = 42/250 = 0.168$. Hence we obtain

$$d = \sqrt{0.168 \cdot 0.832/250} = 0.023.$$

An approximate 95% confidence interval for p is given by

$$I_p = (0.168 \pm 1.96 \cdot 0.023) = (0.168 \pm 0.046) = (0.12, 0.21).$$

In such investigations we often use percentages instead of relative frequencies. As a result of the investigation it can be stated that the interval (12, 21) covers the true percentage with 95% probability. □

Remark 1. How Large a Sample?

It is often desirable to decide on the size of n. Different procedures can then be used, depending on the purpose of the study. We assume here that we want an approximate interval I_p of at most a prescribed length. The example below demonstrates how to proceed in this situation. □

Example 7. Interviews (continued)

The number n of persons should be chosen such that the approximate 95% confidence interval for p has at most length $2 \cdot 0.05$.

(1) Parameter p completely unknown

We have

$$2 \cdot 1.96 \sqrt{\frac{p^*q^*}{n}} < 2 \cdot 0.05.$$

Since $p^*q^* \le 1/4$ for all p^*, it follows that

$$n \gtrsim \frac{1}{4}\left(\frac{1.96}{0.05}\right)^2 = 384.2.$$

Hence 385 persons should be interviewed.

(2) Parameter p at most 0.10

It is assumed known that at most 10% of the population has the opinion H. Then p^*q^* is not much larger than $0.10 \cdot 0.90$. By using this upper limit instead of 1/4 we find

$$n \gtrsim 0.10 \cdot 0.90 \left(\frac{1.96}{0.05}\right)^2 = 138.3.$$

By comparing with the result of (1) we conclude that advance information can be of great importance when planning an investigation. □

Remark 2.

The argument presupposes that the sample is large. If the formulae should give a small value of n, let us say less than 20, the derivations no longer hold, and more exact methods must be used. For example, a table of exact confidence limits can be utilized (see Hald's Table mentioned above.) □

It is common to have two observations, x_1 on X_1 and x_2 on X_2, where $X_1 \sim \text{Bin}(n_1, p_1)$ and $X_2 \sim \text{Bin}(n_2, p_2)$ and p_1, p_2 are unknown. We assume that n_1 and n_2 are large. (Example: throw one die n_1 times, another die n_2 times and count the number of ones in each case.) We shall construct a confidence interval for $p_1 - p_2$ with approximate confidence level $1 - \alpha$.

We start with the point estimates

$$p_1^* = x_1/n_1; \qquad p_2^* = x_2/n_2,$$

and then estimate $p_1 - p_2$ by $p_1^* - p_2^*$. The latter difference is an observation on $Y_1 - Y_2$, where $Y_1 = X_1/n_1$ and $Y_2 = X_2/n_2$. We now make use of the approximate method in §13.5. As is known from the end of §9.2 we have, approximately,

$$Y_1 - Y_2 \sim N(p_1 - p_2, D^2),$$

where

$$D = \sqrt{\frac{p_1q_1}{n_1} + \frac{p_2q_2}{n_2}}.$$

As a standard error we take

$$d = \sqrt{\frac{p_1^*q_1^*}{n_1} + \frac{p_2^*q_2^*}{n_2}} \qquad (q_1^* = 1 - p_1^*, q_2^* = 1 - p_2^*),$$

and obtain the confidence interval

$$I_{p_1-p_2} = (p_1^* - p_2^* - \lambda_{\alpha/2}d, p_1^* - p_2^* + \lambda_{\alpha/2}d). \tag{17}$$

The approximation is better, the larger n_1 and n_2 are.

(b) The Hypergeometric Distribution

All steps in (a) can be modified so as to be valid for the hypergeometric distribution.

As we already know from §12.7, the *LS* estimate of p is $p^* = x/n$. As a confidence interval we obtain

$$I_p = (p^* - \lambda_{\alpha/2} d, p^* + \lambda_{\alpha/2} d) \qquad (d = d_n \sqrt{p^* q^*/n}), \tag{18}$$

where $d_n = \sqrt{(N - n)/(N - 1)}$ is the finite population correction. The confidence level is, approximately, $1 - \alpha$. If n/N is smaller than 0.10 we may take $d_n = 1$ (see §9.3), which does not cause any serious error.

Example 7. Interviews (continued)

Suppose that the population is no longer large in comparison with the sample size $n = 250$; for example, assume that N is 500. Under the same conditions as before we have

$$d_n = \sqrt{\frac{500 - 250}{500 - 1}} = 0.707; \qquad d = d_n \sqrt{p^* q^*/n} = 0.707 \cdot 0.023 = 0.016.$$

Hence an approximate 95% confidence interval for p is given by

$$I_p = (0.168 \pm 1.96 \cdot 0.016) = (0.13, 0.20).$$

This interval is shorter than in the previous version of the example. □

Remark 1 on planning may also be applied to the hypergeometric distribution; see the following example.

Example 8. Planning a Market Survey

In a country there are 500 potential customers for a certain special product. A market researcher wants to take a random sample of these customers and ask them if they like the product or not. No information in advance is available; it is quite possible that interest is divided 50–50, so that $p = 1/2$. The confidence interval is to have level 95% and length $2 \cdot 0.05$. Setting half of the length of the approximate confidence interval equal to 0.05, we obtain the equation

$$1.96 \sqrt{\frac{500 - n}{500 - 1} \cdot \frac{1}{4n}} = 0.05,$$

so that $n = 220$. Hence 220 of the potential customers should be asked. (When the market researcher receives this answer from the company statistician, he will probably give up his request for a short confidence interval and accept a longer one!) □

The example conveys an important message. Let us examine how the answer depends on the size N of the population. The equation in the example is of the form

$$\sqrt{\frac{N - n}{N - 1} \cdot \frac{1}{n}} = \Delta,$$

where Δ is a certain constant. Replacing $N - 1$ by N, which is unimportant if N is reasonable large, we find

$$n = \left(\Delta^2 + \frac{1}{N}\right)^{-1}.$$

By taking numerical examples it is easy to convince oneself that the sample size n does not depend heavily on N. For example, if $\Delta^2 = 0.05$, we get for $N = 20, 50, 100, 1{,}000$ the sample sizes $n = 10, 15, 17, 20$, that is, *n increases only slowly with N*. Also note something that may look paradoxical at first sight: For N larger than 1,000, n is never larger than 20, that is, the size of the population is then of no importance at all! This seems contrary to common sense. People ignorant of statistics often think that the sample size should be a certain percentage of the number of elements in the population. ("Select 10% of the population.") This is entirely wrong, and *such a misconception can have serious economic consequences.*

(c) The Poisson Distribution

In subsection (c) of §12.7 we considered a random sample from Po(m) and showed how to find a point estimate of m. For simplicity, we now assume that we have a single observation x on X. We shall construct an interval estimate of m, assuming it known that m is at least about 15, for then the normal approximation can be used (see the end of §9.4).

As an estimate of m we take (see p. 208)

$$m^* = x.$$

The corresponding standard deviation is (see Theorem 7 in §9.4)

$$D(X) = \sqrt{m},$$

and hence we choose the standard error

$$d = \sqrt{x}.$$

By the approximate method in §13.5 we obtain the confidence interval

$$I_m = (x - \lambda_{\alpha/2}\sqrt{x},\ x + \lambda_{\alpha/2}\sqrt{x}). \tag{19}$$

The confidence level is approximately $1 - \alpha$, and the approximation is better, the larger m is.

Example 9. Accidents

The number of accidents in a factory during time t is assumed to be Po(λt), where λ is the accident rate, which is unknown. During 5 months, 25 accidents were observed. We shall construct a 95% confidence interval for λ, taking a month as a unit.

By (19) we have the interval $(25 - 1.96\sqrt{25},\ 25 + 1.96\sqrt{25}) = (15, 35)$ for the parameter λt. By dividing by the length 5 months, we find that the interval for λ is

(3, 7). Hence with approximately 95% probability, the accident rate is between 3 and 7 accidents per month. The precision of the study is low, because of the short period of observation. □

Remark.

It is often more practical to give the endpoints of the confidence interval in the form of a percentage decrease and increase relative to the observed value x. The interval then becomes simply $(-P, P)$ where

$$P = 100\lambda_{\alpha/2}/\sqrt{x}.$$

We observe here something important: P is inversely proportional to the square root of the observed number x. □

If, quite generally, we have a sample of n values the interval becomes

$$I_m = (\bar{x} - \lambda_{\alpha/2}\sqrt{\bar{x}/n},\ \bar{x} + \lambda_{\alpha/2}\sqrt{\bar{x}/n}). \tag{20}$$

The accuracy is sufficient for most practical purposes if $nm > 15$.

Exercises

1301. A research worker has constructed confidence intervals for 15 different unknown constants. Each confidence interval has confidence level 0.90, and the intervals are based on series of observations which are independent of one another. Some of the confidence intervals, perhaps all of them, are correct, that is, cover the corresponding constant, while others may not be correct, that is, "miss the target".
    (a) Find the probability that each of the 15 intervals contains the corresponding constant.
    (b) Find the most probable value of the number of intervals that "miss the target". (§13.2)

1302. Let the lifetimes of electric bulbs of a certain type be described by a rv with density function $(1/\theta)e^{-x/\theta}$ for $x \geq 0$. Then θ is the mean lifetime. A new bulb is switched on, and it is found that its lifetime is 1,000 hours. Construct a two-sided confidence interval for θ with confidence level 0.95.
    Hint: Choose an interval of the form $(c_1 x, c_2 x)$, where x is the observed lifetime. (§13.3)

1303. (Continuation of Exercise 1302.) Construct instead a one-sided confidence interval of the form (cx, ∞). (§13.3)

1304. Consider a random sample from $N(m, 4)$:

    44.3 45.1 46.1 45.3

    Find a two-sided 95% confidence interval for m. (§13.4)

1305. A solution has the unknown pH value m. Four measurements of m were performed and resulted in

    8.24 8.18 8.15 8.23.

Model: The pH meter has a systematic error Δ and a random error which is $N(0, \sigma^2)$; hence the four measurements form a random sample from $N(m + \Delta, \sigma^2)$. It is known that $\Delta = 0.10$ and $\sigma = 0.05$. Find a 99% confidence interval for m. (§13.4)

1306. (Continuation of Exercise 1305.) Assume that σ is unknown. Find a 99% confidence interval for m. (§13.4)

1307. A person ignorant of statistical theory asserts: If $x_1, \ldots, x_n$ is a random sample from $N(m, \sigma^2)$, where m and σ^2 are unknown, then $(\bar{x} - s, \bar{x} + s)$ is a confidence interval for m with confidence level at least 0.99. For which n is this true? (§13.4)

1308. Using a random sample of eight values from $N(m, \sigma^2)$, where m and σ^2 are unknown, a point estimate of σ has been computed in the usual way. The value obtained is $s = 5.2$. Find a 95% confidence interval for σ^2. Note the considerable length of the interval! (§13.4)

1309. The rv X is $N(m, \sigma^2)$, where m and σ^2 are unknown. We want $(0.9s, 1.1s)$ to be a 95% confidence interval for σ. How many observations of X are then required?
Hint: The approximation in Remark 1, p. 235, can be used. (§13.4)

1310. A chemist performed 10 analyses of the percentage of titanium oxide in a well-homogenized batch of ferric oxide and 50 analyses of the same percentage in another, not so well-homogenized batch. The analyses are considered as random samples from $N(m_1, 0.02^2)$ and $N(m_2, 0.05^2)$, respectively. The chemist found that $\bar{x} = 0.51$ and $\bar{y} = 0.69$. Construct a 90% confidence interval for the difference $m_2 - m_1$ of the true percentages of titanium oxide in the two batches. (§13.4)

1311. Let $x_1, \ldots, x_6$ and $y_1, \ldots, y_{12}$ be random samples from $N(m_1, \sigma^2)$ and $N(m_2, \sigma^2)$, where m_1, m_2 and σ^2 are unknown. Computations gave the values $\bar{x} = 49.2$, $s_x^2 = 8.80$ and $\bar{y} = 37.4$, $s_y^2 = 3.04$, respectively. Find a 90% confidence interval for $m_1 - m_2$. (§13.4)

1312. By means of a random sample of five values from $N(m_1, \sigma_1^2)$, where m_1 is unknown and σ_1^2 known, a 95% confidence interval for m_1 has been constructed in the usual way, resulting in the interval (1.37, 1.53). Similarly, a random sample of seven values from $N(m_2, \sigma_2^2)$, where m_2 is unknown and σ_2^2 known, has resulted in the 95% confidence interval (1.17, 1.29) for m_2. Find a 95% confidence interval for $m_1 - m_2$. (§13.4)

1313. A physician wanted to examine the change of blood pressure (unit: mm Hg) when using a certain medicine. First, the blood pressure was measured for each of 10 persons, then each of them received a certain dose of the medicine and finally after 20 minutes the blood pressure was measured once again. Model: The measured blood pressure of person i before and after is $N(m_i, \sigma_1^2)$ and $N(m_i + \Delta, \sigma_2^2)$, respectively.
(a) Interpret the parameters $m_1, m_2, \ldots, m_{10}$ and Δ.
(b) The result of the measurements was

Person no.	1	2	3	4	5	6	7	8	9	10
Blood pressure before	75	70	75	65	95	70	65	70	65	90
Blood pressure after	85	70	80	80	100	90	80	75	90	100

Construct a 95% confidence interval for Δ. (§13.4)

1314. Let m be the average time required for a certain university degree (unit: month). To estimate m a random sample of 25 persons was selected from the large population of persons studying for this degree. It was found that $\bar{x} = 49.3$, $s = 9.3$. The distribution of study times is nonnormal. In spite of this, construct an approximate 95% confidence interval for m. (§13.5)

1315. Among 600 randomly chosen Swedish citizens it was found that 24 had been foreign citizens. Find a 95% confidence interval for p = relative frequency of such persons among all Swedish citizens. (§13.6)

1316. (Continuation of Exercise 1315.) We want a 95% confidence interval for p of length $2 \cdot 0.005$. Find the required sample size if:
(a) p is totally unknown;
(b) it is known that $p < 0.04$. (§13.6)

1317. A person asserts that, if 1,000 persons are taken at random from a population, it is possible to construct a 90% confidence level for p (where p can have any size) of the form $p^* \pm 0.02$.
(a) If the size N of the population is small enough, the assertion is true; if $N = 1{,}000$ it is trivial. For which values of N is the assertion true?
(b) Replace the sample size 1,000 by a suitable larger number such that the assertion is true for any population size. (§13.6)

1318. Let p be the relative frequency of defective units among 100,000 units in stock. In a random sample of 1,000 units there were 36 defectives.
(a) Give a 95% two-sided confidence interval for p.
(b) Give a 95% one-sided confidence interval for p of the type $(0, a)$.
(c) Give a similar one-sided 95% confidence interval for the total number of defectives in the whole batch. (§13.6)

1319. A research worker R has examined the relative frequency p of a certain property E in a population P comprising 1,000 individuals. He chose 200 individuals at random, counted the number having property E and constructed a 95% confidence interval for p.
(a) His colleague R' plans to perform a similar investigation of population P' consisting of 2,250 individuals. He believes that the relative frequency p' of individuals with E in this population is about the same as in P. How many individuals should he take if he wants to make a 95% interval for p' of the same length as the interval computed by R?
(b) The research worker R'' plans to investigate population P'', which consists of 100,000 individuals, in the same way. How many individuals should he select? (§13.6)

1320. The traffic on a certain road can be described by the model: The number of cars passing a certain point during time t (unit: minute) is $\mathrm{Po}(\lambda t)$. In a traffic count 400 cars passed by during 10 minutes.
(a) Give an approximate 95% confidence interval for 10λ.
(b) Give an approximate 95% confidence interval for λ. (§13.6)

*1321. (a) Let $X_1, \ldots, X_n$ be independent and $N(m, \sigma^2)$ and set

$$X_- = \min_{1 \le i \le n} X_i; \qquad X_+ = \max_{1 \le i \le n} X_i.$$

First find $P(X_- \ge m)$ and $P(X_+ \le m)$, then $P(X_- < m < X_+)$.
(b) Let $x_1, \ldots, x_n$ be a random sample from $N(m, \sigma^2)$, where m and σ^2 are unknown, and let x_- and x_+ be the smallest and the largest value in the sample. What is the confidence level of the interval (x_-, x_+) considered as a confidence interval for m?

*1322. A solution has the unknown pH value m. Four measurements of m were performed and resulted in the values

$$4.32 \qquad 4.22 \qquad 4.23 \qquad 4.37.$$

Another solution has the known pH value 4.84. Six measurements of pH resulted in the values

$$4.71 \qquad 4.63 \qquad 4.69 \qquad 4.76 \qquad 4.58 \qquad 4.83.$$

Model: The pH meter has a systematic error Δ and a random error which is $N(0, \sigma^2)$, where Δ and σ^2 are unknown.
(a) Find a point estimate m^* of m.
(b) Find $D(m^*)$.
(c) Compute the standard error $d(m^*)$ using all values.
(d) Compute a 95% confidence interval for m.

*1323. In a physical experiment two radioactive preparations R_1 and R_2 and a Geiger–Müller counter are used. Particles from R_1 and R_2 are assumed to hit the counter according to Poisson distributions with means $\lambda_1 t$ and $\lambda_2 t$, respectively, where the parameters λ_1 and λ_2 are unknown. During a period of $t = 60$ seconds, the total number of particles registered was 156.

The R_2 was moved to twice its previous distance from the counter; as a result the parameter λ_2 decreases to $\lambda_2/4$. The experiment was repeated, and 130 particles were registered during 120 seconds. Construct an approximate 95% confidence interval for λ_2.

*1324. Let $x_1, \ldots, x_{400}$ be independent observations on $X \sim \mathrm{Exp}(m)$, where m is unknown. It is found that $\sum_1^{400} x_i = 3{,}020$. Construct an approximate 95% confidence interval for the probability $P(X > 10)$.

*1325. The rv X has a uniform distribution in the interval $(0, \theta)$, where θ is unknown. Let $x_1, \ldots, x_n$ be independent observations of X. Show that

$$[\max(x_i), c \max(x_i)] \qquad (c > 1),$$

is a confidence interval for θ. Find the confidence level.

CHAPTER 14

Testing Hypotheses

14.1. Introduction

This chapter deals with the theory of testing hypotheses. Many statisticians have contributed to the development of this theory, R.A. Fisher, J. Neyman, E.S. Pearson, and others. The practical importance of the theory is sometimes questioned, and, as we shall see, care needs to be exercised in using the theory.

In §14.2 a detailed example is given, §14.3 contains the basic ideas, in §14.4 the connection is shown between the testing of hypotheses and interval estimation. In §14.5 applications are made to the normal distribution, in §14.6 the normal approximation is discussed, in §14.7 several examples involving the binomial distribution are given and the so-called χ^2 method is presented, and in §14.8 the practical value of the theory is discussed. Finally, in §14.9 we touch upon the question of repeated tests of significance.

14.2. An Example of Hypothesis Testing

The examples in §10.3 show how problems concerning testing of hypotheses arise in practice. We shall now discuss an example in detail, in order to get acquainted with the basic ideas.

Example 1. ESP

A person claims to have extrasensory perception (ESP). He says that he can tell blindfold whether a coin comes up head or tail. Let p be the unknown probability that his answer is correct when the coin is tossed once. In order to test his ability a psychologist plans to toss a good coin 12 times and use the number x of correct answers to test the null hypothesis H_0 that $p = 1/2$ (which means that the person only guesses).

The model is then a well-known one: For each p the number x is an observation on X, where $X \sim \text{Bin}(12, p)$; in particular, when H_0 is true, we have $X \sim \text{Bin}(12, 1/2)$.

We now formulate the following test of significance: Reject H_0, that is, agree that the person has ESP if x is large enough; say, if $x \geq a$, but not otherwise. The number a is to be determined so as to make the probability small that H_0 is rejected if true. In this way, we guard ourselves fairly well against incorrectly asserting that the person has ESP. The small probability in question is called the level of significance. This level is often chosen to be one of the numbers 0.05, 0.01 and 0.001, but these levels are purely conventional, and other numbers can be chosen.

Let us begin with 0.05. By reasons that will become apparent presently, we require that the level of significance shall be approximately (but not more than) 0.05. Then if $p = 1/2$, $P(X \geq a)$ should be approximately 0.05; for if the number of correct guesses is a or more, we will assert, erroneously, that H_0 is false, that is, that the person has ESP. This leads to the inequality

$$\sum_{i=a}^{12} \binom{12}{i} \left(\frac{1}{2}\right)^{12} \lesssim 0.05.$$

This is not difficult to solve by trial and error. The probabilities appearing in the sum are (for various values of a)

12	11	10	9
0.00024	0.00293	0.01611	0.05371

For $a = 10$ the sum is 0.019, and it is not possible to come closer to 0.05 since this number must not be exceeded. Hence if the person answers correctly at least 10 times, it should be stated that he has ESP, but not otherwise.

The reader can easily see what happens if the level of significance is decreased from 0.05, first to 0.01, then to 0.001. The result is

level of significance		
level of significance $\lesssim$	0.05	$a = 10$,
	0.01	11,
	0.001	12.

In the last case the answers must be all correct. It is clear that the level of significance cannot assume any small value; if we want a level of, say 10^{-6}, then it will be necessary to perform more than 12 tosses.

As we see, the discrete character of the binomial distribution causes some difficulty since the level of significance cannot be attained exactly. When the distribution is continuous, this problem does not arise, as we shall see later. □

14.3. General Method

We shall now give a general description of a test of hypothesis. Consider a random sample $\mathbf{x} = (x_1, \ldots, x_n)$ from a distribution. We want to test a certain *null hypothesis* H_0. The null hypothesis implies that the distribution is specified in some way. A common situation is that the distribution depends on a single unknown parameter θ. Let us assume this for simplicity. (We considered this situation in Example 1, but called the parameter p instead of θ.) H_0 then means that a certain value, or perhaps several values, of θ is specified. We now want to test these values.

If H_0 specifies a single value for θ, the hypothesis is called *simple*, otherwise *composite*. We begin with simple hypotheses.

To test H_0, we first choose a suitable *test quantity* $t = t(\mathbf{x})$. (The ML method is often used to obtain $t(\mathbf{x})$.) The test quantity is an observation on a sample variable $t(\mathbf{X})$. We also choose a *critical region* C. This is a certain portion of the total range of values of t. We then use the following:

Test of significance. If $t \in C$ reject H_0,

$t \notin C$ do not reject H_0.

We choose C so that

$$P(t \in C) = \alpha \qquad \text{when } H_0 \text{ is true.}$$

Here α is selected beforehand and is called the *level of significance* of the test. This is the probability that H_0 is rejected if true. Clearly, α ought to be small. (If the distribution is discrete, it is not always possible to attain α exactly, only $\lesssim \alpha$, see Example 1.)

The test has two possible outcomes: If $t \in C$, the result is said to be *significant* (often with the added phrase "at level α"); if $t \notin C$, the result is said to be *nonsignificant* (at level α). Other wording is possible; for example, we can say that there is a significant (nonsignificant) deviation from the null hypothesis H_0 at level α.

A test of significance is a formalization of ideas from everyday life. Example: You see an animal and formulate the hypothesis H_0: The animal is a horse. As a test quantity you might take t = number of legs. As a critical region we may choose $C = (0, 1, 2, 3, 5, 6, \ldots)$. The test becomes: If $t \leq 3$ or $t \geq 5$ reject H_0, if $t = 4$ do not reject H_0.

The example tells us something important: From a nonsignificant result we cannot infer that H_0 is true, that is, that H_0 should be accepted. If $t = 4$ we do not reject the hypothesis H_0 but cannot therefore accept it as true; the animal can be some other quadruped! The reader is urged to remember this distinction between "not reject" and "accept" H_0. In fact, this argument is well known from science. In science one formulates theories that are used provisionally, that is, they are "not rejected", because they agree well with many observations. Nevertheless, the theories are not necessarily true. If new observations are made, the theories must perhaps be rejected and new hypotheses tested.

In practice, we often use several levels of significance at the same time, typically 0.05, 0.01 and 0.001. We then use the three terms *significant**, *significant*** and *significant**** to state that a result is significant at the level in question (but not for a smaller value of α among these three values). For example, significant* denotes that choice of $\alpha = 0.05$ leads to significance but not $\alpha = 0.01$ and hence not $\alpha = 0.001$.

The reader is warned against using these levels routinely. It often happens that the chosen level is too large. (Example: A significant result may, erroneously, cause a change in an industry with far-reaching consequences for economy and staff, or may condemn a person to a long imprisonment.)

When applying a test of significance, the so-called *P method* is often useful. It will be illustrated by means of an example.

Example 1. ESP (continued)

Suppose that the person answers correctly throughout the experiment, that is, $x = 12$. It is evident that this result is significant***. Similarly, it is found that $x = 11$ is significant** and that $x = 10$ is significant*.

Instead of determining the value of a for all levels we may use the P method, which, in this example, means that we compute

$$P = \sum_{i=x}^{12} \binom{12}{i} \left(\frac{1}{2}\right)^{12},$$

where x is the value of the test quantity. If then $0.01 < P < 0.05$ the result is significant*, and so on. The reader is asked to verify that this procedure works. Note that the P method is only a convenient way of finding the level of significance; the test is unchanged. □

After the reader has studied this example, we shall give an informal, and somewhat imprecise, description of the P method: P is the probability, under H_0, that the test quantity will deviate at least as much as the value observed. (What is imprecise is the word "deviates"; the description is not meaningful unless the points in the critical region have been ordered.)

The critical region C consists sometimes of intervals of type $t \leq a$ or $t \geq b$, where a and b are given quantities. If C consists of one single such interval, the test of significance is said to be *one-sided*. If it consists of two intervals $t \leq a$ and $t \geq b$ where $a < b$ it is said to be *two-sided*. Example 1 contains a one-sided test, the example with the horse a two-sided test.

Example 2. Measurement

In Example 1, p. 224, we considered a research worker attempting to measure a constant θ. In face of errors of measurement, which are $N(0, \sigma^2)$, he obtained the value x. This implies that x is an observation from $N(\theta, \sigma^2)$. We now change the approach. Suppose that the researcher wants to test the hypothesis H_0: $\theta = 2.0$ at level of significance 0.05. As a test quantity he takes the measurement x. It seems reasonable to reject H_0 if x differs much from 2.0. Hence the test of significance ought to be:

$$\text{If} \quad |x - 2.0| \geq \Delta \qquad \text{reject } H_0,$$
$$|x - 2.0| < \Delta \qquad \text{do not reject } H_0.$$

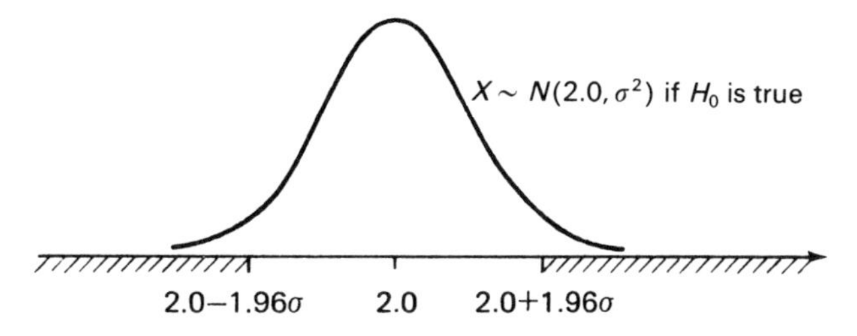

Fig. 14.1. Test of the hypothesis H_0: $m = 2.0$. (Critical region of x shaded in the figure.)

The test is two-sided, for C consists of two intervals $x \leq 2.0 - \Delta$ and $x \geq 2.0 + \Delta$. Here Δ is to be chosen so that the probability is 0.05 that H_0 is rejected if true (that is, if $\theta = 2.0$). Hence we require that if $\theta = 2.0$ then

$$P(|X - 2.0| \geq \Delta) = 0.05,$$

or, taking the complementary probability,

$$P(2.0 - \Delta < X < 2.0 + \Delta) = 0.95.$$

But H_0: $\theta = 2.0$ implies that $X \sim N(2.0, \sigma^2)$ and hence we shall take $\Delta = 1.96\sigma$ (see §8.4, first formula following (13)). Hence if x differs more than 1.96σ from 2.0, we ought to reject H_0, but not otherwise.

A picture is often worth more than many words: Fig. 14.1 contains all the essential information about the test. □

In order to choose a critical region rationally it is necessary to specify not only H_0 but also an *alternative hypothesis* H_1, which may be true if H_0 is false. We do not say that H_1 must necessarily be true if H_0 is not; it often happens that some third hypothesis is the correct one. By a good test we mean one that will reject H_0 with high probability when H_1 is true, which seems reasonable.

The alternative hypothesis may be *simple* (that is, contain one single value θ) or *composite* (that is, consist of several values of θ).

For the moment we suppose that the null hypothesis consists of one value, which we call θ_0.

The argument leads us to introduce the *power function*

$$h(\theta) = P(H_0 \text{ is rejected if } \theta \text{ is the correct value of the parameter}).$$

A test is good if $h(\theta)$ is large for all $\theta \in H_1$. On the other hand, $h(\theta)$ is small for $\theta \in H_0$; it is seen that $h(\theta_0)$ is just α, which is chosen to be a small quantity. We call $h(\theta)$ the *power* of the test at θ.

Example 1. ESP (continued)

We take an example of a simple alternative hypothesis. Suppose that the person boasts: "In nine cases out of ten my answer is correct!" It is then important for the psychologist

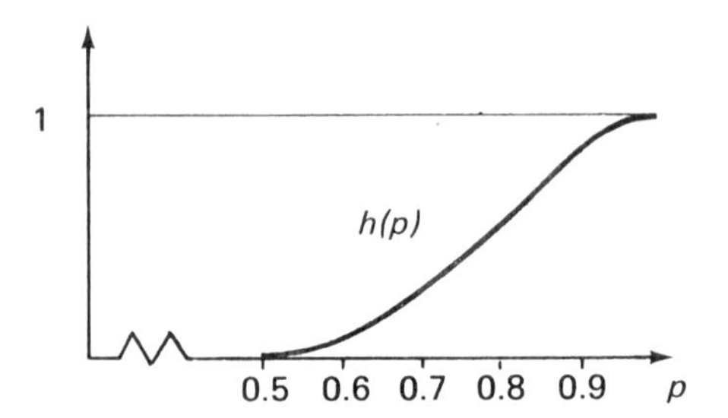

Fig. 14.2. Power function (one-sided test).

to reject H_0 if this assertion is true. Let us examine the situation in detail. As before, the null hypothesis is taken to be H_0: $p = 1/2$, while the alternative hypothesis is H_1: $p = 9/10$. Hence in this case H_1 is a simple hypothesis. We choose $\alpha = 0.05$ and reject H_0 if $x \geq 10$ (see the previous version of the example). It is found that

$$h(0.9) = P(X \geq 10) = \sum_{i=10}^{12} \binom{12}{i} 0.9^i 0.1^{12-i} = 0.89.$$

The power of the test is fairly good: The probability is rather high that the test helps the psychologist to discover the person's ability (if it is as pronounced as required by H_1).

We note in passing that the example illustrates what we have said earlier: H_1 is not necessarily true if H_0 is false. Clearly, some third value of p may be the correct one.

We now change the alternative hypothesis to a composite one: we take H_1: $p > 1/2$. The power function then becomes at each $p > 1/2$

$$h(p) = \sum_{i=10}^{12} \binom{12}{i} p^i (1-p)^{12-i},$$

see Fig. 14.2.

It is seen that the test is good for values of p near 1. However, the power is low for smaller p-values; for $p = 0.7$ it is only about 0.25. This is quite natural, for we cannot expect that as few as 12 tosses will enable us to discriminate well between the "neighbouring" hypotheses $p = 0.5$ and $p = 0.7$. For such a purpose we need a larger experiment.

The power function is in this case a monotone function of p. This is due to the one-sidedness of our test; for such tests the power function is nearly always increasing or decreasing.

If the psychologist does not like the power function, he may increase the number of tosses. The function will then still be (approximately) α at $p = 1/2$, but will be higher than before for values of p greater than 1/2. □

Example 2. Measurement (continued)

Let us assume that $\sigma = 0.04$ (which is a measure of the precision of the experiment). Suppose as before that the researcher wishes to test the hypothesis H_0: $\theta = 2.0$. Let the alternative hypothesis be

$$H_1: \theta \leq 1.9 \qquad \text{or} \qquad \theta \geq 2.1.$$

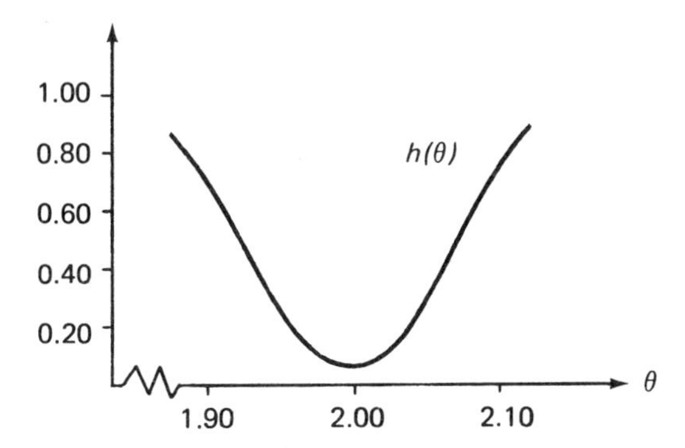

Fig. 14.3. Power function (two-sided test).

The researcher may want to reject H_0 if θ does indeed differ from H_0 as much as H_1 states. Let us evaluate the probability of rejection.

The test is now: Reject H_0 if

$$|x - 2.0| > 1.96 \cdot 0.04.$$

Hence

$$h(\theta) = P(|X - 2.0| > 1.96 \cdot 0.04) \qquad \text{if} \quad X \sim N(\theta, 0.04^2).$$

Formula (8) in §8.4 gives the complementary probability, namely, that X lies between the limits $2.0 \pm 1.96 \cdot 0.04$

$$1 - h(\theta) = \Phi\left(\frac{2.0 + 1.96 \cdot 0.04 - \theta}{0.04}\right) - \Phi\left(\frac{2.0 - 1.96 \cdot 0.04 - \theta}{0.04}\right),$$

so that

$$h(\theta) = 1 - \Phi\left(1.96 + \frac{2.0 - \theta}{0.04}\right) + \Phi\left(-1.96 + \frac{2.0 - \theta}{0.04}\right).$$

This power function has been sketched in Fig. 14.3. It has a minimum (= 0.05, which is the stated level of significance) at $\theta = 2.0$ and is about 0.70 or more for values of θ within H_1; hence the power is not very large. There are two ways out of this undesirable situation: to increase the precision of the measurement (that is, to make σ smaller) or increase the number of observations.

For example, assume that σ is 0.03. It can then be shown that the power function is about 0.90, or more, within H_1, which is a substantial improvement. The probability is therefore quite large that H_0 is rejected if H_1 is true. □

Remark 1. Tests of Composite Hypotheses

We have now discussed at length how to test simple null hypotheses. The reader may perhaps expect an equally detailed discussion of composite null hypotheses. However, this is not needed; only a slight change of the argument is required.

If H_0 consists of several values of θ, the probability $P(H_0$ is rejected) if H_0 is true will depend on which of the values within H_0 is the correct one; hence the old definition of α does not work. Instead we take

$$\alpha = \sup_{\theta \in H_0} h(\theta)$$

as the level of significance. For example, if $\alpha = 0.01$ the risk that H_0 is rejected when true is at most 0.01. Hence we need only determine for which $\theta \in H_0$ the risk is largest that H_0 is rejected by the test; this is usually rather easy. An illustration is given in Example 3 in §14.5. □

Remark 2. Strengthened Test of Significance.

In the example involving the horse, on p. 255, we warned the reader against drawing from a nonsignificant result the general conclusion that H_0 is true. We shall now examine this question in more detail. We want to find out when the conclusion "H_0 is true" is legitimate.

When a null hypothesis is accepted we perform a *strengthened test*.. This is done as follows:

$$\begin{aligned} \text{If} \quad & t \in C \qquad \text{reject } H_0, \\ & t \notin C \qquad \text{accept } H_0 \ (= \text{consider } H_0 \text{ to be true}). \end{aligned}$$

Compared with the usual test we have here replaced "do not reject" by "accept". We distinguish between two alternatives:

(1) The case of two hypotheses

Here we assume that either H_0 or H_1 is true—no third hypothesis enters into the discussion. The strengthening of the test is then possible if and only if $h(\theta)$ is large for all $\theta \in H_1$.

This is seen as follows: We prove that the risk of a wrong conclusion is small whether H_0 or H_1 is true. (The hypotheses may be simple or composite.)

If H_0 is true, the risk of an erroneous conclusion, that is, the probability that H_0 is rejected, is at most

$$\alpha = \sup_{\theta \in H_0} h(\theta)$$

and hence is small. If H_1 is true, the risk of a wrong conclusion, that is, the probability that H_0 is accepted, is equal to

$$P(H_0 \text{ is accepted}) = 1 - P(H_0 \text{ is rejected}) = 1 - h(\theta).$$

The risk is at most equal to

$$\beta = \sup_{\theta \in H_1} (1 - h(\theta)).$$

Since $h(\theta)$ is large within H_1, by assumption, β is small.

We call the error made if H_0 is rejected when H_0 is true the *error of the first kind*, and the error made if H_0 is accepted when H_0 is false the *error of the second kind*. Using this terminology we see that α and β are the largest values of the probability of error of the first kind, and of the second kind, respectively.

(2) The case of more than two hypotheses

Now the alternatives are not only H_0 and H_1; some third hypothesis may be true. More generally, we assume that the hypotheses are $H_0, H_1, \ldots, H_k$, one of which is true. We assume that the power function has not been investigated outside H_0 and H_1.

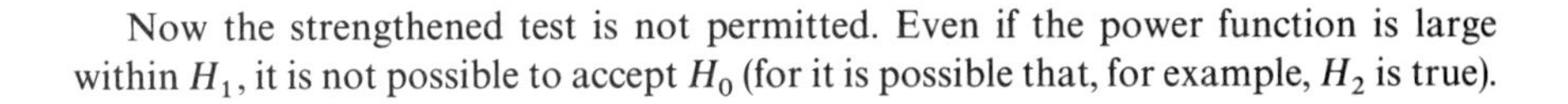

Now the strengthened test is not permitted. Even if the power function is large within H_1, it is not possible to accept H_0 (for it is possible that, for example, H_2 is true). □

Example 2. Measurement (continued)

The example will be used to illustrate the distinction between two hypotheses and several hypotheses. We choose the same H_0 and H_1 as before, and the smaller value of σ, that is, $\sigma = 0.03$.

If we know that H_0 or H_1 is true, we can reject H_0 if $|x - 2.0| \geq 1.96 \cdot 0.03$, that is, if x is larger than 2.06 or smaller than 1.94, and accept H_0 if x lies between these values. The probability of an error of the first kind is then at most 0.05, and the probability of an error of the second kind at most $\beta = 1 - 0.90 = 0.10$ (for we know that the power function within H_1 is 0.90 or more); hence the probability of an error of the second kind is rather small.

On the other hand, if it is possible that H_2: $\theta = 2.05$ holds, it is, of course, impossible to accept H_0 if x lies between 1.94 and 2.06; this third value may then be the correct one. □

14.4. Relation Between Tests of Hypotheses and Interval Estimation

We now present a surprising connection of this chapter with the previous one. It will be demonstrated that there is a close relation between tests of hypotheses and interval estimation. To show this we use an example which is by now familiar to the reader.

Example 2. Measurement (continued)

We describe the situation once more. A research worker measures an unknown constant θ. The error of measurement is $N(0, \sigma^2)$ and hence the observed value x is an observation from $N(\theta, \sigma^2)$.

We found in Example 1, p. 224, that $(x - 1.96\sigma, x + 1.96\sigma)$ is a two-sided 95% confidence interval for θ. On the other hand, we saw on p. 256 that a two-sided test of the hypothesis H_0: $\theta = \theta_0$ consists in rejecting the hypothesis H_0 if $x \leq \theta_0 - 1.96\sigma$ or $x \geq \theta_0 + 1.96\sigma$, that is, if $\theta_0 \leq x - 1.96\sigma$ or $\theta_0 \geq x + 1.96\sigma$, that is, if θ_0 is outside the above confidence interval. Hence the test can be performed by first determining the confidence interval and then finding out whether the hypothetical value θ_0 lies outside or inside this interval. In the former case H_0 is rejected, but not in the latter.

The method also works in the one-sided case. A one-sided test of the hypothesis H_0: $\theta = \theta_0$ against the alternative hypothesis H_1: $\theta > \theta_0$ may be performed by first constructing a one-sided confidence interval with a limit to the left, namely, the interval $(x - 1.64\sigma, \infty)$; H_0 is rejected if θ_0 is outside this interval. Similarly, H_0 is tested against H_1: $\theta < \theta_0$ by determining $(-\infty, x + 1.64\sigma)$ and rejecting H_0 if θ_0 lies outside this interval. □

The relation exemplified above follows quite generally from the argument in §13.3. We showed there, for example, how a two-sided confidence interval can be constructed under rather general conditions. Now suppose that the point estimate θ^* mentioned there is used instead as a test quantity for a two-sided test. According to relation (1) in §13.3, this test can be performed by rejecting H_0: $\theta = \theta_0$ if θ^* lies outside the interval $(b_1(\theta_0), b_2(\theta_0))$. (Clearly, the level of significance is then α.) By relation (2) in §13.3 this is equivalent to saying that θ_0 lies outside the interval $(a_1(\theta^*), a_2(\theta^*))$, and we are through: We have shown that an alternative way of performing the test is to reject H_0 if θ_0 is outside the confidence interval, but not reject H_0 if θ_0 is inside the interval.

An analogous relation exists between one-sided confidence intervals and one-sided tests, and an example of this has been given in Example 2 above.

When the test of hypothesis is performed in the way described in this section we say that it is performed according to the *confidence method.*

14.5. Application to the Normal Distribution

All problems concerning interval estimation in the case of a normal distribution can be reformulated as problems of testing hypotheses. Instead of determining confidence intervals for the mean m, the standard deviation σ and so on, one may test different hypotheses regarding the unknown parameter. All these situations are treated in a similar way. We limit our discussion to two cases:

First, let $x_1, \ldots, x_n$ be a random sample from $N(m, \sigma^2)$. We want to test the simple hypothesis

$$H_0\colon m = m_0.$$

Take as a test quantity

$$u = \begin{cases} (\bar{x} - m_0)/D & \text{if } \sigma \text{ is known} \quad (D = \sigma/\sqrt{n}), \\ (\bar{x} - m_0)/d & \text{if } \sigma \text{ is unknown} \quad (d = s/\sqrt{n}). \end{cases}$$

If H_0 is true, these ratios are observations on rv's which are $N(0, 1)$ and $t(n - 1)$, respectively. Using these facts we can construct appropriate tests.

If σ is *known*, we obtain the following tests:

(a) If H_1 consists of m-values both smaller than m_0 and larger than m_0, the following two-sided test is used: If $|u| \geq \lambda_{\alpha/2}$, reject H_0, but not otherwise. It is convenient to use the following table (see Table 2 at the end of the book):

$\lvert u \rvert$	1.96	2.58	3.29	
α	0.05	0.01	0.001	
Significance	*	**	***	

Hence a value of $|u|$ between 1.96 and 2.58 leads to significance*, and so on.

(b) If H_1 consists only of m-values larger than m_0, a one-sided test is used: If $u \geq \lambda_\alpha$, reject H_0, but not otherwise. The following table is used:

u	1.64	2.33	3.09	
α	0.05	0.01	0.001	
Significance	*	**	***	

(c) If H_1 consists only of m-values smaller than m_0, the following one-sided test is used: If $u \leq -\lambda_\alpha$, reject H_0, but not otherwise. (The same table as in case (b) can be used, with an obvious modification.)

Alternatively, the test of significance can be performed according to the confidence method in §14.4. In all cases, the power function, $h(m) = P(H_0$ is rejected if m is the true value), can be determined according to the principles given earlier.

If σ is *unknown*, the three alternatives are the same, but the two tables above are replaced by a table of the appropriate t distribution. (Determination of the power function is difficult and is beyond the scope of this book.)

Example 3. Sugar Content

This is a continuation of Example 2, p. 234. As before, the sugar content in a sugar-beet is assumed to be an observation from $N(m, \sigma^2)$. The same data on 60 sugar-beets is used.

Let us test the hypothesis

$$H_0: m = 17.0$$

assuming that σ is unknown.

(1) Two-sided test

The alternative hypothesis is taken as all values except 17.0, that is

$$H_1: m \neq 17.0.$$

This is situation (a) (see above) and we reject H_0 if $\bar{x}$ differs sufficiently from 17.0. We found earlier that $\bar{x} = 16.51$, $d = 0.159$, and hence

$$u = \frac{16.51 - 17.0}{0.159} = -3.1.$$

There are $f = 60 - 1 = 59$ degrees of freedom. The test is two-sided and the critical region comprises the dashed parts of Fig. 14.4(a). Since $|u| > t_{0.005}(59) = 2.66$, but $< t_{0.0005}(59) = 3.46$, there is a significant** deviation from the null hypothesis.

(2) One-sided test

The alternative hypothesis is now taken as

$$H_1: m < 17.0.$$

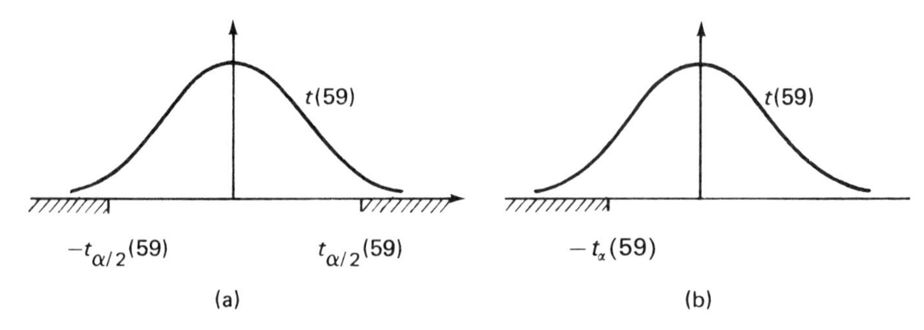

Fig. 14.4. (a) Two-sided test. (b) One-sided test.

This is situation (c). We reject H_0 only if $\bar{x}$ is small enough; that is, we use a one-sided test. The value u is obtained as before, but the ensuing argument is different. The critical region now consists of the part of the t distribution to the left of $-t_\alpha(59)$ (see Fig. 14.4(b)). Since $u < -t_{0.01}(59) = -2.39$ but $> -t_{0.001}(59) = -3.23$, the deviation is still significant**.

If, instead, the null hypothesis had been composite, say

$$H_0\colon m \geq 17.0,$$

and the alternative hypothesis the same as above, the test would have been unchanged. For, the probability of rejecting H_0 when H_0 is true assumes it largest value at $m = 17.0$. (For m-values > 17.0 the probability that $\bar{X}$ is less than a given number is smaller than when $m = 17.0$.) Here is an example showing how a test of a composite hypothesis can sometimes be swiftly transformed into a test of a simple hypothesis (see Remark 1, p. 259). □

The other situation we shall discuss concerns "paired samples". We take an example:

Example 4. Measuring Different Objects

This is a continuation of Example 4 at the end of §13.4. We shall test the hypothesis

$$H_0\colon \Delta = 0$$

which states that there is no systematic difference between B and A. The alternative hypothesis is $H_1\colon \Delta \neq 0$. A two-sided test is then used. Since $\bar{z} = 1.34$, $d = 0.401$, we have

$$u = 1.34/0.401 = 3.34.$$

Since $|u| > t_{0.005}(10) = 3.17$ but $< t_{0.0005}(10) = 4.59$, there is a significant** systematic difference between A and B. □

14.6. Using the Normal Approximation

In §13.5 we showed how the normal approximation can be used for interval estimation. Hypothesis testing can be performed in a similar way. We take a test quantity of the form $(\theta^* - \theta)/D$, or if D is not known, $(\theta^* - \theta)/d$. If the ratio is approximately normally distributed, we can perform tests in the same way as in the foregoing section; the only difference is that the level of significance cannot now be determined exactly.

Example 5. Comparison of Anaesthetics

This is a continuation of Example 6 at the end of §13.5. Let us test the hypothesis

$$H_0: m_1 = m_2,$$

which states that the anaesthetics A and B have the same effect. The alternative hypothesis is assumed to be

$$H_1: m_1 \neq m_2.$$

We found earlier that $\bar{x} = 165.09$, $\bar{y} = 153.10$, $d = 15.19$, and so

$$u = (165.09 - 153.10)/15.19 = 0.66.$$

We reject H_0 if $\bar{x} - \bar{y}$ is large enough positive or large enough negative, that is, the test is two-sided. Since

$$|u| < \lambda_{0.025} = 1.96$$

there is no significant difference between the two anaesthetics. This does *not* prove that there is no difference; it is quite possible that a larger experiment would have given a different result. (Remember the example with the horse!) □

14.7. Application to the Binomial and Related Distributions

(a) The Binomial Distribution

In Example 1 at the beginning of §14.2 we discussed in detail a certain test of a hypothesis for a binomial variable. We shall take two more examples:

Example 6. Tossing a Coin

When a coin is tossed, a head appears with the unknown probability p. We wish to test H_0: $p = 1/2$ against the alternative hypothesis H_1: $p \neq 1/2$. The coin is tossed 30 times and 5 heads are obtained.

Let us first choose the level of significance as 0.01. We then reject H_0 if $x \leq a$ or $x \geq b$, where a and b are such that, if H_0 is true, $P(X \leq a) \lesssim 0.005$, $P(X \geq b) \lesssim 0.005$; hence the test is two-sided. A table of the binomial distribution shows that $a = 7$, $b = 30 - 7 = 23$. (Then $P(X \leq a) = P(X \geq b) = 0.0026$.) Hence the result $x = 5$ is significant, that is, we reject the notion that the coin is fair.

Alternatively, we may use the P method (see §14.3): Compute

$$P/2 = \sum_{i=0}^{5} \binom{30}{i} \left(\frac{1}{2}\right)^i \left(\frac{1}{2}\right)^{30-i} = 0.000162$$

so that $P = 0.00032$. Hence the result is significant***, and we reject the hypothesis that the coin is well balanced. □

Example 7. Taste Testing

A food industry plans to replace the standard production method by a new one. It is then important to know if the flavour of the product is changed.

A so-called triangle test is performed. Each of 500 persons tastes, in random order, three packets of the product, two of which are manufactured according to the standard method and one according to the new method. Each person is asked to select the packet which tastes differently from the two others.

Let x be the number of persons who respond correctly, that is, select the packet manufactured according to the new method. If there is no difference between the tastes, the probability that the new method is selected is 1/3; if there is a difference, we will have $p > 1/3$. Hence our hypotheses are

$$H_0: p = 1/3,$$

$$H_1: p > 1/3.$$

The firm is satisfied with a level of significance of 0.05. Then a should be chosen such that, if X has a binomial distribution with $n = 500$ and $p = 1/3$, then

$$P(X \geq a) = 0.05.$$

Hence the test should be one-sided. The normal approximation shows that

$$P(X \geq a) = 1 - \Phi\left(\frac{a - 500 \cdot 1/3}{\sqrt{500 \cdot 1/3 \cdot 2/3}}\right).$$

The quantity on the right equals 0.05 provided that the expression in parentheses is $\lambda_{0.05} = 1.64$, so that

$$a = 500 \cdot \frac{1}{3} + 1.64 \sqrt{500 \cdot \frac{1}{3} \cdot \frac{2}{3}} = 184.$$

Hence, if at least about 184 persons select the new product, it can be asserted that the new method causes a change of flavour. (However, the test does not disclose whether the flavour is improved!)

In this calculation we did not use u as a test quantity and took the trouble to compute a for a given level of significance. Let us demonstrate how u is computed. Suppose that 200 persons respond correctly. We then obtain

$$u = \left(200 - \frac{500}{3}\right) \Big/ \sqrt{500 \cdot \frac{1}{3} \cdot \frac{2}{3}} = 3.2.$$

We have situation (b) in §14.5, and the critical region is the right tail of the standard normal distribution (see Fig. 14.5). Since $u > 3.09$ for these data, the deviation is significant***, that is, the risk of error is small if we assert that the flavour has changed. □

In §13.6 we discussed an interval estimation problem for two binomial distributions $\text{Bin}(n_1, p_1)$ and $\text{Bin}(n_2, p_2)$, when one observation is available

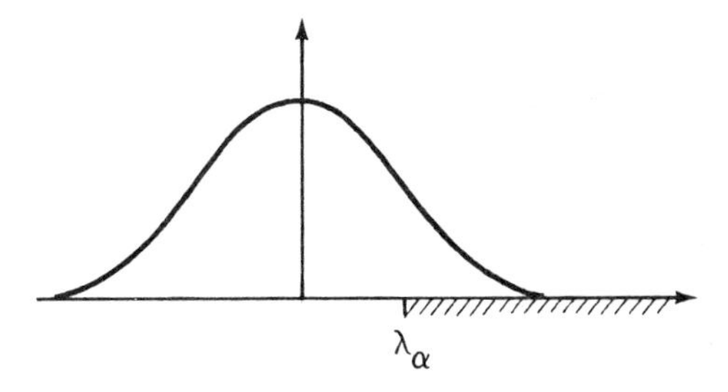

Fig. 14.5. Critical region for one-sided test.

from each distribution. The corresponding hypothesis testing problem can be treated as follows:

The null hypothesis is assumed to be

$$H_0\colon p_1 = p_2.$$

When H_0 is true, an estimate of p (the common value of p_1 and p_2) is given by

$$p^* = (x_1 + x_2)/(n_1 + n_2).$$

(By the way, this is the ML estimate of p in this situation.) Using the same symbols as before if H_0 is true, we obtain

$$D = \sqrt{\frac{pq}{n_1} + \frac{pq}{n_2}} = \sqrt{pq\left(\frac{1}{n_1} + \frac{1}{n_2}\right)}.$$

As a standard error we take

$$d = \sqrt{p^*q^*\left(\frac{1}{n_1} + \frac{1}{n_2}\right)}.$$

Form the ratio $u = (p_1^* - p_2^*)/d$ in the usual way and perform an approximate test. The approximation is better the larger n_1 and n_2.

The test can also be used when we want to compare observations from two distributions which themselves can be approximated by binomial ones. This happens, for example, when samples are taken from finite populations and the sample sizes are small compared to the population sizes. (As we have mentioned in §9.3 the hypergeometric distribution can then be approximated by the binomial distribution.) The following example refers to this situation.

Example 8. Seeking Housing

From a large population of Swedish married couples, 556 couples with children and 260 couples without children were selected at random. Three hundred and twenty-four of the former and ninety-eight of the latter had to wait at most one year for housing, and the rest more than one year. Let p_1 be the proportion in the whole population of couples with children who had to wait at most one year, and let p_2 be the corresponding proportion for couples without children. The estimates are

$$p_1^* = 324/556 = 0.583; \qquad p_2^* = 98/260 = 0.377.$$

To perform the test we compute

$$p^* = (324 + 98)/(556 + 260) = 0.517,$$

$$d = \sqrt{0.517 \cdot 0.483\left(\frac{1}{556} + \frac{1}{260}\right)} = 0.0376,$$

$$u = (0.583 - 0.377)/0.0376 = 5.48.$$

Since $|u|$ exceeds 3.29, the difference is significant***. Hence, with little risk of committing an error, we assert that, in the population studied, the proportion of families with children who obtained housing within one year is different from the corresponding proportion of families without children.

We have chosen here the alternative hypothesis $p_1 \neq p_2$ and have used a two-sided test. If it is assumed known that $p_1 \geq p_2$ (the case $p_1 < p_2$ seems unrealistic), then a one-sided test should be used: reject H_0 if u is large. Since u is larger than $\lambda_{0.001} = 3.09$, we again conclude that the difference is significant***. The interpretation is now stronger than before: With little risk of error we can state that, in the population, the proportion of families with children who have obtained housing is larger than the corresponding proportion of families without children. □

We shall now use the binomial distribution in a somewhat unexpected context. At the end of §14.5 we analysed "paired samples", assuming a normal distribution. Sometimes it is possible to use a less sensitive method, the so-called *sign test*, which has the advantage that the assumption of normality is not necessary.

We have n pairs of values $(x_1, y_1), \ldots, (x_n, y_n)$ obtained, for example, by examining n objects according to two methods A and B. The values x_i, y_i are regarded as independent observations on the rv's X_i, Y_i. We wish to decide whether, besides random variation, there is a systematic difference between X_i and Y_i.

The sign test is performed as follows: Determine the number of pairs for which the x-value is larger than the y-value (or, if this is more convenient, smaller than the y-value). This number is called z. If there is no systematic difference between X_i and Y_i, the number z should be about $n/2$.

Let us describe this in more detail: The hypotheses are

H_0: X_i and Y_i have the same distribution for all i,
H_1: For all i, the distribution of Y_i is shifted in the same direction relative to the distribution of X_i,

(see Fig. 14.6; the figure refers to a single pair of rv's).

If H_0 is true, z is an observation from Bin(n, 1/2). This is due to the fact that, when H_0 is true, then $P(X_i > Y_i) = P(X_i < Y_i) = 1/2$ (show this!). Hence each comparison within a pair may be regarded as a trial of the type "toss a symmetric coin", where $x > y$ and $x < y$ correspond to head and tail, respectively.

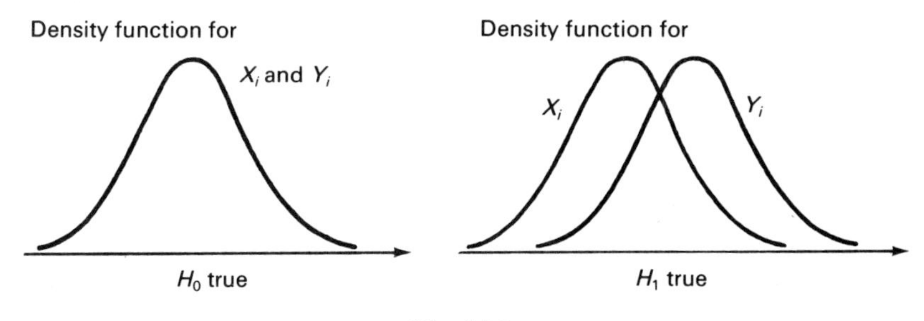

Fig. 14.6.

Hence the original problem has been replaced by one discussed previously in this section. As usual, we have to choose between a one-sided and two-sided test, and this depends on the practical situation.

Example 9. Measuring Different Objects

We have discussed this problem twice before, the last time as a problem of testing hypotheses, assuming a normal distribution (see Example 4 at the end of §13.4 and Example 4 at the end of §14.5). Now we shall use the sign test. We reproduce the data once more:

	Object										
	1	2	3	4	5	6	7	8	9	10	11
A	20.2	22.0	19.7	21.4	16.3	17.0	24.5	15.6	16.0	13.2	19.0
B	21.6	22.9	20.0	23.6	15.7	17.5	27.8	19.8	16.4	14.7	20.1
sign $(B - A)$	+	+	+	+	−	+	+	+	+	+	+

(If two values in a pair are equal, this pair is deleted.) We use a two-sided test. Since $n = 11$, $z = 10$ we find, using the P method,

$$P/2 = \sum_{j=10}^{11} \binom{11}{j} \left(\frac{1}{2}\right)^{11} = \left(\frac{1}{2}\right)^{11} + 11\left(\frac{1}{2}\right)^{11} = 3/512,$$

or $P = 0.012$. The result is significant*; that is, at the level 0.05 we assert that there is a systematic difference between A and B. □

In the earlier version of the example we had significance**, but there we used a more sensitive test, assuming normality. This is an instructive situation: The same data were analysed by different statistical methods, and the conclusions were different in degree of significance. This does not at all mean that the methods are wrong, only that they differ in power.

We have now discussed the binomial distribution and its many applications at such length that we have no space left for the hypergeometric distribution and the Poisson distribution. The loss is not very great: With the information now available it is easy to transform the interval estimation problems for these distributions to hypothesis testing problems, for almost the identical steps as for the binomial are required.

Instead, we shall discuss something new.

(b) The χ^2 Method

As in §9.5 we perform n independent trials, each of which has r outcomes $A_1, \ldots, A_r$ with probabilities $p_1, \ldots, p_r$ ($\sum p_i = 1$). Further, let $x_1, \ldots, x_r$ be the frequencies of the outcomes $A_1, \ldots, A_r$ obtained in such a sequence of trials. These frequencies are regarded as an observation on a rv with a multinomial distribution.

Assume that the probabilities $p_1, \ldots, p_r$ are unknown and that we wish to test a certain hypothesis H_0 concerning their values. The χ^2 method is used as follows: As a test quantity we take the expression

$$Q = \sum_{i=1}^{r} \frac{(x_i - np_i)^2}{np_i},$$

after inserting the values of $p_1, \ldots, p_r$ that we wish to test. The reason why we choose this test quantity is that it looks reasonable (the more x_i deviates from the expectation np_i, the larger is Q) and also that it is not very difficult to handle this expression mathematically.

We introduce the following

Test of significance. Reject H_0 if $Q > \chi^2_\alpha(f)$, where $f = r - 1$, otherwise do not reject H_0.

For large n, this test has approximate level of significance α. (The proof is difficult; see, for example, H. Cramér, *Mathematical Methods of Statistics*, Almqvist and Wiksell, Uppsala, 1945; Princeton University Press, Princeton, NJ, 1946, Chapter 30.)

It is not possible to answer precisely how large n should be to make the approximation acceptable. The following rule of thumb is often used: All values of np_i should be at least 5. (However, this rule is unnecessarily restrictive, if the numbers p_i are fairly alike; if they are exactly alike, np_i may be allowed to be as small as 1.)

Example 10. Throwing a Die

A die was thrown 96 times, and the following frequencies of ones, twos, etc., were obtained: 15, 7, 9, 20, 26, 19. We wish to test the hypothesis that the die is well-balanced, that is, we formulate the null hypothesis

$$H_0: p_1 = p_2 = \cdots = p_6 = 1/6.$$

We may use the χ^2 method, for all np_i equal $96 \cdot 1/6 = 16$ and hence the sample is large enough. We obtain

$$Q = \frac{(15-16)^2}{16} + \frac{(7-16)^2}{16} + \frac{(9-16)^2}{16} + \frac{(20-16)^2}{16} + \frac{(26-16)^2}{16} + \frac{(19-16)^2}{16} = 16.0.$$

There are $f = r - 1 = 6 - 1 = 5$ degrees of freedom. Since Q is larger than $\chi^2_{0.01}(5) = 15.1$ (see Table 4 at the end of the book), we assert that the die is not symmetric: the result is significant**. □

The χ^2 method can also be applied when we have performed several sequences of independent trials and want to examine whether the sequences are *homogeneous*. By this term we mean that the same set of probabilities $p_1, \ldots, p_r$ appear in all sequences. These numbers $p_1, \ldots, p_r$ are not specified.

Assume that s sequences have been performed:

Sequence	Frequency of A_1	A_2	$\ldots$	A_r	Number of trials
1	x_{11}	x_{12}	$\ldots$	x_{1r}	n_1
2	x_{21}	x_{22}	$\ldots$	x_{2r}	n_2
$\vdots$	$\vdots$	$\vdots$		$\vdots$	$\vdots$
s	x_{s1}	x_{s2}	$\ldots$	x_{sr}	n_s
Sum	S_1	S_2	$\ldots$	S_r	N

In order to test the homogeneity, we take

$$Q = \sum_{i,j} \frac{(x_{ij} - n_i p_j^*)^2}{n_i p_j^*},$$

where $p_j^* = S_j/N$. (See the above table for the meaning of S_j and N.) It is seen that p_j^* is the best estimate of the common value p_j that can be formed from the combined observations. We now use the following

Test of homogeneity. Reject the hypothesis of homogeneity if

$$Q > \chi^2_\alpha(f) \qquad (f = (r-1)(s-1)),$$

but not otherwise.

Example 11. Throwing Two Dice

Besides the sample in Example 10 we have a sample of 80 throws performed with another die, resulting in the frequencies 10, 11, 7, 15, 21, 16. We wish to test the hypothesis that the probabilities of a "one" are the same, the probabilities of a "two" are the same etc., on both dice. (We are not interested in knowing whether the dice are well-balanced.) Hence we have the following data:

	Frequencies						Number of trials
Sample 1	15	7	9	20	26	19	96
Sample 2	10	11	7	15	21	16	80
Sum	25	18	16	35	47	35	176
p_j^*	0.142	0.102	0.091	0.199	0.267	0.199	

The numbers p_j^* in the last row have been obtained from those in the preceding row by dividing by the total number, 176, of throws. We find

$$Q = \frac{(15 - 96 \cdot 0.142)^2}{96 \cdot 0.142} + \frac{(7 - 96 \cdot 0.102)^2}{96 \cdot 0.102} + \cdots + \frac{(16 - 80 \cdot 0.199)^2}{80 \cdot 0.199} = 2.21.$$

There are $(6 - 1)(2 - 1) = 5$ degrees of freedom. Since Q does not exceed $\chi^2_{0.05}(5) = 11.1$, the result is not significant. Hence the data are compatible with the hypothesis that the dice are identical, that is, both show "one" with the same probability "two" with the same probability, and so on. □

Example 8. Seeking Housing (continued)

The observations on p. 267 can now be analysed in a new way. Since the population is assumed to be very large, the random selection of the 556 married couples with children may be said to correspond to a sequence of 556 approximately independent trials with the outcome A_1 (waiting-time at most one year) and A_2 (waiting-time more than one year). As before, the corresponding probabilities are denoted by p_1 and $1 - p_1$, respectively. Analogously, we regard the 260 couples without children as a sequence of 260 trials with the same outcomes A_1 and A_2 and probabilities p_2 and $1 - p_2$, respectively. (The notation differs somewhat from that used above in the description of the χ^2 method.)

As before, we want to test the hypothesis that A_1 occurs with the same probability in both sequences (and hence also A_2), in other words, that the sequences are homogeneous. The material is now arranged in a so called 2×2 *table*:

	Frequency A_1	A_2	Sum
With children	324	232	556
Without children	98	162	260
Sum	422	394	816

Using the χ^2 method we obtain, on noting that $422/816 = 0.517$ and $394/816 = 0.483$,

$$Q = \frac{(324 - 556 \cdot 0.517)^2}{556 \cdot 0.517} + \frac{(232 - 556 \cdot 0.483)^2}{556 \cdot 0.483} + \frac{(98 - 260 \cdot 0.517)^2}{260 \cdot 0.517}$$

$$+ \frac{(162 - 260 \cdot 0.483)^2}{260 \cdot 0.483} = 30.0.$$

We have $r = s = 2$ and hence $f = (2-1)(2-1) = 1$. Since Q exceeds $\chi^2_{0.001}(1) = 10.8$, the result is significant***. Hence we reject the assumption of homogeneity; p_1 is different from p_2.

It should be added that the difference between the two statistical methods we have used here and on p. 267 is only apparent; it can be shown that they always lead to the same result. □

For friends of simple special formulae it may be added that, in the case of a 2×2 table, Q can be determined in the following simple manner. Let the table be

Frequency	
A_1	A_2
a	b
c	d

Then Q reduces to

$$Q = \frac{(a+b+c+d)(ad-bc)^2}{(a+b)(c+d)(a+c)(b+d)}.$$

A reader who likes algebra can derive this expression from the general expression for Q. Let us check it in the example we just gave:

$$Q = \frac{(324+232+98+162)(324 \cdot 162 - 98 \cdot 232)^2}{(324+232)(98+162)(324+98)(232+162)}$$

$$= \frac{816 \cdot 29752^2}{556 \cdot 260 \cdot 422 \cdot 394} = 30.0.$$

Remark. Beware of Misusing χ^2

We warn the reader against misusing the χ^2 method. Let us suggest an incorrect application. An election had the following result: In town X, 39.9% of the 19,359 persons voted for party S. In town Y, 48.7% of the 68,075 voters voted for S. Is there a significant difference between these percentages? A routine application of the χ^2 method would give high significance.

However, in order to qualify as a statistical method such as the χ^2 method, some sort of random selection from a population must be involved. This is not possible in the above example, for we have performed a complete investigation of town X and Y and know the situation in full. The situation would have been different it we had taken random samples from the two towns (with sample-sizes small compared to the numbers of inhabitants) and compared the number of votes. It would then have been possible to use the χ^2 method to compare the proportions voting for S in the two towns. In the present case, we already know these figures, and by simply looking at them, we find that they differ considerably.

14.8. The Practical Value of Tests of Significance

It was indicated at the beginning of this chapter that the practical value of a test of significance is sometimes questioned. We shall now try to explain why this is so, and base our discussion on the following situation:

We wish to examine the systematic difference between two ways A and B of making measurements. For this purpose, we perform two equally long series of measurements $x_1, x_2, \ldots, x_n$ and $y_1, y_2, \ldots, y_n$ corresponding to method A and B, respectively. These series are considered to be random samples from $N(m + \Delta, \sigma^2)$ and $N(m, \sigma^2)$, respectively, where σ is known. Hence the x-values vary about an expectation $m + \Delta$ and the y-values about m. The systematic difference, Δ, is the quantity of interest.

Choose the null hypothesis $H_0\colon \Delta = 0$ and the alternative hypothesis $H_1\colon \Delta \neq 0$. As explained earlier, we should then use a two-sided test.

Determine

$$u = (\bar{x} - \bar{y})/D,$$

where

$$D = D(\bar{x} - \bar{y}) = \sigma\sqrt{\frac{1}{n} + \frac{1}{n}} = \sigma\sqrt{\frac{2}{n}},$$

and reject H_0 if $|u| \geq \lambda_{\alpha/2}$.

Now assume that H_1 is true. Let us distinguish between two situations:

(a) The Difference Δ is Large Compared to σ

We indicate this situation in Fig. 14.7(a). In this case, the numerator in u will probably be large compared to the denominator, and significance is obtained with high probability.

(b) The Difference Δ is Small Compared to σ

The situation is illustrated in Fig. 14.7(b). Now significance is generally not obtained if the sample size n is small. However, remarkably enough,

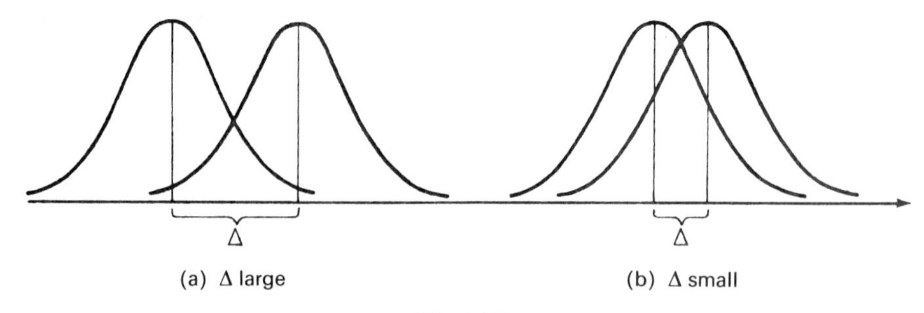

Fig. 14.7.

significance will practically always be obtained *as soon as n is large enough.* A strict proof can be given by means of the power function, but we provide only a vague explanation here: If n increases, the denominator D becomes smaller, but the numerator will stay near Δ. Hence the ratio will be large with high probability so that significance is obtained.

The argument shows that for large n the test of significance acts in the same way in the two cases (a) and (b). In both cases, significance is obtained if only n is large enough, and hence *even if the difference* Δ *is quite small.* This implies that for large n a test of significance is of practical value only if the researcher wishes to discover a difference *regardless of whether it is large or small.* Such a situation is rather uncommon, at least in industrial applications of statistics; however, it occurs in more basic research.

An interval estimate is not subject to the same disadvantage. Confidence intervals provide all the information furnished by a test of significance, and, in addition, other information. The interval covers the unknown parameter with certain probability. A test of significance tells us only whether a certain null hypothesis should be rejected or not.

14.9. Repeated Tests of Significance

One sometimes wants to perform whole sequences of tests of significance. Certain complications then arise, as we shall now see.

First, assume that we have several independent samples and that one null hypothesis (not necessarily the same) is tested on each sample, always at the same level of significance α. Let us imagine that all null hypotheses are true. The probability that a certain test will lead to a rejection of the corresponding null hypothesis is as usual equal to α. The probability that at least one of the tests will result in significance is equal to $1 - (1 - \alpha)^r$, where r is the number of tests. This probability is always greater than α, and, if r is large, considerably greater. For example, if $\alpha = 0.01$ and $r = 20$, it is about 0.18. This implies a fairly large risk that at least one of the null hypotheses is rejected, although all are true.

The argument shows that the total risk of reaching a wrong conclusion can be considerably larger than α, if several tests are performed. In order to compensate for this, we can choose a very small value of α. However, it is generally better to add a remark to the presented results mentioning the increased risk of erroneous rejection.

Second, assume that several different hypotheses are tested on the same data. Similar difficulties arise as before. In this situation it is, in general, complicated to ascertain the total risk of error, since the tests are not independent.

Let us point out that a null hypothesis and the alternative hypothesis should both be formulated *before* the samples are collected. Hypotheses put forward because of some finding suggested by the data cannot be tested by the methods discussed in this chapter.

Although it is outside the scope of the present chapter, we want to point out, in order to be fair, that interval estimation leads to similar problems. If we have constructed a confidence interval for a parameter θ based on certain data and repeat the construction using new data, then the probability that *both* intervals cover the parameter is equal to $(1 - \alpha)^2$; this is less than the confidence level $1 - \alpha$ for each of them. Hence the combined confidence level depends on the number of intervals. Similar difficulties occur if the same data are used for constructing confidence intervals for several unknown parameters. In addition, the endpoints of the intervals can then be dependent, which complicates the determination of the combined confidence level.

Exercises

The first three exercises are related to Example 1, p. 253.

1401. By performing 15 trials, A wants to investigate whether B has ESP. He decides to believe that B has ESP if B responds correctly in at least 11 trials. Assume that B guesses.
(a) Find the distribution of X = number of correct answers.
(b) What is the probability that A will believe that B has ESP? (§14.2)

1402. Find the smallest number of trials required for testing the hypothesis H_0: B guesses, with a probability of error that is at most 10^{-6}. In the test, it is required that all answers should be correct. (§14.2)

1403. Person C has a certain degree of ESP and gives a correct response with probability 0.8. A requires all answers in 20 throws to be correct, in order to be convinced. What is the probability that C will convince A about his ESP? (§14.2)

1404. The lifetime (unit: hours) of a certain type of light bulb has an exponential distribution with expectation θ, that is,

$$f_X(x) = (1/\theta)e^{-x/\theta} \qquad \text{for} \quad x \geq 0.$$

The manufacturer states that $\theta = 1{,}000$. A person doubts that θ is so large. He plans to test the null hypothesis H_0: $\theta = 1{,}000$ by buying a new bulb and determining its lifetime x_1. If x_1 is small, say $x_1 < a$, he will reject H_0. In order to find a he considers the equation $P(X < a \text{ if } \theta = 1{,}000) = \alpha$, where α is the level of significance.
(a) Determine a as a function of α.
(b) Suppose that $x_1 = 75$. Is this result significant at the 5% level?
(c) Answer the same question for $x_1 = 50$. (§14.3)

1405. (Continuation of Exercise 1404.) Assume that $x_1 = 45$. Test H_0 by means of the P method. (§14.3)

1406. A one-armed bandit gives a win with unknown probability p. The number of rounds required up to and including the first win then has probability function

$$p_X(k) = p(1-p)^{k-1} \qquad \text{for} \quad k = 1, 2, 3, \ldots.$$

It is said that $p = 0.2$ but a person A doubts that p is so large. He wishes to test the hypothesis H_0: $p = 0.2$ against H_1: $p < 0.2$. Suppose that he loses ten times and then wins. Can A then reject H_0 at level of significance 0.10 or less? Use the P method. (§14.3)

1407. If a person in Sweden is suspected of drunken driving, three determinations of the concentration of alcohol in his blood are made. The results x_1, x_2 and x_3 are assumed to be a random sample from $N(m, 0.05^2)$, where m is the true concentration (unit: per mille). If $m > 0.5$ the person has transgressed the law. Suppose that the court declares the person guilty if

$$\bar{x} > 0.5 + \lambda_{0.01} \cdot 0.05/\sqrt{3}$$

but not otherwise. In statistical parlance, the court tests H_0: $m = 0.5$ against H_1: $m > 0.5$ at level of significance 0.01. Which of the following statements give a correct description of what will happen in the long run?
1: at most 1% of all those acquitted are guilty;
2: at most 1% of all those innocent are declared guilty;
3: at most 1% of all those guilty are acquitted;
4: at most 1% of all those declared guilty are innocent. (§14.3)

1408. A pharmaceutical firm sometimes adds a certain colour to a product. It is required to know how the colour affects the appearance of the product. From the factory, 10 packages are chosen at random, and the turbidity of the content is measured after some storing time. Results:

3.9 4.1 4.4 4.0 3.8 4.0 3.9 4.3 4.2 4.4.

Without addition of colour, the turbidity is on the average 4.0. Do the results indicate that the turbidity has increased? Model: The measurements are assumed to be a random sample from $N(m, 0.2^2)$. Test the hypothesis H_0: $m = 4.0$ against H_1: $m > 4.0$ with a test at level 0.05. (§14.5)

1409. (Continuation of Exercise 1408.) If m is the correct value, what is the distribution of the rv that the test quantity is an observation of? Determine the power function of the test, that is, determine $P(H_0$ is rejected if m is the true value). Also determine the power of the test for $m = 3.8$ and $m = 4.3$. (§14.5)

1410. Using the arithmetic mean $\bar{x}$ of n independent observations from $N(m, 2^2)$, the hypothesis H_0: $m = 1$ is tested against H_1: $m < 1$ at level of significance 0.05 by means of the test

$$\text{reject } H_0 \text{ if } \bar{x} < 1 - 2\lambda_{0.05}/\sqrt{n}.$$

Determine the smallest value of n such that, for $m = 0$, the power of the test is at least 0.99. (§14.5)

1411. To investigate the content of mercury in pike in a certain lake, ten pike were examined. Result (unit: mg/kg):

0.8 1.6 0.9 0.8 1.2 0.4 0.7 1.0 1.2 1.1.

Model: Mercury content in a pike has a $N(m, \sigma^2)$ distribution.

(a) Is it possible to reject H_0: $m = 0.9$ against H_1: $m > 0.9$ at level of significance 0.05?

(b) Is it possible to reject H_0: $m = 1.1$ against H_1: $m < 1.1$ at level of significance 0.05? (§14.5)

1412. In a laboratory, 18 determinations were made of the gravitational constant g. The arithmetic mean was $\bar{x} = 972$ and the standard deviation $s = 6.0$. Model: $N(g, \sigma^2)$, where the parameters are both unknown. Test at level of significance 0.05 the hypothesis H_0: $g = 981$ against H_1: $g \neq 981$. (§14.5)

1413. Eight persons measure their own height in the morning and at night (unit: cm). Results:

person	1	2	3	4	5	6	7	8
morning	172	168	180	181	160	163	165	177
night	172	167	177	179	159	161	166	175

The differences between the morning and the night values are assumed to be a random sample from $N(m, \sigma^2)$. Test the hypothesis H_0: $m = 0$ against H_1: $m \neq 0$ at level of significance 0.05. (§14.5)

1414. The weights of certain tablets vary as a rv with mean m and standard deviation $\sigma = 0.02$. In order to check the weight, 35 tablets are weighed. As a point estimate of m the value $\bar{x} = 0.69$ is taken. Test the hypothesis H_0: $m = 0.65$ against H_1: $m \neq 0.65$ by a test with approximate level of significance 0.05. The weights do not necessarily have a normal distribution. (§14.6)

1415. (Continuation of Exercise 1414.) Suppose that σ is unknown and that $s = 0.018$ is taken as an estimate of σ.

(a) Find the standard error of $\bar{x}$.

(b) Perform a test similar to that in Exercise 1414. (§14.6)

1416. A die is suspected of producing too few sixes. It is thrown 12 times, and no six appears. Does this indicate that the suspicion is well-founded? (§14.7)

1417. A die is suspected of producing too few sixes. It is thrown 120 times, and a six appears 8 times. Does this indicate that the suspicion is well-founded? (§14.7)

1418. A rv X is Bin(n, p), where p is unknown. Let x be an observation on X. An amateur statistician wishes to test H_0: $p = 0.5$ against H_1: $p \neq 0.5$, using the estimate $p^* = x/n$. He proposes the following test: Reject H_0 if $|p^* - 0.5| > 0.1$. Find the level of significance of this test if:

(a) $n = 10$;

(b) $n = 100$. (§14.7)

1419. A person asserting that he could find water with a divining rod was tested in the following way: He was led into a big room with 10 containers, well apart from each other, and was told that five were filled with water and five were empty. He identified four of the containers with water correctly and one erroneously. Test the hypothesis that he guessed. Find the P value. (§14.7)

1420. The rv X is Po(m). By means of 50 independent observations on X one wants to test the hypothesis H_0: $m = 0.2$ against the hypothesis H_1: $m > 0.2$. The sum of the observations is 19. Should H_0 be rejected? Use the P method.
Hint: The sum of n observations from Po(m) is distributed as Po(nm). (§14.7)

1421. The rv X assumes the values 0, 1, 2, 3. The following data consist of 4,096 independent observations on X:

Observation	0	1	2	3
Frequency	1,764	1,692	552	88

Test the hypothesis that $X \sim \text{Bin}(3, 1/4)$ at level of significance 1%. (§14.7)

1422. From each of three large populations P_1, P_2 and P_3 a random sample was taken and classified according to sex. Result

	male	female
P_1	46	54
P_2	78	72
P_3	143	107

Test at the 5% level the hypothesis that the sex distribution is the same in the three populations. (§14.7)

*1423. We have two random samples, each of size ten, from $N(m_1, 0.3^2)$ and $N(m_2, 0.4^2)$, where m_1 and m_2 are unknown parameters. We want to test the hypothesis $m_1 = m_2$ with a suitable two-sided test at level of significance 0.01.
(a) Find the power of the test when $m_1 - m_2 = 0.6$.
(b) Increase each sample by the same number of values so that the power becomes 0.99 for $m_1 - m_2 = 0.6$. How many more values are required approximately?

*1424. A scientist plans to collect a random sample of n values from Po(m), where m is unknown. He then wants to test the null hypothesis H_0: $m = 4$ against the alternative H_1: $m = 5$ in such a way that the probabilities of error of the first and second kind are 0.001 and 0.01, respectively. How shall n be chosen and how shall the test be performed?

*1425. Let

2.75 2.48 2.16 2.13 1.59 2.61 2.34

be a random sample from $N(m_1, \sigma^2)$ and 2.95 a single observation from $N(m_2, 4\sigma^2)$. The parameters m_1, m_2 and σ are unknown. Test the hypothesis H_0: $m_1 = m_2$ against the alternative hypothesis H_1: $m_1 < m_2$.

*1426. You want to collect a random sample from $N(m, \sigma^2)$, where the parameters are unknown. Your purpose is to test H_0: $\sigma = 0.1$ against H_1: $\sigma = 0.2$. The probabilities of errors of the first and second kind should both be at most 0.05. What is the smallest possible sample size?

CHAPTER 15

Linear Regression

15.1. Introduction

In many areas of the natural and social sciences the relations between two or more variables are studied. What is the relation between the number of children in a family and the need for housing? Between smoking and longevity? Between intelligence quotient and income? Such questions can be both important and difficult to answer. We cannot discuss here the variety of problems arising in this area; only one model will be considered. We call it a model for simple linear regression.

In §15.2 the model is presented, in §15.3 we derive point estimates and in §15.4 the corresponding interval estimates. In §15.5 we discuss a special case of the model. §15.6 contains various remarks.

15.2. A Model for Simple Linear Regression

Example 1. Calibration of a Balance

Known weights $x_1, x_2, \ldots, x_n$ are placed on a balance, one at a time. On the scale one reads corresponding values $y_1, y_2, \ldots, y_n$. The scale is constructed in such a way that, if we disregard the random errors for a moment, each pair (x_i, y_i) satisfies the linear relation ("calibration curve")

$$y = \alpha' + \beta x.$$

The equation shows the relation between weight and scale reading, apart from weighing errors. It can, in principle, be used for calibration of the balance: For any y read on the scale it is possible to compute the weight x; see the dashed curve in Fig. 15.1.

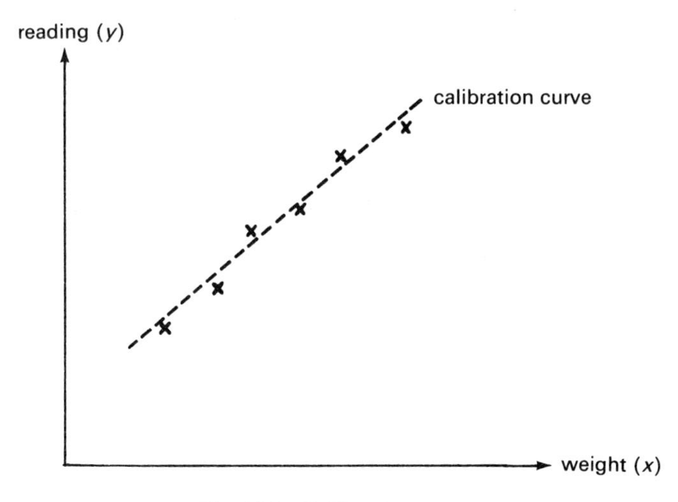

Fig. 15.1. Calibration curve.

A model that includes random errors is

$$y_i = \alpha' + \beta x_i + e_i,$$

where $e_1, \ldots, e_n$ denote these errors; they measure the vertical deviations of the points from the calibration curve. □

This is an example of *simple linear regression*. A model for this situation is the following: We have n pairs of values

$$(x_1, y_1), \ldots, (x_n, y_n),$$

where $x_1, \ldots, x_n$ are given quantities and $y_1, \ldots, y_n$ observations on independent rv's $Y_1, \ldots, Y_n$, where $Y_i \sim N(m_i, \sigma^2)$. Each expectation m_i depends linearly on x_i, that is

$$m_i = \alpha' + \beta x_i \qquad (i = 1, \ldots, n).$$

The line

$$y = \alpha' + \beta x \tag{1}$$

is called the *theoretical regression line*. It shows how the expectation depends on the *regression variable* x. Note that x is a mathematical variable. The coefficient β is called the *regression coefficient*. This coefficient shows how much the expectation changes when x is increased by one unit. If β should be 0, the expectation is constant; that is, it does not depend on x. By means of the regression line we can, for each given value x, determine the corresponding expectation.

The model is illustrated in Fig. 15.2. The theoretical regression line is dashed in the figure. Each pair of values has been plotted: x_i is the abscissa and y_i the ordinate. The deviations in a vertical direction from the points to

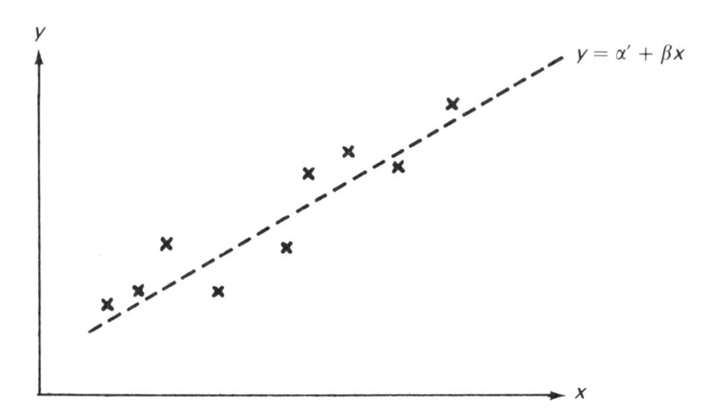

Fig. 15.2. Simple linear regression.

the line, that is, $e_i = y_i - m_i = y_i - \alpha' - \beta x_i$, are observations from $N(0, \sigma^2)$. The smaller σ, the smaller do these distances tend to be; that is, the better do the points conform to the line.

15.3. Point Estimates

In practice, the theoretical regression line is usually unknown, that is, the parameters α' and β are unknown. One of the purposes of regression analysis is then to estimate the equation of the line. It is, of course, possible to do this simply by drawing a straight line through the plotted points, as well as one can. If σ is small, the result may be quite good. However, it is more satisfactory to use an objective numerical method. We shall use the *LS* method for determining point estimates of α' and β.

It is convenient to rewrite equation (1):

$$y = \alpha + \beta(x - \bar{x}) \qquad (\bar{x} = \sum x_i/n). \tag{1'}$$

This means that we replace α' by $\alpha - \beta\bar{x}$; hence α and β are now the unknown parameters. According to the *LS* method we estimate α and β by seeking the minimum of

$$Q(\alpha, \beta) = \sum_{i=1}^{n} (y_i - m_i)^2,$$

where

$$m_i = \alpha + \beta(x_i - \bar{x}).$$

The quantity Q is the sum of the squares of the vertical distances e_i from the points to the line. We find

$$\frac{\partial Q}{\partial \alpha} = -2\sum_{i=1}^{n} (y_i - m_i); \qquad \frac{\partial Q}{\partial \beta} = -2\sum_{i=1}^{n} (x_i - \bar{x})(y_i - m_i).$$

If we set the derivatives equal to zero and solve the resulting system of equations, we find after some simplification the *LS* estimates

$$\alpha^* = \bar{y}; \qquad \beta^* = \frac{\sum (x_i - \bar{x})(y_i - \bar{y})}{\sum (x_i - \bar{x})^2}. \tag{2}$$

We now understand why writing the model in the form (1′) is advantageous: The estimate of α has a simple and well-known expression. Note that both estimates are linear in the observations y_i. In fact, we have

$$\alpha^* = \frac{1}{n} \sum y_i; \qquad \beta^* = \sum c_i y_i,$$

where

$$c_i = (x_i - \bar{x}) / \sum (x_j - \bar{x})^2. \tag{3}$$

We have here used the fact that $\sum (x_i - \bar{x}) = 0$.

By inserting α^* and β^* into (1′) we obtain the *estimated regression line*

$$y = \alpha^* + \beta^*(x - \bar{x}).$$

For any given value $x = x_0$ we can use this line for estimating the corresponding expectation $m_0 = \alpha + \beta(x_0 - \bar{x})$. We call this estimate m_0^*, and so

$$m_0^* = \alpha^* + \beta^*(x_0 - \bar{x}).$$

This procedure is termed *prediction of a point on the theoretical regression line* or, more briefly, *prediction of expectation* (see Fig. 15.3).

By means of probability theory we shall now examine the estimates α^*, β^* and m_0^*. According to known theorems for expectations and variances we find

$$\begin{aligned} E(\alpha^*) &= \sum E(Y_i)/n = \sum m_i/n = \sum [\alpha + \beta(x_i - \bar{x})]/n = \alpha, \\ V(\alpha^*) &= \sum V(Y_i)/n^2 = n\sigma^2/n^2 = \sigma^2/n. \end{aligned} \tag{4}$$

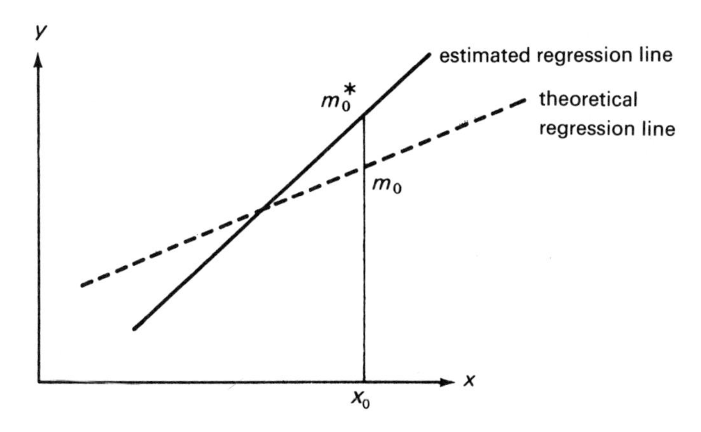

Fig. 15.3. Prediction of expectation.

Hence the estimate α^* is unbiased and its variance is small when n is large. By using (3) we further find

$$E(\beta^*) = E(\sum c_i Y_i) = \sum c_i m_i,$$
$$V(\beta^*) = V(\sum c_i Y_i) = \sum c_i^2 V(Y_i) = \sum c_i^2 \sigma^2.$$

Inserting here the expressions for c_i and m_i, we find after reduction

$$\begin{aligned} E(\beta^*) &= \beta, \\ V(\beta^*) &= \sigma^2 / \sum (x_i - \bar{x})^2. \end{aligned} \tag{5}$$

Hence also β^* is unbiased, and its variance is small when the sum of squares $\sum (x_i - \bar{x})^2$ is large.

This proves that the *LS* estimates of α and β have good properties. It is possible to show one more important property (the proof of which is outside the scope of this book): The estimates are more efficient than all other unbiased estimates of α and β.

The readers would perhaps like to know how the *ML* estimates look. The answer is very brief: They coincide with the *LS* estimates. (To show this, is a good but rather lengthy exercise.)

Let us add: It is possible to show that α^* and β^* are independent. No proof of this is given.

We now study the properties of the predicted value m_0^*. By using earlier expressions together with the property of independence just mentioned we find

$$\begin{aligned} E(m_0^*) &= E[\alpha^* + \beta^*(x_0 - \bar{x})] = \alpha + \beta(x_0 - \bar{x}) = m_0, \\ V(m_0^*) &= V[\alpha^* + \beta^*(x_0 - \bar{x})] = V(\alpha^*) + V(\beta^*)(x_0 - \bar{x})^2 \\ &= \sigma^2 \left[\frac{1}{n} + \frac{(x_0 - \bar{x})^2}{\sum (x_i - \bar{x})^2} \right]. \end{aligned} \tag{6}$$

We see that m_0^* is unbiased and that its variance depends on the value x_0 whose m_0-value we want to estimate: The more x_0 deviates from $\bar{x}$, the larger is the variance of m_0^*. Hence it is "more difficult" to accurately predict m_0 when x_0 lies far away from $\bar{x}$ than when it lies near $\bar{x}$.

Remark 1. Minimum Value of $Q(\alpha, \beta)$

The minimum value of $Q(\alpha, \beta)$ is

$$Q_0 = \sum [y_i - \alpha^* - \beta^*(x_i - \bar{x})]^2. \tag{7}$$

This so-called *residual sum of squares* is the sum of squares of the vertical distances from the plotted points to the estimated regression line.

The conscientious reader may have noted that we have not proved that Q_0 in (7) is really the least possible value of $Q(\alpha, \beta)$. This follows easily from the relation

$$Q(\alpha, \beta) = n(\alpha^* - \alpha)^2 + (\beta^* - \beta)^2 \sum (x_i - \bar{x})^2 + Q_0.$$

The reader may derive this expression as an exercise. □

Remark 2. Estimate of σ

It is often desired to estimate the standard deviation σ (see the model at the beginning of §15.2). The ML estimate of the variance can be shown to be Q_0/n, where Q_0 is given by (7); however, this estimate is not unbiased. The corrected ML estimate is

$$s^2 = Q_0/(n - 2).$$

As an estimate of σ we take

$$s = \sqrt{Q_0/(n - 2)}.$$

□

Remark 3. Numerical Computations

When performing numerical calculations, the estimates in (2) and Q_0 in (7) are usually rewritten as

$$\alpha^* = \sum x_i/n; \qquad \beta^* = S_{xy}/S_{xx}; \qquad Q_0 = S_{yy} - S_{xy}^2/S_{xx},$$

where

$$S_{xx} = \sum (x_i - \bar{x})^2 = \sum x_i^2 - \frac{1}{n}(\sum x_i)^2,$$

$$S_{xy} = \sum (x_i - \bar{x})(y_i - \bar{y}) = \sum x_i y_i - \frac{1}{n}(\sum x_i)(\sum y_i),$$

$$S_{yy} = \sum (y_i - \bar{y})^2 = \sum y_i^2 - \frac{1}{n}(\sum y_i)^2.$$

□

Example 2. Bilirubin and Protein

In an investigation of $n = 20$ newborn children suffering from a certain disease a doctor determined the amount of bilirubin (x) and the protein concentration (y) in the spinal fluid. Result (in appropriate units)

x	0.14	0.08	0.07	0.26	0.08	0.02	0.03	0.22	0.06	0.23
y	83	65	71	140	135	30	30	128	80	168

x	0.29	0.04	0.13	0.14	0.07	0.05	0.13	0.06	0.05	0.08
y	139	88	121	125	56	98	101	96	73	116

By means of simple linear regression we shall derive an equation which, for a given amount of bilirubin x, predicts the expected value of the protein concentration y. We obtain (see Remark 3)

$$\sum x_i = 2.23; \qquad \sum x_i^2 = 0.3701; \qquad S_{xx} = 0.3701 - \frac{1}{20}2.23^2 = 0.12146,$$

$$\sum y_i = 1{,}943; \qquad \sum y_i^2 = 215{,}061; \qquad S_{yy} = 215{,}061 - \frac{1}{20}1{,}943^2 = 26{,}299,$$

$$\sum x_i y_i = 259.79; \qquad S_{xy} = 259.79 - \frac{1}{20}2.23 \cdot 1{,}943 = 43.1455,$$

$$\alpha^* = 1{,}943/20 = 97.15; \qquad \beta^* = 43.1455/0.12146 = 355.2,$$

$$\bar{x} = 2.23/20 = 0.1115.$$

Estimated regression line:

$$y = 97.15 + 355.2(x - 0.1115) = 57.5 + 355.2x.$$

For a given value $x = x_0$ of the amount of bilirubin it is now possible to predict the expected value $m = m_0$ of the protein concentration of the child. For example, when $x_0 = 0.20$ we find $m_0^* = 129$. □

15.4. Interval Estimates

We shall derive interval estimates of α and β. As mentioned earlier, the point estimates (2) are linear functions of the observations y_i. It follows from Theorem 3′ in §8.5 that, since the y_i's are observations on normally distributed rv's, so are the point estimates. Means and variances are given in (4) and (5) and the distributions are therefore completely specified.

It is now evident how to proceed if *the standard deviation σ is known*. By the usual argument we find that

$$\begin{aligned} I_\alpha &= (\alpha^* - \lambda_{p/2} D, \alpha^* + \lambda_{p/2} D) \qquad (D = \sigma/\sqrt{n}), \\ I_\beta &= (\beta^* - \lambda_{p/2} D, \beta^* + \lambda_{p/2} D) \qquad (D = \sigma/\sqrt{\textstyle\sum (x_i - \bar{x})^2}), \end{aligned} \tag{8}$$

is a confidence interval for α and β, respectively, with confidence level $1 - p$. (We use p in order to avoid a clash of symbols.)

We shall also briefly and without proof describe the corresponding procedure when *the standard deviation σ is unknown*. It resembles very much the procedure used in the case of one sample from $N(m, \sigma^2)$.

As an estimate of σ we take (see Remark 2, p. 285)

$$s = \sqrt{Q_0/(n-2)},$$

where Q_0 is given by (7).

It can be proved that

$$\begin{aligned} I_\alpha &= (\alpha^* - t_{p/2}(f)d, \alpha^* + t_{p/2}(f)d) \qquad (d = s/\sqrt{n}, f = n - 2), \\ I_\beta &= (\beta^* - t_{p/2}(f)d, \beta^* + t_{p/2}(f)d) \qquad (d = s/\sqrt{\textstyle\sum (x_i - \bar{x})^2}, f = n - 2), \end{aligned} \tag{9}$$

are confidence intervals for α and β, respectively, with confidence level $1 - p$. The result looks familiar: One adds and subtracts the standard error, multiplied by a suitably chosen constant.

A confidence interval for a predicted expectation is constructed analogously (see the following example).

Example 2. Bilirubin and Protein (continued)

Let us construct a 95% confidence interval for the coefficients in the estimated regression line assuming that σ is unknown. We then compute, in addition to what was found earlier (see again Remark 3 at the end of §15.3)

$$Q_0 = 26{,}299 - \frac{43.1455^2}{0.12146} = 10{,}973,$$

$$s = \sqrt{10{,}973/(20 - 2)} = 24.7.$$

Since $t_{0.025}(18) = 2.10$, we obtain the following result:

$$\text{Parameter } \alpha\text{:} \quad d = 24.7/\sqrt{20} = 5.52,$$

$$I_\alpha = (97.15 \pm 2.10 \cdot 5.52) = (86, 109).$$

$$\text{Parameter } \beta\text{:} \quad d = 24.7/\sqrt{0.12146} = 70.9,$$

$$I_\beta = (355.2 \pm 2.10 \cdot 70.9) = (210, 500).$$

We shall also construct a confidence interval for the expectation $m_0 = \alpha + \beta(0.20 - \bar{x})$ corresponding to $x_0 = 0.20$. We have already found the estimate $m_0^* = 129$. We find (see the second formula (6))

$$d = 24.7\sqrt{\frac{1}{20} + \frac{(0.20 - 0.1115)^2}{0.12146}} = 8.38,$$

and hence

$$I_{m_0} = (129 \pm 2.10 \cdot 8.38) = (110, 150).$$

The intervals are large, due to the high variability of the biological data. □

15.5. Several y-Values for Each x-Value

The model for simple linear regression can, of course, be used even if, among the pairs $(x_1, y_1), \ldots, (x_n, y_n)$, there are several equal x-values. This situation is common in planned experiments. We shall give an example.

Example 3. Injection of Androsterone in Capons

This example has been taken (with permission) from C.W. Emmens, *Principles of Biological Assay* (Chapman and Hall, London, 1948).

Twenty-five capons were divided into 5 groups of 5 birds in each, and each group was injected with a different dose of androsterone. The standard deviation is assumed

Table 15.1. Increase in comb size of 25 capons.

Dose (x)				
1	2	3	4	5
8	5	13	17	17
1	6	7	14	17
1	9	12	14	20
3	7	10	19	18
1	4	11	13	15

to be $\sigma = 2.5$. The result is given in Table 15.1 (doses in suitable unit; y is the increase in size of the comb expressed as mm increases in length plus height).

We arrange the data as follows:

x_i	1	1	1	1	1	2	2	2	2	2	...	5	5	5	5	5
y_i	8	1	1	3	1	5	6	9	7	4	...	17	17	20	18	15.

We can now immediately use the usual formulae, taking $n = 25$:

$$\sum x_i = 5 \cdot 1 + 5 \cdot 2 + 5 \cdot 3 + 5 \cdot 4 + 5 \cdot 5 = 75; \qquad \bar{x} = 75/25 = 3.0,$$

$$\sum x_i^2 = 5 \cdot 1^2 + 5 \cdot 2^2 + 5 \cdot 3^2 + 5 \cdot 4^2 + 5 \cdot 5^2 = 275,$$

$$S_{xx} = \sum x_i^2 - \frac{1}{n}(\sum x_i)^2 = 275 - \frac{1}{25} \cdot 75^2 = 50.0,$$

$$\sum y_i = 8 + 1 + 1 + \cdots + 18 + 15 = 262,$$

$$\sum x_i y_i = 1 \cdot 8 + 1 \cdot 1 + 1 \cdot 1 + \cdots + 5 \cdot 18 + 5 \cdot 15 = 978,$$

$$S_{xy} = \sum x_i y_i - \frac{1}{n}(\sum x_i)(\sum y_i) = 978 - \frac{1}{25} \cdot 75 \cdot 262 = 192.0.$$

We find

$$\alpha^* = 262/25 = 10.48; \qquad \beta^* = 192.0/50.0 = 3.84.$$

The estimated regression line is

$$y = 10.48 + 3.84(x - 3.0)$$

or, in reduced form,

$$y = -1.04 + 3.84x.$$

A 95% confidence interval for β is, according to the second formula of (9)

$$I_\beta = (3.84 \pm 1.96 \cdot 2.5 \cdot /\sqrt{50.0}) = (3.1, 4.5).$$

Suppose that we want to estimate the mean increase in comb size for a bird which is given the dose $x_0 = 6$. Assuming that the regression continues to be linear in this part of the dose interval, we get

$$m_0^* = 10.48 + 3.84(6 - 3.0) = 22.00.$$

A 95% confidence interval is given by

$$I_{m_0} = \left(22.00 \pm 1.96 \cdot 2.5\sqrt{\frac{1}{25} + \frac{(6-3.0)^2}{50.0}}\right) = (19.7, 24.3).$$ □

15.6. Various Remarks

(a) The Danger of Extrapolation

Two variables are generally linearly related only within a certain limited region (see Fig. 15.4). It is then dangerous to extrapolate outside the region (the dashed line indicates such an erroneous extrapolation).

(b) Nonlinear Regression

If a linear model is not applicable, the analysis given in this chapter cannot be used. Sometimes it is possible to change the scale along the x-axis, and in this way transform a nonlinear relation into a linear one.

Suppose, for example, that, apart from random variation,

$$y = cx^{\beta}.$$

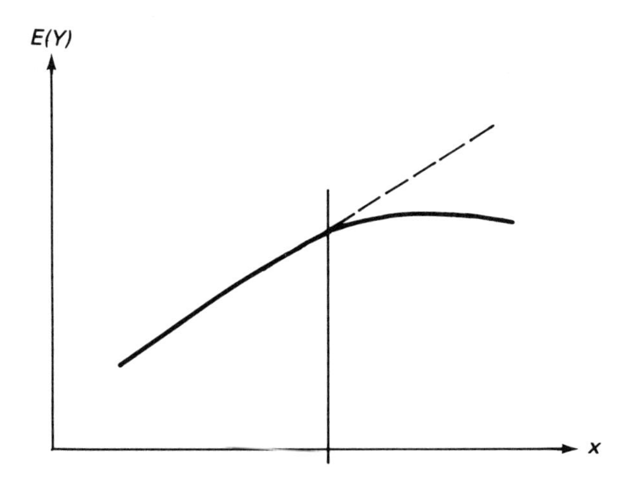

Fig. 15.4. The danger of extrapolation.

On taking logarithms we obtain the linear relation

$$\ln y = \ln c + \beta \ln x$$

between $\ln y$ and $\ln x$. We may then use the earlier theory provided that the deviations after taking logarithms are normally distributed with constant standard deviation.

(c) Nonconstant Standard Deviation

It is rather common that the standard deviation measuring the variation about the theoretical regression line is not constant; for instance, it may increase with x. Such nonconstancy often becomes apparent from visual inspection of the usual plot diagram. In such situations special methods have to be resorted to, for example, weighted regression analysis.

(d) Time Series

When the regression variable x is a time variable, it is common to study the dependence of Y on x by time series analysis (see the list of references at the end of the book).

(e) Statistical Versus Causal Relations

We emphasized earlier that if two variables depend on each other in a statistical sense, this does not necessarily mean that there is a causal relation between them. If we find that Y, on the average, increases with x, it cannot be logically inferred that x causes the increase in Y; a third variable may affect both x and Y. (Example: let x be the length of a child's foot and Y the child's ability to read. Quite certainly, it will be found that, on the average, Y increases with x, for older children will, as a rule, read better than younger ones!)

(f) Multiple Regression

A variable Y often depends on several regression variables. For example, the strength of steel may depend on several ingredients used in the manufacture of the steel. In such situations we may use *multiple regression analysis*, which is a very powerful technique for many practical problems (see the list of references).

Exercises

1501. In a solution of a certain substance the light extinction y is a linear function of the concentration x:

$$y = \alpha' + \beta x.$$

A technician has seven solutions with known concentrations $x_1, \ldots, x_7$ and in each case determines the light extinction. Each of the measurements $y_1, \ldots, y_7$ is subject to an error that is an observation from $N(0, \sigma^2)$. Results:

x_i	0.40	0.70	1.00	1.20	1.40	1.70	2.00
y_i	0.23	0.34	0.42	0.55	0.61	0.77	0.84

(a) How much does the light extinction increase (according to the model) when the concentration is increased by one unit? Estimate this quantity.
(b) Estimate the light extinction when the concentration is 1.20 and also find the standard deviation of the estimate expressed in σ.
(c) The same problem when the concentration is 0.
(d) Estimate σ. (§15.3)

1502. From the numerical data $(x_1, y_1), \ldots, (x_{10}, y_{10})$ the following has been computed:

$$\sum x_i = 12.0; \quad \sum x_i^2 = 18.40; \quad \sum y_i = 15.0; \quad \sum y_i^2 = 27.86; \quad \sum x_i y_i = 20.40.$$

Model: $y_i = \alpha' + \beta x_i + e_i$, where $e_1, \ldots, e_{10}$ are independent observations from $N(0, \sigma^2)$. Construct a 95% confidence interval for
(a) β,
(b) α'. (§15.4)

1503. Consider the following statistical data:

x_i	1	2	3	4	5	6	7	8
y_i	1.5	2.3	1.7	2.0	2.5	1.9	2.2	2.4

Suppose that y_i is an observation from $N(\alpha + \beta(x_i - \bar{x}), \sigma^2)$, $i = 1, 2, \ldots, 8$. Test the null hypothesis that $\beta = 0$ by a two-sided test at the level of significance 5%. (§15.4)

1504. The anode current I_a (unit: mA) was determined for a triode for some grid voltages V_g (unit: volt); the anode voltage was in each case 100. Two independent measurements were made for each grid voltage. Result:

V_g	-4.00	-3.50	-3.00	-2.50	-2.00
I_a	1.0	1.5	3.0	3.9	4.9
	1.3	1.7	2.7	3.4	5.9

Linear regression is assumed with normally distributed errors of measurements. Construct a 95% confidence interval for the transconductance of the tube (the slope of the regression line). (§15.4)

1505. Two random samples were collected, one, $y_1, \ldots, y_{20}$, from $N(m_i, 0.04)$, where $m_i = \alpha + \beta(x_i - \bar{x})$, and one, $y'_1, \ldots, y'_{30}$, from $N(m'_i, 0.09)$, where $m'_i = \alpha' + \beta'(x'_i - \bar{x}')$. The statistical analysis gave the result:

$$\sum_1^{20} (x_i - \bar{x})^2 = 3.240; \qquad \sum_1^{20} (x_i - \bar{x})(y_i - \bar{y}) = 1.381,$$

$$\sum_1^{30} (x'_i - \bar{x}')^2 = 4.012; \qquad \sum_1^{30} (x'_i - \bar{x}')(y'_i - \bar{y}') = 3.257.$$

Test the hypothesis $\beta = \beta'$ by a two-sided test of significance. (§15.4).

*1506. When a certain type of particle moves in an accelerator, it covers a distance $vt + at^2/2$ in time t, where v is the initial velocity and a an unknown quantity characteristic of the type of particle and accelerator. On one occasion, five measurements were made of the distance for particles with initial velocity 0. Result:

i	1	2	3	4	5
t_i	5.00	5.00	6.00	7.00	7.00
y_i	6.22	6.20	8.93	12.27	12.39

Further, a measurement was made of the distance for a particle with the unknown initial velocity v_0, which resulted in

$$(t_0, y_0) = (7.00, 14.46).$$

Model: $y_0, y_1, \ldots, y_5$ are independent observations on $Y_0, Y_1, \ldots, Y_5$ where

$$Y_0 \sim N(v_0 t_0 + a t_0^2/2, 0.070^2),$$

$$Y_i \sim N(a t_i^2/2, 0.070^2) \qquad \text{for} \quad i = 1, 2, \ldots, 5.$$

Find the ML estimate of v_0 and a 95% confidence interval for v_0.

CHAPTER 16

Planning Statistical Investigations

16.1. Introduction

We have now discussed probability theory and statistical theory long enough, and it is time to change pace. In this chapter we shall go out into the real world and treat problems arising when planning statistical investigations. After a general discussion in §16.2, we devote §16.3 to noncomparative investigations and §16.4 to comparative investigations. §16.5 contains some final remarks. In an Appendix, it is shown how to handle random numbers.

16.2. General Remarks About Planning

That the planning of an investigation is extremely important needs, in a way, no justification. It is one of the most painful experiences of the professional statistician, on receiving data from a client, to be forced to say that, because of bad planning, it is impossible do draw any sensible conclusions from the data, and that it is therefore necessary to begin anew. Planning is very important because of an often neglected fact: When planning the investigation, the investigator can actively affect the choice of theoretical model for the data to be produced. If the planning is good, it is often possible to use a model leading to clear conclusions, while, if the planning is bad, no model at all can be constructed.

It should not be thought that size in itself determines whether an investigation is good or bad. Even quite large investigations can be worthless.

An important part of planning is to formulate a plan which is the best possible within the range of available resources. We will show that such problems can sometimes be solved by rather simple considerations. However, the planning of a large investigation is often very complicated, and to handle the tools of the trade one needs both proper statistical training and expert knowledge in the field studied. Planning activities should therefore often be organized by a team of specialists.

There is, of course, no general method of planning the immense number of investigations that are performed in practice. However, certain common features apply. In the following list, we give some steps in planning that are relevant in an investigation. The list is meant for general information only, and is not used in the sequel. (Always remember that the planning details should be *written down* and be agreed upon by all participants.)

Check list for planning statistical investigations:

Stage 1. Practical preliminaries

- Writing down information available to the investigation
- Allocation of resources (time, personnel, money, and so on)
- Formulation of practical problem
- Desired properties of the solution (population studied, precision wanted and so on)
- Kind of investigation

Stage 2. Theoretical issues

- Construction of model
- Formulation of theoretical problem
- Desired properties of the solution
- Construction of plan for the investigation
- Choice of statistical method

Stage 3. Final practical steps

- Detailed plan for data collection
- Plan for collection of data
- Plan for data processing
- Plan for statistical analysis
- Plan for presentation of results

The division into stages could be done in other ways. The stages are seldom performed in strict time order. Stage 1 and Stage 2 are often discussed at the same time. When working with Stage 3, the other stages may have to be elaborated, and so on.

16.3. Noncomparative Investigations

In a noncomparative investigation, information is sought about some unknown quantity, for example, a physical constant, the percentage of some substance in a chemical compound, consumer interest in a commodity, the precision in steering a spacecraft.

(a) A Single Sample

The investigator takes n elements from the population of interest and obtains the sample

$$\boxed{x_1}\ \boxed{x_2}\ \cdots\ \boxed{x_n}.$$

When the population is physical, the squares may symbolize the actual elements and the values x_i the corresponding observations. In the case of a physical population this type of sampling is often called *simple sampling.*

Example 1. Harvest Damage

To get information about the total area of harvest damaged in some way on N Swedish farms, one takes a sample of n farms and determines the damaged area for each of them. □

Example 2. Iron Ore

To get information about the percentage of iron in a quantity of ore, one takes n small amounts and determines the percentage in each. □

Example 3. Determination of Physical Constant

To estimate a physical constant, n measurements are performed. □

Example 4. Investigation of Precision of Measurement

To get information of the precision of a method, n measurements are performed. □

Example 5. Investigation of Disease

The unknown quantity is a certain measurable property of patients who suffer from a certain disease. The sample is a number of patients in a hospital with this ailment. □

In order to be able to construct a useful model for a noncomparative investigation, we have several requirements on the sampling procedure. Two of the most important requirements are the following: The sample must be random, and there must be no disturbing factors.

Let us begin with the *requirement of randomness.* All models studied in this book assume that the elements are taken from the population at random, with or without replacement. This procedure is called *simple random sampling.* In order to apply such models successfully, we must take suitable measures.

Let us first assume that the investigation concerns a physical population as in Examples 1 or 2. In the case of a finite population (which is most common), we use a table of *random numbers* or somc equivalent method (see Appendix to this chapter). Such a procedure assumes that the elements in the population can be numbered. (A written list of the members is ideal.) In industrial sampling, numbering is usually impractical or impossible, and other methods have to be devised. When material is in motion, for example, on a conveyor belt or in a pipe, sampling methods can often be devised by the use of scoops or ladles. It is much more difficult to take random samples from heterogeneous material arranged in heaps, such as stone or coal. The sampling can then easily become biased.

Biased sampling often arises through taking samples from the wrong population. For example, suppose that in some region we want a random sample of families with children. Suppose that we have a list of all children in the area, select a random sample of children from this list and note the corresponding families. This will not give us a random sample of families since families with many children have a greater chance of entering the sample than those with few children.

When studying physical populations, the term *representative selection* is often used, especially by statistical amateurs. This term is dangerous because it may convey the false idea that a sample, in order to be useful, must "represent" the population in the sense that it is a faithful miniature copy of it. If the population consists of human beings, one then means that the sample should have, approximately, the same proportions of males and females, of old and young, etc., as the whole population. We know that such requirements are not necessary. Simple random sampling presupposes only that the elements are selected at random. (However, see the discussion of stratification on p. 299 where extra conditions are used.)

In many investigations—the population may be physical or not—it is impossible to select the elements at random in the way discussed here. Then the question arises whether it is even possible to introduce a probability model. We here encounter one of the greatest difficulties in practical statistical work, and the decision demands much experience and good judgement.

Let us consider a typical and reasonably simple situation of this kind (see Examples 3 and 4). When determining, say, a physical constant, the population is fictitious—it consists of the infinite set of values obtained if the investigation is performed "forever". The values can be regarded as randomly drawn from this population provided that the experiment is *reproducible*. We mean by that, somewhat vaguely, that it can be repeated over and over again under similar conditions.

Scientifically, a new era began when research workers such as Newton, Galileo, and many others, started to perform reproducible experiments, which made it possible to test hypotheses objectively and to estimate unknown parameters—the reader is asked to excuse the unhistorical use of the terminology of modern statistics!

Let us now consider a more difficult situation (see Example 5). A doctor examines patients who suffer from a certain disease, perhaps a rare one. The data collected refer to all patients treated in a certain US hospital during a five-year period. It is doubtful whether this can be regarded as a random sample from a population. From which population? From the population of all persons in the USA having the disease during the period? Of course not, for those in hospital are, on the whole, more seriously ill than the others affected. From the population of all persons treated in a hospital? We must then assume that patients from different hospitals do not differ, which is uncertain. Such difficulties do not apply only to medical statistics. Similar situations occur in most areas where statistics is used. It is therefore very important to clarify in each particular case what is meant in practice by the statement that a set of data constitute a random sample from a given population.

The topic is extremely important, but we limit ourselves to this brief discussion, which stresses the importance of randomness and also illustrates the difficulties involved. If the requirement of randomness is not fulfilled, long experience and good judgement are needed in deciding whether the conclusions based on the theoretical model are valid at all in practice. To use statistical methods correctly is an art, not only a question of theoretical knowledge. If an investigation is badly planned or if the practical situation is such that the condition of randomness is not satisfied, application of statistical methodology may do more harm than good. The investigation may then get a quite undeserved air of respectability.

We shall now discuss the requirement that *disturbing factors do not appear* in the investigation. We shall show by means of some examples what is meant by this. We mentioned earlier that an experiment should be reproducible. This is, of course, a necessary but not a sufficient condition for the validity of an experiment. Also, systematic errors must be absent, for if they occur, the experiment can be misleading. If the instrument is poorly calibrated and gives, say, too high values, the results are incorrect even if the measurement is repeated many times. The erroneous calibration is then a disturbing factor.

Another common disturbing factor is caused by *missing values*; for example, they may occur when human populations are studied. In an opinion survey, the interviewer visiting the households reaches only those at home. Also, a selected person may refuse to participate. Similar difficulties arise when letters are sent to people, firms or organizations. In particular, sensitive questions can cause many missing replies. It is often difficult to know how serious the effect of such values is. Sometimes 5% missing values can be unimportant, sometimes they can spoil the investigation altogether. As an example of the latter, we can take post-examination of people who have undergone an operation for some disease. If the doctor then gets information only about the persons who have survived but not about those who have died, the investigation will certainly be spoiled!

Missing values may also be troublesome in industrial investigations. Let us take a situation that should never arise in practice. When sampling a heap of material of, say, coal or ore, the sampling device must be constructed in such a way that large lumps are not excluded from the sampling process.

Strictly speaking, the requirement that there be no disturbing factors is too severe. Indeed, it presupposes that an ideal situation can be attained that exists only in the model world. Naturally, it is not possible altogether to avoid systematic errors, missing values and so forth. It is more realistic to require that disturbing factors shall be eliminated to such an extent that they do not spoil the conclusions of the investigation. Even here, long experience and good judgement are required to attain a satisfactory result.

What has been said here is only part of the work of planning the investigation. The investigator also has to choose an appropriate theoretical model and to fix the sample size. Another book would be needed for a detailed discussion of these important questions. In fact, the procedure varies in practice, for it depends to a large extent on the prior information. Sometimes it is known beforehand how the model should be chosen, sometimes it is necessary to work in stages. For instance, it is very common that a small pilot investigation is first performed before the final one is planned (see the scheme on p. 294).

(b) Several Samples

Many noncomparative investigations result in samples from several populations, say from k populations:

Sample 1	x_{11}	x_{12}	...	x_{1n_1},
2	x_{21}	x_{22}	...	x_{2n_2},
...	...	...	...	...
k	x_{k1}	x_{k2}	...	x_{kn_k}.

As before, we want to use the data in order to get information on some unknown parameter, or several unknown parameters.

We illustrate this situation with modifications to Examples 1–4, p. 295.

Example 1. Harvest Damage

The farms are divided into k classes according to magnitude. From each class a sample of farms is taken. □

Example 2. Iron Ore

The iron ore is stored in K railway cars. Of these, k cars are chosen. From each, a number of samples are taken. □

Example 3. Determination of Physical Constant

Several measurements are performed at each of k laboratories. □

Example 4. Investigation of Precision of Measurement

During each of k days, a number of measurements are made. □

The planning is the same as before in principle, but there is a new stage: the total number of values should be allocated to the samples in an appropriate way.

We sometimes use a *stratified sample survey*. We mean by this an investigation of a population that has been divided into k smaller populations or *strata*.

Example 1 elucidates the idea behind stratification. Instead of selecting a single sample from the whole population of farms, we stratify the farms according to size and examine a number of farms in each class (= stratum). It may be advantageous to examine all farms above a certain size, for these farms weigh heavily in the determination of total harvest damage.

For example, assume that we want to select a total of 160 farms from 2,000. We may then choose the following numbers, where N_i = total number of farms in the ith stratum, n_i = sample size in the ith stratum (1 hectare = 2.5 acres):

Stratum	Size of farm in hectares	N_i	n_i
1	– 10	1,600	100
2	10– 50	300	25
3	50–100	90	25
4	100–	10	10
		2,000	160

We shall now give a more detailed explanation of why a stratified survey may be desirable. At the same time we will see how a model is constructed in a somewhat more complicated situation than usual.

(The model we shall give presupposes that the number of units taken from each stratum is very small compared to the total number of units in the stratum. Hence it cannot be applied to the example with the farms. For a model which can be used in that example, see the references at the end of the book.)

Let us assume that, in each stratum, there is an unknown parameter, say the parameter m_i in the ith stratum ($i = 1, \ldots, k$). Also suppose that the value x_{ij} in the sample $x_{i1}, \ldots, x_{in_i}$ from this stratum is an observation on a rv X_{ij} with mean m_i and standard deviation σ_i. However, these parameters are only of secondary interest, because we want information about the total population. In fact, we assume that the unknown quantity to be estimated is a weighted mean

$$m = \sum_1^k c_i m_i \qquad \left(\sum_1^k c_i = 1\right)$$

of the parameters m_i (with known weights c_i).

As an estimate *of* m we take the weighted mean

$$m^* = \sum_1^k c_i \bar{x}_i$$

of the sample means in the different samples. We shall examine this estimate.

Using easily understood notation and known theorems about means and variances, we find

$$E(m^*) = \sum c_i E(\bar{X}_i) = \sum c_i m_i = m,$$

$$V(m^*) = \sum c_i^2 V(\bar{X}_i) = \sum c_i^2 \sigma_i^2 / n_i.$$

The first relation states that m^* is unbiased. The second relation shows that the variance of m^* depends on the standard deviations *within* strata but not on differences between the m_i-values. Hence stratification is advantageous when the population of interest can be divided into homogeneous parts, that is, into strata within which the standard deviations are small.

We shall now briefly discuss the choice of the numbers n_i in a stratified survey. One possibility is to use *proportional allocation*: n observations in all are made, distributed among the strata in the same proportions as the weights c_i; that is, we take

$$n_i = n \cdot c_i.$$

It is better still to use *optimal allocation.* As before let n be fixed. We choose n_i such that the variance $V(m^*)$ becomes as small as possible. Mathematically, this means that we minimize

$$V(m^*) = \sum_1^k c_i^2 \sigma_i^2 / n_i$$

subject to the side condition $n_1 + \cdots + n_k = n$. The solution can be shown to be

$$n_i = h \cdot c_i \cdot \sigma_i,$$

where h is selected so that the sum is n. (The answer is then adjusted so that each n_i becomes an integer.) It is seen that strata with large standard deviation are allotted more values than those with small standard deviation, which is quite reasonable. A difficulty with this method is that the standard deviations are not always known beforehand. One then has to guess intelligently or, better, make a pilot survey.

Instead of giving n a fixed value, it is possible to start from a given total cost $\sum d_i n_i$, where d_i is the cost per item in the ith sample. The variance of m^* is then minimized (subject to the cost constraint) if the numbers n_i are proportional to $c_i \sigma_i / \sqrt{d_i}$.

There are many other possibilities for planning noncomparative investigations. When large-scale sampling surveys are planned, it is, of course, important to keep the costs down, and statisticians have devised many procedures adapted to different practical circumstances.

16.4. Comparative Investigations

In a *comparative statistical investigation* we want to use samples in order to compare two or more *treatments* A, B, C, ... We limit the discussion to two treatments. (The term treatment is sometimes somewhat artificial, but provides a convenient general terminology.)

Example 6. Medicines

In order to compare two medicines A and B, a doctor gives A to one group of patients, B to another group and compares the result. □

Example 7. Manufacturing Methods

An industry plans to replace the present manufacturing method A by a new one, B. For comparison, a number of products are manufactured according to each method, and the difference is studied. □

Example 8. High or Low Houses?

A sociologist wishes to compare how people like high-rise housing, A, and housing in smaller buildings, B. For this purpose, he wants to interview a number of families in each type of house and make a comparison of their opinions. □

Comparative investigations are of two kinds. In a *comparative experimental investigation* (more briefly: comparative experiment) the investigator himself decides by chance how to distribute the treatments to the elements. If this is not done, we have a *comparative nonexperimental investigation* (more briefly: comparative survey).

Examples 6 and 7 describe comparative experiments and Example 8 describes a comparative survey.

Ethical questions concerning experimental investigations in more sensitive situations have often been discussed. For example, can medicines be administered to cancer patients in this way?

A good comparative experiment admits more reliable conclusions than a nonexperimental comparative investigation. Since the investigator himself allots the treatments, disturbing factors can be eliminated so that a correct comparison can be made.

The interpretation of data from a comparative nonexperimental investigation is generally more difficult, since the comparison can be spoiled by disturbing factors. Nevertheless, such investigations should not be condemned as useless. In much scientific activity, especially in the social sciences but also in natural sciences such as astronomy, geophysics and geology, experiments cannot be performed, and inferences must be based on nonexperimental data.

In the rest of this chapter we discuss only planning of experimental investigations.

(a) Completely Randomized Experiment

Let us consider a *completely randomized experiment*. It is obtained by allotting the treatments at random to the elements. For example, suppose that equally many elements, say n, are to receive treatment A and treatment B. It is then decided by the use of random numbers how the treatments should be allocated. This procedure is called *randomization*. It has the great advantage that the comparison of the treatments is not affected systematically by disturbing factors, and hence can be done correctly.

In Example 6 there is a risk that the doctor, consciously or unconsciously, gives one of the medicines to patients of a certain type, for example, to those who are the most sick. For that reason, randomization is a safeguard.

In Example 7 it may happen that the raw material varies from time to time. It would then be objectionable to first produce units according to method A and later according to method B. It might then happen that an observed difference does not depend on the methods but on changes in the raw material. The randomization precludes such faulty conclusions.

In an Appendix to this chapter it is shown how randomization is performed in practice.

We shall briefly discuss the choice of model when the experiment is completely randomized. As a result of such an experiment we obtain two samples

$$\begin{array}{lllll} A: & x_1 & x_2 & \dots & x_n, \\ B: & y_1 & y_2 & \dots & y_n, \end{array}$$

from the two populations corresponding to the treatments. Several models are possible; we shall give two:

First, assume that the values can be regarded as independent samples from $N(m_1, \sigma_1^2)$ and $N(m_2, \sigma_2^2)$. Then $m_1 - m_2$ measures the difference between the treatment effects. As we know from Chapters 13 and 14, we can construct a confidence interval for the difference, or test some null hypothesis concerning its size. If the distributions are not normal, we can use the normal approximation when n is large, as described in §13.5 and §14.5.

An important question concerns the choice of sample size n. Even on this point earlier chapters can serve as guides. For example, we may plan the experiment so that the confidence interval for $m_1 - m_2$ attains a given length. It is not necessary to collect equally many observations for each treatment. If A is more expensive than B, one may have fewer values in treatment group A than in group B so as to reduce the cost of experiment.

Second, assume that, in the experiment, it is noted whether or not an element has a certain property E. (Example: In Example 6 it may be of interest to see if a patient recovers or not after a certain period following treatment.) Then the values x_i and y_i in the above schemes are replaced by E's and not-E's, depending on the outcome. The number of elements with property E in the two samples can often be regarded as observations from Bin(n, p_1) and Bin(n, p_2). The analysis can be carried out according to the methods described in §13.6 and §14.7. Even in this case the samples may comprise different numbers of elements. The advantage of the completely randomized experiment is clear. However, the method sometimes has a drawback: If the elements differ very much, the precision of the comparison is not very high.

In Example 6 there may be substantial variations between the patients, for they may react very differently to the same treatment. In Example 7 the units produced may differ considerably because of variations of the raw material. If the first of the two models above is then applied, it may happen that σ_1 and σ_2 become large. However, we may compensate for this by taking larger samples.

(b) Randomized Block Experiment

Let us assume that the differences between the elements are so large that it is desirable to get rid of them, or, at least, to reduce their influence. It is then possible to use a *randomized block experiment*:

Elements that are similar are brought together into *blocks* with two elements in each. Within each block, the treatments are distributed at random so that one element is given treatment A and the other treatment B. In total, n blocks are used.

In Example 6 the blocks can be constructed in different ways. One possibility is to choose block = patient, that is, give the same patient first one treatment, then the other, the order being determined by chance. Another possibility is to choose block = pair of twins, that is, give one twin treatment A and the other twin treatment B. (A pair of twins may either be taken literally or may mean two persons chosen in such a way that they are matched approximately with respect to age, social background, and so on.) In Example 7 one may produce two units at a time, one according to method A, the other according to method B.

An excellent feature of the plan just described is that comparisons may be made *within blocks*. The comparisons are not affected by differences that may exist between the blocks.

The result of a randomized block experiment can be written:

	Block			
	1	2		n
A:	x_1	x_2	...	x_n
B:	y_1	y_2	...	y_n
Diff. $B - A$:	z_1	z_2	...	z_n

(Superficially, this resembles data from a completely randomized experiment, but carefully note the difference: x_1 and y_1 belong to Block 1, and so on.) Several different models have been proposed, and we shall mention two:

If normal distributions can be assumed, the analysis can often be performed according to the method for paired samples (see subsection (d) in §13.4 and the end of §14.5). The values for A and B in the jth block are then assumed to be observations from $N(m_j, \sigma_1^2)$ and $N(m_j + \Delta, \sigma_2^2)$. Here m_j measures the effect caused by the block. Hence differences between the numbers $m_1, \ldots, m_n$ reflect differences between the blocks used in the experiment. The interesting quantity is Δ, which measures the difference between the treatments B and A, in this order. We know that paired samples are analysed statistically by taking differences, which eliminates $m_1, \ldots, m_n$. Because of the similarities of the elements paired in the blocks, the differences $z_1, \ldots, z_n$ may be expected to have a comparatively small variation. Hence the experiment will be more precise than the completely randomized experiment.

The model for paired samples can be used even if the distributions are not normal, if only n is large enough. The approximate methods in §13.5 and §14.6 can then be employed.

Another common method for analysing randomized block experiments with two treatments is the sign test, which does not assume normal distributions, and can be applied for all values of n.

In this section we have only discussed how to plan investigations involving two treatments. In practice it is very common that more than two treatments are compared at the same time. The planning can then be quite involved, and a whole branch of statistics is devoted to such questions: the design of experiments and other investigations; see, further, the references at the end of the book.

16.5. Final Remarks

After reading about statistical methods, many readers are perhaps caught up by enthusiasm. The procedures look simple and there seem to be many possibilities for applying them in practice. This is true to a large extent, but is not the whole truth. Life is more complicated than appears from the streamlined examples in books. Furthermore, we have simplified our task by refraining from a discussion of the important subject of *testing a model*. Only extensive experience makes the model builder a master. Let us also mention once more that statistical investigations should often be planned and handled by specialists working in collaboration.

We also stress again the importance of computers in statistical investigations. With the help of computers it is much easier nowadays to test the usefulness of a proposed model and to perform complicated statistical analyses rapidly.

Appendix: How to Handle Random Numbers

When taking random samples from a population or randomizing the treatments in an experimental investigation it is often convenient to use random numbers. We limit the discussion to finite populations.

Consider a population of N elements from which we want to draw a subset of n elements at random.

Definition. If all possible subsets have the same probability of being drawn, the subset is said to be chosen at random. □

For example, if the possible subsets are A_1A_2, A_1A_3 and A_2A_3, the probability should be 1/3 for each.

When selecting such a subset, various devices can be used. Of special value is a *table of random numbers*. Such a table contains randomly chosen numbers 0, 1, ..., 9. (More precisely: At each position in the table, these numbers appear with the same probability, and the numbers at different positions are independent.) Here follow some random numbers; more can be taken from Table 9 at the end of the book:

91 98 23 15 83 37 00 22 75 19 32 49 22 67 36 81 18 61 34 78
38 73 54 84 85 18 77 87 37 26 25 50 59 07 69 34 28 37 10 56

Computers can also be used for generating random numbers.

(a) Drawing with Replacement

First, we consider drawing random numbers with replacement, which is performed as follows:

We begin by assuming that N is larger than 10, but not larger than 100, let us say $N = 69$, and take $n = 10$. A starting point is chosen in the table of random numbers. One two-digit number is noted at a time with 00 interpreted as 100, and numbers greater than N are skipped. If the above random numbers are used, starting from the beginning, we obtain 23 15 37 22 19 32 49 22 67 36.

We now assume that N is larger than 100 but not larger than 1,000, let us say 356, and that $n = 5$. We then group the random numbers as follows: 9198 2315 8337 .... We now use, say, the three last digits in each group and obtain in a similar way as before: 198 315 337 022 249.

(b) Drawing without Replacement

Drawing without replacement is the most common procedure. We then proceed as in (a) with the single difference that, if a previously obtained random number appears, it is skipped, and a new number is taken. Hence more than n operations may be needed for collecting n numbers.

If no table of random numbers is available, it is, of course, possible to write all numbers from 1 to N on slips of paper, place them in a hat, mix well and draw one slip at a time until n have been obtained. But it is difficult to mix well, which makes it safer to use tables of random numbers.

It is now also clear how to perform randomization in a completely randomized experiment with two treatments A and B (see p. 302). In fact, it can be seen as a special form of drawing without replacement. Assume that $n = 10$ elements shall be given treatment A and 10 treatment B. We then number the elements in the investigation from 1 to 20, draw 10 numbers at random without replacement and give the corresponding elements treatment A, the other treatment B.

Remark. Systematic Selection

Commonly, elements in a set have a certain order from the beginning. For example, they may be recorded on a list or on cards in a card index. If it is really known that this order is random, one needs no table of random numbers; it is then possible to take the n first values. Alternatively, we may take every kth element counted from some starting point, where k is such that the elements are spread over the whole population. This procedure is called *systematic selection* or systematic sampling (see Fig. 16.1).

If the order in the set is not random from the start, the result can, of course, be catastrophic, for if the list is periodic with period k, a nonrandom sample will be obtained. (Example: Select n houses in some area by choosing every kth one in a street. Perhaps only corner houses will be obtained!) □

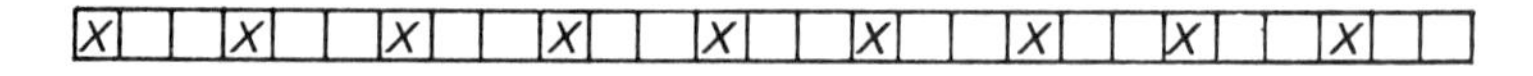

Fig. 16.1. Systematic selection with $k = 3$, $n = 9$. Selected elements marked by X.

Selected Exercises

(a) Probability

P1. The probability that a letter lies somewhere in a desk with seven drawers is p. It is equally likely to be in any one of the drawers. If six of them have been inspected without the letter having been found, find the probability that the letter lies in the seventh drawer.

P2. Find the mean and variance of $X \sim \chi^2(f)$.

P3. Let X_1, X_2, X_3, X_4, Y_1, Y_2, Y_3, Y_4 be independent rv's such that $X_i \sim N(4, 1)$, $i = 1, 2, 3, 4$ and $Y_i \sim N(3, 4)$, $i = 1, 2, 3, 4$. Determine

$$P(X_1 + 2X_2 + 3X_3 + 4X_4 < Y_1 + 2Y_2 + 3Y_3 + 4Y_4).$$

P4. (a) Throw two fair dice once and let X be the sum of the points. Find the distribution of X.
(b) Mark two fair dice with 1, 2, 2, 3, 3, 4 and 1, 3, 4, 5, 6, 8, respectively. Throw them once and let Y be the sum of the points. Find the distribution of Y.

P5. "A chain is no stronger than its weakest link." The load X when a link breaks is assumed to be Exp(50) (unit: newton). How many links can a chain at most consist of, if the probability that the chain breaks when the load is 10 newton must be at most 0.5?

P6. The two-dimensional rv (X, Y) has density function

$$f_{X,Y}(x, y) = \begin{cases} c \cdot x^2 y & 0 \le y \le x \le 1, \\ 0 & \text{otherwise.} \end{cases}$$

(a) Determine c.
(b) Determine the marginal densities.

P7. A factory produces 25 chairs of a certain type during a working-day. The probability is 0.95 that a chair is nondefective. Each chair is inspected before it is delivered, and defective chairs are later sold at a lower price. Find, approximately, the probability that more than 25 working-days are required to produce 600 nondefective chairs.

P8. Select two cards at random without replacement from an ordinary deck of cards. Find the probability that at least one of the cards is a heart.

P9. The two-dimensional rv (X, Y) has density function

$$f_{X,Y}(x, y) = c(y - x)^{\alpha}, \qquad 0 \le x < y < 1,$$

where $\alpha > -1$, $c = (\alpha + 1)(\alpha + 2)$. Find the means of X and Y.

P10. A box contains two defective and three nondefective transistors. To find the latter, the transistors are tested, one at a time, until either the two defective ones or the three nondefective ones have been found. Find the probability function of the required number, X, of tests.

P11. A test used to diagnose a certain disease has the following properties. It is positive with probability 0.99 if a person has the disease, and with probability 0.05 if a person does not have the disease. In the whole population, 1% has the disease. Determine the conditional probability that a person with a positive test has the disease.

P12. Let U_1 and U_2 be two urns containing in all two red balls and one white ball. Initially the balls are in U_1. First, one ball is drawn at random from U_1 and placed in U_2. Second, another drawing of one of the three balls is made at random; this ball is taken from the urn it is in and placed in the other urn. This procedure is repeated and stops as soon as a red ball is placed in U_2. Determine the probability that the white ball also then lies in U_2.

P13. A fair die is thrown until the same side turns up twice in succession. Determine the probability that this happens in throws $k - 1$ and k.

P14. In a food factory marmalade is filled into cardboard containers each of which weighs exactly 30 grams. The container is placed on a scale and filled with marmalade until the scale shows the weight m. The container then contains a total of Z grams of marmalade. The scale is subject to a random error $X \sim N(0, 7.5^2)$.

(a) Find the relation between Z, X and m.

(b) Find the distribution of Z.

(c) Choose m so that 95% of all containers hold at least 450 grams of marmalade.

P15. (Continuation of Exercise P14.) Assume that the weight of an empty container is no longer constant, but a rv $Y \sim N(30, 3^2)$. Assume also that X and Y are independent.

(a) Find the distribution of Z.

(b) How should m now be chosen?

P16. The rv X is Po(m). Determine, approximately, the mean and standard deviation of $Y = \sqrt{X}$.

Hint: Use Gauss's approximation formulae.

P17. The two-dimensional rv (X, Y) has density function

$$f_{X,Y}(x, y) = \tfrac{1}{2} \qquad \text{for} \quad |x| + |y| \le 1.$$

(a) Are X and Y uncorrelated?
(b) Are X and Y independent?

P18. The rv X has a Rayleigh distribution with density function $f_X(x) = 2\beta x e^{-\beta x^2}$, $x \ge 0$. Find $E(X)$ and $V(X)$.

P19. A fair coin is tossed three times. Let X be the number of heads at the first two throws and Y the total number of heads. Find the joint probability function of X and Y and the marginal probability functions.

P20. At a hamburger shop, three kinds of hamburgers are sold. The price (in English pounds (£)) of a hamburger is assumed to be a rv X which takes on the values 1, 1.20, 1.50 with probability 0.3, 0.2, 0.5. One day 300 hamburgers were sold. Find approximately the probability that the total sale that day amounted to at least £400.

P21. The weight, X, of a certain product is $N(10, 3)$. The products are divided into four classes as follows:

weight	class
$X \le 8$	A
$8 < X \le 9$	B
$9 < X \le 11$	C
$X > 11$	D

Compute the probability that, out of 10 randomly chosen items, 1, 3, 4 and 2 are in class A, B, C and D, respectively.

P22. The rv's X and Y are independent and uniformly distributed over the interval (0, 1). Determine the density function of the rv $Z = (X + Y)/2$.

P23. A bar of unit length is divided at random into two parts. Find the mean length of the smaller part.

P24. The two-dimensional rv (X, Y) has density function

$$f_{X,Y}(x, y) = 120xy(1 - x - y), \qquad x \ge 0, \quad y \ge 0, \quad x + y \le 1.$$

Find $C(X, Y)$.

P25. The rv's X and Y are independent and $\text{Bin}(n_1, p)$ and $\text{Bin}(n_2, p)$. Set $Z = \arcsin\sqrt{X/n_1} - \arcsin\sqrt{Y/n_2}$. Find approximately the mean and variance of Z.
Hint: Use Gauss's approximation formulae.

P26. A and B play the following game. A throws a fair die. If one or six turns up, he wins. Otherwise B throws the die repeatedly until either one of two events occurs: E = "B throws a onc or a six" or F = "B throws the same outcome as A". B wins if E occurs and A wins if F occurs. Find the probability that A wins the game.

P27. A and B have bought ten fruits. Neither knows that three of the fruits are poisonous. A eats four fruits and B six. Find the probability that:
(a) A becomes poisoned;
(b) B becomes poisoned;
(c) both A and B become poisoned.

P28. The rv's X and Y are independent and uniformly distributed over the interval (0, 1). Derive the probability that either one of the variables is at least twice as large as the other.

P29. The triangle RST is equilateral with side 1. A point Q is chosen at random in the triangle. Derive the distribution function for the distance from Q to the side ST.

P30. A person buys three ropes of equal length. The strengths (unit: newton) of the ropes are independent with the same density function $10^{-6}xe^{-x/1000}$ $(x \geq 0)$.
(a) Find the probability that each rope breaks if the load is 1,500 newton.
(b) If the three ropes are tied into one long rope, what is the probability that it sustains a load of 1,500 newton without breaking?

P31. A person goes from one place to another, first by bus, then by taxi. The respective waiting-times (unit: minutes) are X and Y. The rv X has a uniform distribution in the interval (0, 10) and the rv Y has an exponential distribution with density function $(1/8)e^{-y/8}$, $y \geq 0$. The rv's X and Y are independent. Find the probability that the total waiting-time exceeds 16 minutes.

P32. A fair die is thrown n times. Find the probability that each side turns up at least once if:
(a) $n = 6$;
(b) $n = 7$.

P33. From an ordinary deck of cards 13 cards are drawn at random without replacement. Consider the events "six hearts occur" and "six diamonds occur". Determine the probability that at least one of these two events occurs.

P34. A person has two coins A and B. The coin A is an ordinary fair coin, and B is a false coin with heads on both sides. The person selects one coin at random.
(a) When the selected coin is tossed, it shows head. Find the probability that it is coin A.
(b) The selected coin is tossed once more, and again shows head. Find the probability that it is coin A.

P35. The rv's $X_1, X_2, \ldots, X_{100}$ are independent and uniformly distributed in the interval (0, 1). Let $Y = X_1 X_2 \cdot \ldots \cdot X_{100}$. Determine approximately the probability $P(Y < 10^{-40})$.

P36. A coin is such that $P(\text{head}) = p$, $P(\text{tail}) = q = 1 - p$. The coin is tossed until head has turned up r times. Let X be the number of trials performed.
(a) Find the probability function of X.
(b) Find the mean and variance of X.
Hint: In (b) you may if you like use the fact that $X = U_1 + \cdots + U_r$, where U_1 is the number of trials until the first head appears, and so on.

P37. The r-dimensional rv $(X_1, X_2, \ldots, X_r)$ has a multinomial distribution. (The usual symbols are used.) Show that

$$C(X_i, X_j) = -np_ip_j.$$

Hint: Consider $V(X_i + X_j)$.

P38. The rv's X_1 and X_2 are independent and $N(0, 1)$. Determine the density function of the rv $Y = (X_1^2 + X_2^2)^{1/2}$.

P39. Select n points at random on the circumference of a circle. Find the probability that all points lie on some semicircle.
Hint: Let $A_1, A_2, \ldots, A_n$ be the points counted in, say, the positive direction. To find the required probability, it suffices to consider each semicircle C_i beginning in A_i, $i = 1, 2, \ldots, n$.

P40. A person starts from A in the figure below and moves each time between the corners by going to one of the three other corners, selected at random with the same probability 1/3. Determine the mean number of moves until he has visited B, C and D.

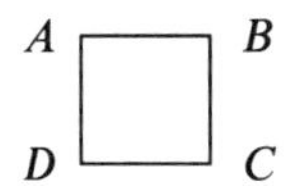

(b) Statistics

S1. A very large batch of units has a relative frequency p of defectives where p is unknown but ≤ 0.04. It is required to take a random sample of n units from the batch and use this sample to construct an estimate of p with standard error of at most 0.02. How large should n be?

S2. Let $x_1, \ldots, x_n$ be independent observations on a rv with density function

$$(1/a)x^{(1/a)-1} \qquad (0 < x < 1; a > 0),$$

where the parameter a is unknown. Find the ML estimate of a and show that it is unbiased.

S3. Let

11.3 2.1 1.1 8.9 4.6 5.7 13.5 24.5 16.4

be independent observations on a rv X such that $\ln X \sim N(m, \sigma^2)$, where m and σ are unknown parameters. Construct a two-sided confidence interval with confidence level 95% for:
(a) m;
(b) σ.

S4. The data set $x_1, \ldots, x_{20}$ has arithmetic mean 2.0 and standard deviation 0.36. The data are transformed to new values $y_1, \ldots, y_{20}$ by the transformation $y_i = 1.0 + 0.4x_i$, $i = 1, \ldots, 20$. Compute the coefficient of variation of $y_1, \ldots, y_{20}$.

S5. The lifetime of certain components has density function

$$x^2e^{-x/a}/2a^3 \qquad (x > 0),$$

where a is an unknown parameter. The lifetimes of a random sample of n components are $x_1, \ldots, x_n$. Find the *ML* estimate of a and prove that it is unbiased.

S6. A method of measurement was studied by measuring two distances A and B, 10 times and 15 times, respectively. Results:

$$A: \quad \sum_1^{10} x_i = 1{,}216; \qquad \sum_1^{10} x_i^2 = 168{,}731,$$

$$B: \quad \sum_1^{15} y_i = 1{,}731; \qquad \sum_1^{15} y_i^2 = 232{,}654.$$

All errors of measurement are independent and $N(0, \sigma^2)$, where σ is unknown. Construct a two-sided confidence interval for σ with confidence level 0.95.

S7. An urn contains 7 balls, a of which are red, and the others blue. In order to test H_0: $a \leq 2$ against H_1: $a > 2$, two balls are drawn at random without replacement. The null hypothesis H_0 is rejected if both balls are red. Find the power of the test for all possible values of a.

S8. Let

1.21 0.20 0.39 0.68 2.00 1.07 0.96 0.55 0.80

2.45 0.67 0.34 0.22 0.77 1.24 0.39 0.29 0.59

be a random sample of 18 values from a distribution with density function $\exp[-(x - a)]$, $x > a$, where a is unknown. Test the hypothesis $a = 0$ against the hypothesis $a > 0$ at the level of significance 5%, using the smallest value in the sample as a test quantity.

S9. Consider a regression model with n pairs (x_i, y_i) where y_i is an observation on $Y_i \sim N(\beta x_i, 1)$ and β is unknown. Find the *ML* estimate of β and its variance.

S10. The rv Y has a geometric distribution such that $P(Y = k) = p(1 - p)^k$, $k = 0, 1, \ldots$. The parameter p is unknown. We have the following random sample of 50 values from this distribution:

0 0 0 0 0 0 0 0 0 0 0 0 0 0 0 0 0 0 0 1 0 1 1 0 1 0 1

0 1 0 0 2 1 1 0 0 1 0 0 1 0 0 0 1 0 0 0 0 0 0 0 0.

Find:
(a) the *ML* estimate of p;
(b) the *LS* estimate of p.

S11. Let

1.8807	0.1251	2.0899	2.1059	1.8722	0.1346	2.2767	1.5771
−0.7184	−0.2791	2.1957	2.2535	3.0155	0.9356	−0.4501	0.7535
2.0769	1.1986	0.3574	2.5372	2.6740	1.3706	1.0940	2.8718
0.2142	2.0177	−0.5822	0.7159	2.7182	0.9693		

be a random sample of 30 observations from $N(m, m)$, where m is unknown. Find the *ML* estimate of m.

S12. Let

$$12 \quad 0 \quad 6 \quad 2 \quad 9 \qquad \text{and} \qquad 14 \quad 11 \quad 7 \quad 12 \quad 16$$

be random samples from $\text{Po}(m_1)$ and $\text{Po}(m_2)$, respectively. Test the hypothesis $m_1 = m_2$ with an approximate two-sided test.

S13. A person counts the number of trucks passing his bedroom window from 11 p.m. to 7 a.m. during three nights. The number passing during one night is supposed to be $\text{Po}(\lambda)$, where λ is unknown. He obtained the numbers 136, 154 and 127.

(a) Prove that the *ML* estimate and the *LS* estimate of λ are identical.

(b) Construct a two-sided confidence interval for λ with approximate confidence level 99%.

S14. In a factory the method of production has been changed. In order to find out whether the quality of a certain product has been affected, a random sample of 100 units produced before the change was compared with a similar sample produced after the change. It appeared that 36 units produced before the change and 27 units produced after the change did not meet specifications. Test the hypothesis H_0 that the quality has not been affected.

S15. Let

$$3.24 \quad 3.37 \quad 3.29 \quad 3.18 \quad 3.51$$

$$3.20 \quad 3.35 \quad 3.38 \quad 3.29 \quad 3.42$$

be a random sample from $N(\ln(M + 1), \sigma^2)$ where M and σ are unknown. Construct a two-sided confidence interval for M with confidence level 0.95.

S16. Let $x_1, \ldots, x_{15}$ be a random sample from $\text{Exp}(a)$, where a is unknown. It is found that $\sum x_i = 27$. Test H_0: $a = 1$ against H_1: $a > 1$ with an exact test at the level of significance 0.05.

S17. Let

$$2 \quad 10 \quad 6 \quad 3 \quad 6 \quad 3$$

be a random sample from $\text{Po}(4\lambda)$, where λ is unknown. Construct a one-sided confidence interval $(0, a)$ for λ with the approximate confidence level 99%.

S18. When analysing n samples according to two methods A and B, a chemist obtained n pairs of values (x_i, y_i), $i = 1, \ldots, n$. Model: x_i is an observation on $X_i \sim N(m_i, \sigma_1^2)$ and y_i is an observation on $Y_i \sim N(m_i, \sigma_2^2)$. The parameter σ_1 is known, but the parameters $m_1, \ldots, m_n$ and σ_2 are unknown. Describe how a two-sided 95% confidence interval for σ_2 is found.

S19. The rv X assumes the values 1, 2 and 3 with probabilities θ, θ and $1 - 2\theta$, where θ is an unknown parameter. Let $x_1, \ldots, x_n$ be a random sample from this distribution.

(a) Derive the *ML* estimate of θ.

(b) Describe how the *ML* estimate can be used for testing hypotheses concerning θ both when n is small and when n is large.

S20. A chemist wants to compare two pH meters, called type A and type B. He believes that there exists a systematic difference d, such that type B delivers, on the average, smaller values than type A. Therefore the pH value was determined for five different solutions, using both instruments. Results:

Solution	1	2	3	4	5
Type A	6.23	4.16	8.79	10.11	3.56
Type B	6.10	3.96	8.82	9.83	3.50

Test the chemist's belief, using a suitable test of significance at the level of significance 5%. Normal distributions may be assumed.

S21. Let $x_1, \ldots, x_n$ be independent observations on a rv which has a uniform distribution over the interval $(-a, a)$. The constant a is unknown. Show that $c\sum_1^n x_i^2$ is an unbiased estimate of the parameter a^2 if c is suitably chosen. Prove that the estimate is consistent.

S22. An object consists of two parts A and B which have been weighed three and four times, respectively. Further, the whole object has been weighed twice, using the same balance. Results:

A	12.07	12.01	12.04	
B	18.34	18.36	18.35	18.32
$A + B$	30.35	30.39		

Each result is subject to a random error with mean 0 and standard deviation σ. Find the LS estimate of the total weight of the object.

S23. Let $x_1, \ldots, x_n$ be a random sample from $N(m, \sigma^2)$ where m is known and $\theta = \sigma^2$ is unknown. The estimates

$$\theta_1^* = \sum_1^n (x_i - m)^2/n,$$

$$\theta_2^* = \sum_1^n (x_i - \bar{x})^2/(n - 1),$$

are both unbiased. Which of them has the smallest variance?

S24. A machine produces units with weights (in grams) distributed as $N(50.0, \sigma^2)$, where σ is unknown. A unit is considered to be nondefective if its weight is between 47.0 and 53.0. To control σ, units are weighed, one at a time, until a defective unit is obtained.

On one occasion, the first defective unit was found when the tenth unit was weighed. Determine the ML estimate of σ.

S25. A firm asserts that a certain measuring device is such that the absolute error (that is, the absolute value of the difference between the measurement and the correct value) exceeds 0.05 in at most one case in 10. A person assumes that the error is $N(0, \sigma^2)$ but does not believe σ to be so small as to make the assertion correct. He therefore performs 15 independent measurements and obtains the errors

-0.027 0.012 0.039 -0.062 -0.032 0.074 -0.006 0.013

-0.019 0.010 0.047 0.039 0.005 -0.048 0.016

Test the assertion of the firm.

S26. Let $x_1, \ldots, x_{16}$ be a random sample from $N(m, 1)$. Reject the hypothesis H_0: $m = 0$ if $|\bar{x}| > 0.5$. Derive the power function $h(m)$ of the test and draw its graph.

S27. Consider the following data:

x_i	1	2	3	4	5	6
y_i	3.0	3.9	5.6	6.7	8.8	10.1

Model: y_i is an observation from $N(\alpha + \beta x_i, \sigma^2)$ where α, β and σ^2 are unknown parameters. Test the hypothesis H_0: $\alpha = 0$ with a two-sided test of significance.

S28. The numbers of dots (defects) in two types, A and B, of laminates of a certain size are assumed to be $\text{Po}(m_A)$ and $\text{Po}(m_B)$, respectively. On 30 laminates of type A and 45 of type B, the following numbers of dots were found:

A	1	3	1	0	0	0	2	1	1	0	2	0	0	2	0
	1	0	2	0	0	2	0	0	1	1	0	0	1	0	0
B	0	0	0	0	0	2	0	0	1	1	1	0	0	0	0
	0	1	0	0	1	0	0	1	0	1	0	0	0	0	0
	0	0	0	1	0	0	0	0	2	1	0	1	1	1	0

Estimate the difference $m_A - m_B$ and find a standard error of the difference.

S29. Let $x_1, \ldots, x_8$ and $y_1, \ldots, y_{12}$ be random samples from $\text{Exp}(\theta)$ and $\text{Exp}(3\theta)$, respectively, where θ is an unknown parameter. It is found that $\sum x_i = 19.0$, $\sum y_i = 63.6$. Derive the ML estimate of θ and find its variance.

S30. The maximum height H of the waves during one year at a certain place is assumed to be a rv having a Rayleigh distribution with density function

$$f_H(x) = (x/a)e^{-x^2/2a} \qquad (x > 0),$$

where a is an unknown parameter. During 8 years the following heights (in meters) have been observed:

2.5 2.9 1.8 0.9 1.7 2.1 2.2 2.8.

Use the ML estimate of a to find an estimate of the "1,000 year wave", that is, a wave so high that it occurs, on average, only once in a thousand years.

S31. A physicist has performed five measurements of a physical constant m. The measurements are assumed to be $N(m, \sigma^2)$ with known variance σ^2. The phys-

icist obtained the 90% confidence interval (7.02, 7.14). How many additional measurements are required if the physicist wants:
(a) a confidence interval with half the length;
(b) a confidence interval with the same length but 95% confidence level?

S32. A person has an observation x on $X \sim \text{Bin}(10, p)$, where p is unknown. He used the normal approximation to construct the confidence interval

$$I_p = (x/10 \pm 1.96\sqrt{x(10-x)/1{,}000}).$$

The desired confidence level was 95%. The true confidence level is a function $K(p)$ of p. Find $K(0.50)$.

S33. A market research institute wants to estimate the fraction p in a population of 1,200 families who would be willing to buy a certain product. The institute therefore plans to ask n persons, drawn at random without replacement, if they want or do not want to buy the product. It is known beforehand that at most 1/4 will be positive. Find n so that a confidence interval for p with approximate confidence level 95% and total length 0.05 can be constructed.

S34. Let $x_1, \ldots, x_9$ be a random sample from $N(m, 1)$. Construct a test of $H_0\colon m = 2$ against $H_1\colon m \neq 2$ at the level of significance 5%.
(a) Derive the power function of the test.
(b) Suppose that the power is not good enough. How many values are needed in order to give the test a power of 0.99 for $m = 1$?

S35. A dentist wants to compare two anaesthetics A and B by testing them on each of 15 patients. He is interested to know whether an injection results in anaesthesia (event H) or does not result in any anaesthesia (event H^*). Results:

	Patient no.															
	1	2	3	4	5	6	7	8	9	10	11	12	13	14	15	16
A	H	H	H	H	H	H^*	H	H	H	H	H	H	H^*	H	H	H
B	H	H	H^*	H	H^*	H^*	H^*	H^*	H	H	H^*	H^*	H	H	H^*	H^*

Test whether A and B differ with respect to the incidence of anaesthesia. Determine the P value (two-sided test).

S36. A pond contains N fish, where N is unknown. In order to estimate N, one catches 100 fish, marks them and returns them to the pond. A day later, 400 fish are caught, 42 of which are marked. Construct an approximate two-sided 95% confidence interval for N, under suitable assumptions.

S37. When studying the velocity of a chemical reaction at constant temperature, a chemist determined at different times t_i the concentration y_i of a substance participating in the reaction. Results:

t_i	0	184	319	400	526	575
y_i	2.33	2.08	1.91	1.82	1.67	1.62

As a model the chemist chose $y_i = c \cdot \exp(\beta t_i) \cdot \varepsilon_i$ where $\ln \varepsilon_i$ is an observation from $N(0, \sigma^2)$ and the parameters c, β and σ are unknown. Determine a 95% confidence interval for $m = c \cdot \exp(600\beta)$.

S38. A product is manufactured in units. If the length of a unit exceeds 10 (mm), it is defective. The lengths are assumed to be $N(m, 0.1^2)$. In order to control a large batch, the manufacturer selects n units and will accept the batch if and only if the arithmetic mean of the lengths of these units is less than a.

Find a and n so that the following conditions are fulfilled:

(a) The probability is 90% that a batch with 0.1% defective units is accepted.

(b) The probability is 5% that a batch with 1% defective units is accepted.

S39. Consider the usual regression model with n pairs (x_i, y_i) where y_i is an observation on $Y_i \sim N(\alpha + \beta(x_i - \bar{x}), \sigma^2)$. The parameters α, β and σ are unknown. Show that the ML estimates of α and β coincide with the LS estimates.

S40. A circle with known centre has the unknown radius R. In the circle n points are chosen at random. The distances from the points to the centre are $x_1, \ldots, x_n$. Find an unbiased estimate of R based on $\max(x_i)$.

References

This list of references is longer than is usual in elementary books. It is intended to demonstrate the breadth of probability and statistics as reflected in the literature. Therefore, theoretical and applied areas not treated in the present book are also represented in the list. Most books are on an elementary or intermediate level, but there are also some advanced titles.

Popular Books

R.J. Brook, G.C. Arnold, Th.H. Hassard and R.M. Pringle (Eds.). *The Fascination of Statistics.* Marcel Dekker, New York, 1986.

S.K. Campbell. *Flaws and Fallacies in Statistical Thinking.* Prentice-Hall, Englewood Cliffs, NJ, 1974.

M. Hollander and F. Proschan. *The Statistical Exorcist. Dispelling Statistics Anxiety.* Marcel Dekker, New York, 1984.

D.S. Moore. *Statistics, Concepts and Controversies*, 2nd ed. Freeman, New York, 1985.

J.M. Tanur *et al. Statistics: A Guide to the Unknown.* Holden-Day, San Francisco, 1972.

General Textbooks

Elementary

J.L. Hodges and E.L. Lehmann. *Basic Concepts of Probability and Statistics*, 2nd ed. Holden-Day, San Francisco, 1970.

P.G. Hoel, S.C. Port and C.J. Stone. *Introduction to Probability Theory.* Houghton Mifflin, Boston, 1971.

P.G. Hoel, S.C. Port and C.J. Stone. *Introduction to Statistical Theory.* Houghton Mifflin, Boston, 1971.

S. Ross. *A First Course in Probability.* Macmillan, New York, 1976.

Less Elementary or More Comprehensive

K.L. Chung. *Elementary Probability Theory with Stochastic Processes.* Springer-Verlag, Berlin, 1974.
W. Feller. *An Introduction to Probability Theory and Its Applications*, Vol. 1, 3rd ed. John Wiley and Sons, New York, 1968.
J.G. Kalbfleisch. *Probability and Statistical Inference.* Vol. 1: *Probability*, Vol. 2: *Statistical Inference*, 2nd ed. Springer-Verlag, New York, 1985.
H.J. Larsen. *Introduction to Probability Theory and Statistical Inference*, 3rd ed. John Wiley and Sons, New York, 1982.
R.J. Larsen and M.L. Marx. *An Introduction to Mathematical Statistics and Its Applications.* Prentice-Hall, Englewood Cliffs, NJ, 1983.
B.W. Lindgren. *Statistical Theory*, 3rd ed. Macmillan, New York, 1976.
A.M. Mood, F.A. Graybill and D.C. Boes. *Introduction to the Theory of Statistics*, 3rd ed. McGraw-Hill, New York, 1974.
M. Woodroofe. *Probability with Applications.* McGraw-Hill, New York, 1975.

Advanced

L. Breiman. *Probability.* Addison-Wesley, Reading, MA, 1968.
K.L. Chung. *A Course in Probability Theory*, 2nd ed. Academic Press, New York, 1974.
W. Feller. *An Introduction to Probability Theory and Its Applications*, Vol. 2, 2nd ed. John Wiley and Sons, New York, 1971.

Inference Theory

E.L. Lehmann. *Testing Statistical Hypotheses*, 2nd ed. John Wiley and Sons, New York, 1986.
E.L. Lehmann. *Theory of Point Estimation.* John Wiley and Sons, New York, 1983.
S.D. Silvey. *Statistical Inference.* Penguin Books, Harmondsworth, 1970.

Inventory Models

E. Naddor. *Inventory Systems.* John Wiley and Sons, New York, 1966.

Lifetimes. Reliability

D.R. Cox and D. Oakes. *Analysis of Survival Data.* Chapman and Hall, London, 1984.
J.D. Kalbfleisch and R.L. Prentice. *The Statistical Analysis of Failure Time Data.* John Wiley and Sons, New York, 1980.
K.C. Kapur and L.R. Lamberson. *Reliability in Engineering Design.* John Wiley and Sons, New York, 1977.

Linear Models

F.A. Graybill. *Theory and Application of the Linear Model.* Duxbury Press, North Scituate, MA, 1976.
W. Mendenhall. *Introduction to Linear Models and the Design and Analysis of Experiments.* Wadsworth, Belmont, 1968.

Multivariate Analysis

T.W. Anderson. *Introduction to Multivariate Statistical Analysis*, 2nd ed. John Wiley and Sons, New York, 1984.

Nonparametric Methods

M. Hollander and D.A. Wolfe. *Nonparametrical Statistical Methods.* John Wiley and Sons, New York, 1973.
Ch.H. Kraft and C. van Eeden. *A Nonparametric Introduction to Statistics.* Macmillan, New York, 1968.
E.L. Lehmann. *Nonparametrics: Statistical Methods Based on Ranks.* Holden-Day, San Francisco, 1975.

Operations Research

F.S. Hillier and G.J. Lieberman. *Introduction to Operations Research*, 4th ed. Holden-Day, San Francisco, 1986.

Pattern Recognition

V.A. Kovalevsky. *Image Pattern Recognition.* Springer-Verlag, New York, 1984.

Planning of Experiments

G.E.P. Box, W.G. Hunter and J.S. Hunter. *Statistics for Experiments.* John Wiley and Sons, New York, 1978.
W.G. Cochran and G.M. Cox. *Experimental Designs*, 2nd ed. John Wiley and Sons, New York, 1957.
D.C. Montgomery. *Design and Analysis of Experiments*, 2nd ed. John Wiley and Sons, New York, 1984.

Quality Control

D.C. Montgomery. *Introduction to Statistical Quality Control.* John Wiley and Sons, New York, 1985.

Queueing Theory

R.B. Cooper. *Introduction to Queueing Theory*, 2nd ed. Pergamon Press, London, 1981.
D. Gross and C.M. Harris. *Fundamentals of Queueing Theory.* John Wiley and Sons, New York, 1974.

Regression

N.R. Draper and H. Smith. *Applied Regression Analysis*, 2nd ed. John Wiley and Sons, New York, 1981.
D.C. Montgomery and E.A. Peck. *Introduction to Linear Regression Analysis.* John Wiley and Sons, New York, 1982.

Sampling Methods

W.G. Cochran. *Sampling Techniques*, 3rd ed. John Wiley and Sons, New York, 1977.
W. Mendenhall, L. Ott and R.L. Scheaffer. *Elementary Survey Sampling.* Wadsworth, Belmont, 1971.

Simulation

B.J.T. Morgan. *Elements of Simulation.* Chapman and Hall, London, 1984.
R.Y. Rubenstein. *Simulation and the Monte Carlo Method.* John Wiley and Sons, New York, 1981.

Stochastic Processes

G.R. Grimmett and D.R. Stirzaker. *Probability and Random Processes.* Oxford University Press, Oxford, 1982.
P.G. Hoel, S.C. Port and C.J. Stone. *Introduction to Stochastic Processes.* Houghton Mifflin, Boston, 1971.
S.M. Ross. *Introduction to Probability Models*, 3rd ed. Academic Press, New York, 1985.
S.M. Ross. *Stochastic Processes.* John Wiley and Sons, New York, 1983.

Time Series

G.E.P. Box and G.M. Jenkins. *Time Series Analysis: Forecasting and Control*, rev. ed. Holden Day, Oakland, 1976.
C. Chatfield. *The Analysis of Time Series: Theory and Practice.* Chapman and Hall, London, 1975.

Tables

General

Biometrika Tables for Statisticians, Vol. 1, 3rd ed; Vol. 2. Cambridge University Press, Cambridge, 1966, 1972.
R.A. Fisher and F. Yates. *Statistical Tables for Biological, Agricultural and Medical Research*, 6th ed. Oliver and Boyd, London and Edinburgh, 1963.
A. Hald. *Statistical Tables and Formulas.* John Wiley and Sons, New York, 1952.
D.V. Lindley and W.F. Scott. *New Cambridge Elementary Statistical Tables.* Cambridge University Press, Cambridge, 1984.
D.B. Owen. *Handbook of Statistical Tables.* Addison-Wesley, Reading, MA, 1962.

Special

A Million Random Digits with 100 000 *Normal Deviates.* Rand Corporation, Glencoe, 1955.
G.J. Lieberman and D.B. Owen. *Tables of Hypergeometric Probability Function.* Stanford University Press, Stanford, 1961.
Tables of the Binomial Probability Distribution. National Bureau of Standards, New York, 1950.
Tables of the Cumulative Binomial Probabilities. Ordnance Corps Pamphlet 20-1, Washington, DC, 1956.
Tables of the Individual and Cumulative Terms of Poisson Distribution. Van Nostrand, Princeton, NJ, 1962.

Tables[1]

[1] The tables are taken from a Swedish publication (see Preface), and the same numbering of the tables has been used as in that publication.

Table 1. Normal distribution. $\Phi(x) = P(X \leq x)$ where $X \sim N(0, 1)$. When x is negative, use the fact that $\Phi(-x) = 1 - \Phi(x)$.

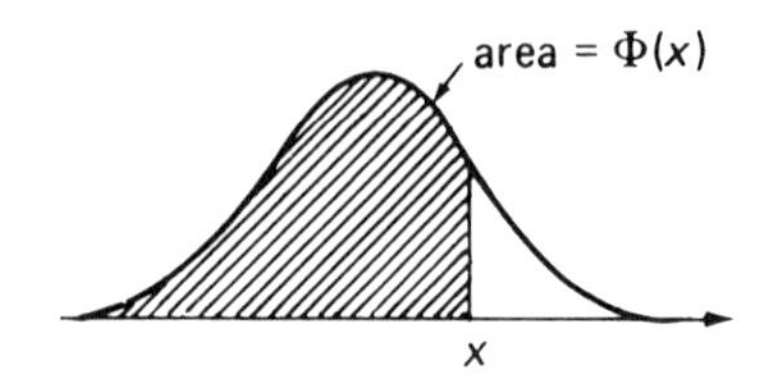

x	.00	.01	.02	.03	.04	.05	.06	.07	.08	.09
0.0	.5000	.5040	.5080	.5120	.5160	.5199	.5239	.5279	.5319	.5359
0.1	.5398	.5438	.5478	.5517	.5557	.5596	.5636	.5675	.5714	.5753
0.2	.5793	.5832	.5871	.5910	.5948	.5987	.6026	.6064	.6103	.6141
0.3	.6179	.6217	.6255	.6293	.6331	.6368	.6406	.6443	.6480	.6517
0.4	.6554	.6591	.6628	.6664	.6700	.6736	.6772	.6808	.6844	.6879
0.5	.6915	.6950	.6985	.7019	.7054	.7088	.7123	.7157	.7190	.7224
0.6	.7257	.7291	.7324	.7357	.7389	.7422	.7454	.7486	.7517	.7549
0.7	.7580	.7611	.7642	.7673	.7704	.7734	.7764	.7794	.7823	.7852
0.8	.7881	.7910	.7939	.7967	.7995	.8023	.8051	.8078	.8106	.8133
0.9	.8159	.8186	.8212	.8238	.8264	.8289	.8315	.8340	.8365	.8389
1.0	.8413	.8438	.8461	.8485	.8508	.8531	.8554	.8577	.8599	.8621
1.1	.8643	.8665	.8686	.8708	.8729	.8749	.8770	.8790	.8810	.8830
1.2	.8849	.8869	.8888	.8907	.8925	.8944	.8962	.8980	.8997	.9015
1.3	.9032	.9049	.9066	.9082	.9099	.9115	.9131	.9147	.9162	.9177
1.4	.9192	.9207	.9222	.9236	.9251	.9265	.9279	.9292	.9306	.9319
1.5	.9332	.9345	.9357	.9370	.9382	.9394	.9406	.9418	.9429	.9441
1.6	.9452	.9463	.9474	.9484	.9495	.9505	.9515	.9525	.9535	.9545
1.7	.9554	.9564	.9573	.9582	.9591	.9599	.9608	.9616	.9625	.9633
1.8	.9641	.9649	.9656	.9664	.9671	.9678	.9686	.9693	.9699	.9706
1.9	.9713	.9719	.9726	.9732	.9738	.9744	.9750	.9756	.9761	.9767
2.0	.97725	.97778	.97831	.97882	.97932	.97982	.98030	.98077	.98124	.98169
2.1	.98214	.98257	.98300	.98341	.98382	.98422	.98461	.98500	.98537	.98574
2.2	.98610	.98645	.98679	.98713	.98745	.98778	.98809	.98840	.98870	.98899
2.3	.98928	.98956	.98983	.99010	.99036	.99061	.99086	.99111	.99134	.99158
2.4	.99180	.99202	.99224	.99245	.99266	.99286	.99305	.99324	.99343	.99361
2.5	.99379	.99396	.99413	.99430	.99446	.99461	.99477	.99492	.99506	.99520
2.6	.99534	.99547	.99560	.99573	.99585	.99598	.99609	.99621	.99632	.99643
2.7	.99653	.99664	.99674	.99683	.99693	.99702	.99711	.99720	.99728	.99736
2.8	.99744	.99752	.99760	.99767	.99774	.99781	.99788	.99795	.99801	.99807
2.9	.99813	.99819	.99825	.99831	.99836	.99841	.99846	.99851	.99856	.99861
3.0	.99865									
3.1	.99903									
3.2	.99931									
3.3	.99952									
3.4	.99966									
3.5	.99977									
3.6	.99984									
3.7	.99989									
3.8	.99993									
3.9	.99995									
4.0	.99997									

Table 2. Quantiles of normal distribution. $P(X > \lambda_\alpha) = \alpha$ where $X \sim N(0, 1)$.

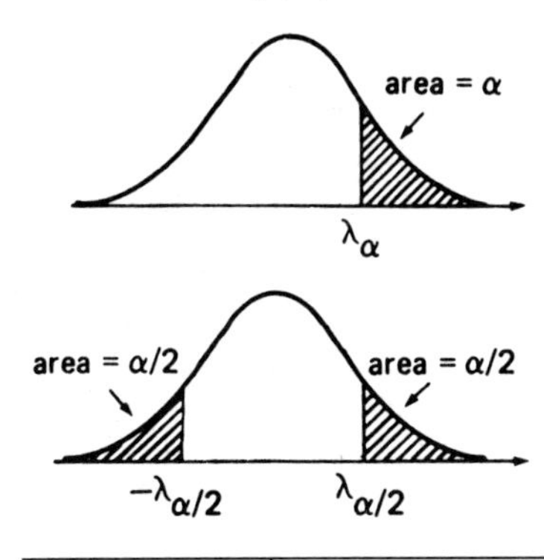

α	λ_α	α	λ_α
0.10	1.2816	0.001	3.0902
0.05	1.6449	0.0005	3.2905
0.025	1.9600	0.0001	3.7190
0.010	2.3263	0.00005	3.8906
0.005	2.5758	0.00001	4.2649

Table 3. t Distribution. $P(X > t_\alpha(f)) = \alpha$ where $X \sim t(f)$.

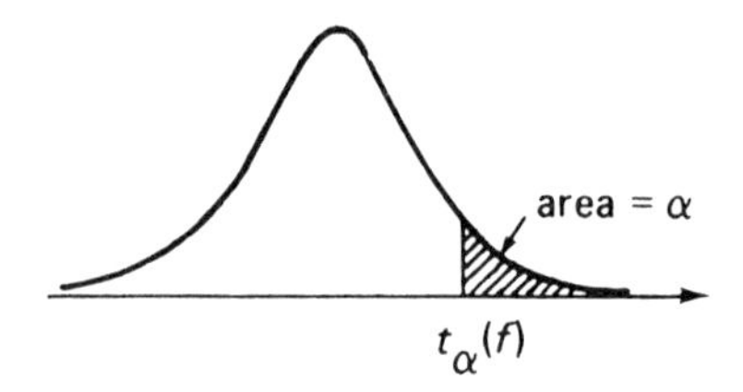

f \ α	0.10	0.05	0.025	0.01	0.005	0.001	0.0005
1	3.08	6.31	12.71	31.82	63.66	318.31	636.61
2	1.89	2.92	4.30	6.96	9.92	22.33	31.60
3	1.64	2.35	3.18	4.54	5.84	10.21	12.92
4	1.53	2.13	2.78	3.75	4.60	7.17	8.61
5	1.48	2.02	2.57	3.36	4.03	5.89	6.87
6	1.44	1.94	2.45	3.14	3.71	5.21	5.96
7	1.41	1.89	2.36	3.00	3.50	4.79	5.41
8	1.40	1.86	2.31	2.90	3.36	4.50	5.04
9	1.38	1.83	2.26	2.82	3.25	4.30	4.78
10	1.37	1.81	2.23	2.76	3.17	4.14	4.59
11	1.36	1.80	2.20	2.72	3.11	4.02	4.44
12	1.36	1.78	2.18	2.68	3.05	3.93	4.32
13	1.35	1.77	2.16	2.65	3.01	3.85	4.22
14	1.34	1.76	2.14	2.62	2.98	3.79	4.14
15	1.34	1.75	2.13	2.60	2.95	3.73	4.07
16	1.34	1.75	2.12	2.58	2.92	3.69	4.02
17	1.33	1.74	2.11	2.57	2.90	3.65	3.97
18	1.33	1.73	2.10	2.55	2.88	3.61	3.92
19	1.33	1.73	2.09	2.54	2.86	3.58	3.88
20	1.33	1.72	2.09	2.53	2.85	3.55	3.85
21	1.32	1.72	2.08	2.52	2.83	3.53	3.82
22	1.32	1.72	2.07	2.51	2.82	3.51	3.79
23	1.32	1.71	2.07	2.50	2.81	3.48	3.77
24	1.32	1.71	2.06	2.49	2.80	3.47	3.75
25	1.32	1.71	2.06	2.49	2.79	3.45	3.73
26	1.32	1.71	2.06	2.48	2.78	3.44	3.71
27	1.31	1.70	2.05	2.47	2.77	3.42	3.69
28	1.31	1.70	2.05	2.47	2.76	3.41	3.67
29	1.31	1.70	2.05	2.46	2.76	3.40	3.66
30	1.31	1.70	2.04	2.46	2.75	3.39	3.65
40	1.30	1.68	2.02	2.42	2.70	3.31	3.55
60	1.30	1.67	2.00	2.39	2.66	3.23	3.46
120	1.29	1.66	1.98	2.36	2.62	3.16	3.37
∞	1.28	1.64	1.96	2.33	2.58	3.09	3.29

Table 4. χ^2 Distribution. $P(X > \chi^2_\alpha(f)) = \alpha$ where $X \sim \chi^2(f)$.

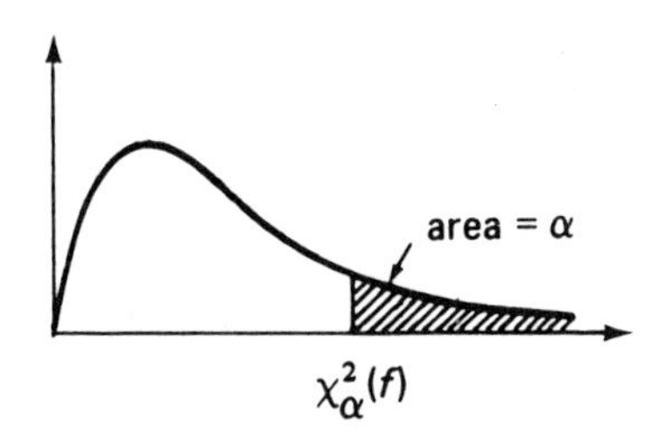

f \ α	0.9995	0.999	0.995	0.99	0.975	0.95	0.05	0.025	0.01	0.005	0.001	0.0005
1	0.00	0.00	0.00	0.00	0.00	0.00	3.84	5.02	6.63	7.88	10.8	12.1
2	0.00	0.00	0.01	0.02	0.05	0.10	5.99	7.38	9.21	10.6	13.8	15.2
3	0.02	0.02	0.07	0.12	0.22	0.35	7.81	9.35	11.3	12.8	16.3	17.7
4	0.06	0.09	0.21	0.30	0.48	0.71	9.49	11.1	13.3	14.9	18.5	20.0
5	0.16	0.21	0.41	0.55	0.83	1.15	11.1	12.8	15.1	16.7	20.5	22.1
6	0.30	0.38	0.68	0.87	1.24	1.64	12.6	14.4	16.8	18.5	22.5	24.1
7	0.48	0.60	0.99	1.24	1.69	2.17	14.1	16.0	18.5	20.3	24.3	26.0
8	0.71	0.86	1.34	1.65	2.18	2.73	15.5	17.5	20.1	22.0	26.1	27.9
9	0.97	1.15	1.73	2.09	2.70	3.33	16.9	19.0	21.7	23.6	27.9	29.7
10	1.26	1.48	2.16	2.56	3.25	3.94	18.3	20.5	23.2	25.2	29.6	31.4
11	1.59	1.83	2.60	3.05	3.82	4.57	19.7	21.9	24.7	26.8	31.3	33.1
12	1.93	2.21	3.07	3.57	4.40	5.23	21.0	23.3	26.2	28.3	32.9	34.8
13	2.31	2.62	3.57	4.11	5.01	5.89	22.4	24.7	27.7	29.8	34.5	36.5
14	2.70	3.04	4.07	4.66	5.63	6.57	23.7	26.1	29.1	31.3	36.1	38.1
15	3.11	3.48	4.60	5.23	6.26	7.26	25.0	27.5	30.6	32.8	37.7	39.7
16	3.54	3.94	5.14	5.81	6.91	7.96	26.3	28.8	32.0	34.3	39.3	41.3
17	3.98	4.42	5.70	6.41	7.56	8.67	27.6	30.2	33.4	35.7	40.8	42.9
18	4.44	4.90	6.26	7.01	8.23	9.39	28.9	31.5	34.8	37.2	42.3	44.4
19	4.91	5.41	6.84	7.63	8.91	10.1	30.1	32.9	36.2	38.6	43.8	46.0
20	5.40	5.92	7.43	8.26	9.59	10.9	31.4	34.2	37.6	40.0	45.3	47.5
21	5.90	6.45	8.03	8.90	10.3	11.6	32.7	35.5	38.9	41.4	46.8	49.0
22	6.40	6.98	8.64	9.54	11.0	12.3	33.9	36.8	40.3	42.8	48.3	50.5
23	6.92	7.53	9.26	10.2	11.7	13.1	35.2	38.1	41.6	44.2	49.7	52.0
24	7.45	8.08	9.89	10.9	12.4	13.8	36.4	39.4	43.0	45.6	51.2	53.5
25	7.99	8.65	10.5	11.5	13.1	14.6	37.7	40.6	44.3	46.9	52.6	54.9
26	8.54	9.22	11.2	12.2	13.8	15.4	38.9	41.9	45.6	48.3	54.1	56.4
27	9.09	9.80	11.8	12.9	14.6	16.2	40.1	43.2	47.0	49.6	55.5	57.9
28	9.66	10.4	12.5	13.6	15.3	16.9	41.3	44.5	48.3	51.0	56.9	59.3
29	10.2	11.0	13.1	14.3	16.0	17.7	42.6	45.7	49.6	52.3	58.3	60.7
30	10.8	11.6	13.8	15.0	16.8	18.5	43.8	47.0	50.9	53.7	59.7	62.2
40	16.9	17.9	20.7	22.2	24.4	26.5	55.8	59.3	63.7	66.8	73.4	76.1
50	23.5	24.7	28.0	29.7	32.4	34.8	67.5	71.4	76.2	79.5	86.7	89.6
60	30.3	31.7	35.5	37.5	40.5	43.2	79.1	83.3	88.4	92.0	99.6	103.
70	37.5	39.0	43.3	45.4	48.8	51.7	90.5	95.0	100.	104.	112.	116.
80	44.8	46.5	51.2	53.5	57.2	60.4	102.	107.	112.	116.	125.	128.
90	52.3	54.2	59.2	61.8	65.6	69.1	113.	118.	124.	128.	137.	141.
100	59.9	61.9	67.3	70.1	74.2	77.9	124.	130.	136.	140.	149.	153.

Table 7. Poisson distribution. $P(X \leq x)$ where $X \sim \text{Po}(m)$.

x \ m	0.1	0.2	0.3	0.4	0.5	0.6	0.7	0.8	0.9
0	.90484	.81873	.74082	.67032	.60653	.54881	.49659	.44933	.40657
1	.99532	.98248	.96306	.93845	.90980	.87810	.84420	.80879	.77248
2	.99985	.99885	.99640	.99207	.98561	.97688	.96586	.95258	.93714
3	1.00000	.99994	.99973	.99922	.99825	.99664	.99425	.99092	.98654
4		1.00000	.99998	.99994	.99983	.99961	.99921	.99859	.99766
5			1.00000	1.00000	.99999	.99996	.99991	.99982	.99966
6					1.00000	1.00000	.99999	.99998	.99996
7							1.00000	1.00000	1.00000

x \ m	1.0	1.2	1.4	1.6	1.8	2.0	2.2	2.4	2.6
0	.36788	.30119	.24660	.20190	.16530	.13534	.11080	.09072	.07427
1	.73576	.66263	.59183	.52493	.46284	.40601	.35457	.30844	.26738
2	.91970	.87949	.83350	.78336	.73062	.67668	.62271	.56971	.51843
3	.98101	.96623	.94627	.92119	.89129	.85712	.81935	.77872	.73600
4	.99634	.99225	.98575	.97632	.96359	.94735	.92750	.90413	.87742
5	.99941	.99850	.99680	.99396	.98962	.98344	.97509	.96433	.95096
6	.99992	.99975	.99938	.99866	.99743	.99547	.99254	.98841	.98283
7	.99999	.99996	.99989	.99974	.99944	.99890	.99802	.99666	.99467
8	1.00000	1.00000	.99998	.99995	.99989	.99976	.99953	.99914	.99851
9			1.00000	.99999	.99998	.99995	.99990	.99980	.99962
10				1.00000	1.00000	.99999	.99998	.99996	.99991
11						1.00000	1.00000	.99999	.99998
12								1.00000	1.00000

x \ m	2.8	3.0	3.2	3.4	3.6	3.8	4.0	4.2	4.4
0	.06081	.04979	.04076	.03337	.02732	.02237	.01832	.01500	.01228
1	.23108	.19915	.17120	.14684	.12569	.10738	.09158	.07798	.06630
2	.46945	.42319	.37990	.33974	.30275	.26890	.23810	.21024	.18514
3	.69194	.64723	.60252	.55836	.51522	.47348	.43347	.39540	.35945
4	.84768	.81526	.78061	.74418	.70644	.66784	.62884	.58983	.55118
5	.93489	.91608	.89459	.87054	.84412	.81556	.78513	.75314	.71991
6	.97559	.96649	.95538	.94215	.92673	.90911	.88933	.86746	.84365
7	.99187	.98810	.98317	.97693	.96921	.95989	.94887	.93606	.92142
8	.99757	.99620	.99429	.99171	.98833	.98402	.97864	.97207	.96420
9	.99934	.99890	.99824	.99729	.99598	.99420	.99187	.98887	.98511
10	.99984	.99971	.99950	.99919	.99873	.99807	.99716	.99593	.99431
11	.99996	.99993	.99987	.99978	.99963	.99941	.99908	.99863	.99799
12	.99999	.99998	.99997	.99994	.99990	.99983	.99973	.99957	.99934
13	1.00000	1.00000	.99999	.99999	.99997	.99996	.99992	.99987	.99980
14			1.00000	1.00000	.99999	.99999	.99998	.99997	.99994
15					1.00000	1.00000	1.00000	.99999	.99998
16								1.00000	1.00000

Table 7 (*continued*)

x \ m	4.6	4.8	5.0	5.5	6.0	6.5	7.0	7.5	8.0
0	.01005	.00823	.00674	.00409	.00248	.00150	.00091	.00055	.00034
1	.05629	.04773	.04043	.02656	.01735	.01128	.00730	.00470	.00302
2	.16264	.14254	.12465	.08838	.06197	.04304	.02964	.02026	.01375
3	.32571	.29423	.26503	.20170	.15120	.11185	.08177	.05915	.04238
4	.51323	.47626	.44049	.35752	.28506	.22367	.17299	.13206	.09963
5	.68576	.65101	.61596	.52892	.44568	.36904	.30071	.24144	.19124
6	.81803	.79080	.76218	.68604	.60630	.52652	.44971	.37815	.31337
7	.90495	.88667	.86663	.80949	.74398	.67276	.59871	.52464	.45296
8	.95493	.94418	.93191	.89436	.84724	.79157	.72909	.66197	.59255
9	.98047	.97486	.96817	.94622	.91608	.87738	.83050	.77641	.71662
10	.99222	.98958	.98630	.97475	.95738	.93316	.90148	.86224	.81589
11	.99714	.99601	.99455	.98901	.97991	.96612	.94665	.92076	.88808
12	.99902	.99858	.99798	.99555	.99117	.98397	.97300	.95733	.93620
13	.99969	.99953	.99930	.99831	.99637	.99290	.98719	.97844	.96582
14	.99991	.99985	.99977	.99940	.99860	.99704	.99428	.98974	.98274
15	.99997	.99996	.99993	.99980	.99949	.99884	.99759	.99539	.99177
16	.99999	.99999	.99998	.99994	.99983	.99957	.99904	.99804	.99628
17	1.00000	1.00000	.99999	.99998	.99994	.99985	.99964	.99921	.99841
18			1.00000	.99999	.99998	.99995	.99987	.99970	.99935
19				1.00000	.99999	.99998	.99996	.99989	.99975
20					1.00000	1.00000	.99999	.99996	.99991
21							1.00000	.99999	.99997
22								1.00000	.99999
23									1.00000

Table 7 (*continued*)

x \ m	8.5	9.0	9.5	10.0	11.0	12.0	13.0	14.0	15.0
0	.00020	.00012	.00007	.00005	.00002	.00001	.00000	.00000	.00000
1	.00193	.00123	.00079	.00050	.00020	.00008	.00003	.00001	.00000
2	.00928	.00623	.00416	.00277	.00121	.00052	.00022	.00009	.00004
3	.03011	.02123	.01486	.01034	.00492	.00229	.00105	.00047	.00021
4	.07436	.05496	.04026	.02925	.01510	.00760	.00374	.00181	.00086
5	.14960	.11569	.08853	.06709	.03752	.02034	.01073	.00553	.00279
6	.25618	.20678	.16495	.13014	.07861	.04582	.02589	.01423	.00763
7	.38560	.32390	.26866	.22022	.14319	.08950	.05403	.03162	.01800
8	.52311	.45565	.39182	.33282	.23199	.15503	.09976	.06206	.03745
9	.65297	.58741	.52183	.45793	.34051	.24239	.16581	.10940	.06985
10	.76336	.70599	.64533	.58304	.45989	.34723	.25168	.17568	.11846
11	.84866	.80301	.75199	.69678	.57927	.46160	.35316	.26004	.18475
12	.90908	.87577	.83643	.79156	.68870	.57597	.46310	.35846	.26761
13	.94859	.92615	.89814	.86446	.78129	.68154	.57304	.46445	.36322
14	.97257	.95853	.94001	.91654	.85404	.77202	.67513	.57044	.46565
15	.98617	.97796	.96653	.95126	.90740	.84442	.76361	.66936	.56809
16	.99339	.98889	.98227	.97296	.94408	.89871	.83549	.75592	.66412
17	.99700	.99468	.99107	.98572	.96781	.93703	.89046	.82720	.74886
18	.99870	.99757	.99572	.99281	.98231	.96258	.93017	.88264	.81947
19	.99947	.99894	.99804	.99655	.99071	.97872	.95733	.92350	.87522
20	.99979	.99956	.99914	.99841	.99533	.98840	.97499	.95209	.91703
21	.99992	.99983	.99964	.99930	.99775	.99393	.98592	.97116	.94689
22	.99997	.99993	.99985	.99970	.99896	.99695	.99238	.98329	.96726
23	.99999	.99998	.99994	.99988	.99954	.99853	.99603	.99067	.98054
24	1.00000	.99999	.99998	.99995	.99980	.99931	.99801	.99498	.98884
25		1.00000	.99999	.99998	.99992	.99969	.99903	.99739	.99382
26			1.00000	.99999	.99997	.99987	.99955	.99869	.99669
27				1.00000	.99999	.99994	.99980	.99936	.99828
28					1.00000	.99998	.99991	.99970	.99914
29					1.00000	.99999	.99996	.99986	.99958
30						1.00000	.99998	.99994	.99980
31						1.00000	.99999	.99997	.99991
32							1.00000	.99999	.99996
33								1.00000	.99998
34								1.00000	.99999
35									1.00000

Table 8. Binomial distribution. $P(X \leq x)$ where $X \sim \text{Bin}(n, p)$. When $p > 1/2$, use the fact that $P(X \leq x) = P(Y \geq n - x)$ where $Y \sim \text{Bin}(n, 1 - p)$.

n	x \ p	0.05	0.10	0.15	0.20	0.25	0.30	0.40	0.50
2	0	.90250	.81000	.72250	.64000	.56250	.49000	.36000	.25000
	1	.99750	.99000	.97750	.96000	.93750	.91000	.84000	.75000
3	0	.85738	.72900	.61413	.51200	.42188	.34300	.21600	.12500
	1	.99275	.97200	.93925	.89600	.84375	.78400	.64800	.50000
	2	.99988	.99900	.99663	.99200	.98438	.97300	.93600	.87500
4	0	.81451	.65610	.52201	.40960	.31641	.24010	.12960	.06250
	1	.98598	.94770	.89048	.81920	.73828	.65170	.47520	.31250
	2	.99952	.99630	.98802	.97280	.94922	.91630	.82080	.68750
	3	.99999	.99990	.99949	.99840	.99609	.99190	.97440	.93750
5	0	.77378	.59049	.44371	.32768	.23730	.16807	.07776	.03125
	1	.97741	.91854	.83521	.73728	.63281	.52822	.33696	.18750
	2	.99884	.99144	.97339	.94208	.89648	.83692	.68256	.50000
	3	.99997	.99954	.99777	.99328	.98438	.96922	.91296	.81250
	4	1.00000	.99999	.99992	.99968	.99902	.99757	.98976	.96875
6	0	.73509	.53144	.37715	.26214	.17798	.11765	.04666	.01563
	1	.96723	.88573	.77648	.65536	.53394	.42018	.23328	.10938
	2	.99777	.98415	.95266	.90112	.83057	.74431	.54432	.34375
	3	.99991	.99873	.99411	.98304	.96240	.92953	.82080	.65625
	4	1.00000	.99994	.99960	.99840	.99536	.98907	.95904	.89063
	5	1.00000	1.00000	.99999	.99994	.99976	.99927	.99590	.98438
7	0	.69834	.47830	.32058	.20972	.13348	.08235	.02799	.00781
	1	.95562	.85031	.71658	.57672	.44495	.32942	.15863	.06250
	2	.99624	.97431	.92623	.85197	.75641	.64707	.41990	.22656
	3	.99981	.99727	.98790	.96666	.92944	.87396	.71021	.50000
	4	.99999	.99982	.99878	.99533	.98712	.97120	.90374	.77344
	5	1.00000	.99999	.99993	.99963	.99866	.99621	.98116	.93750
	6	1.00000	1.00000	1.00000	.99999	.99994	.99978	.99836	.99219
8	0	.66342	.43047	.27249	.16777	.10011	.05765	.01680	.00391
	1	.94276	.81310	.65718	.50332	.36708	.25530	.10638	.03516
	2	.99421	.96191	.89479	.79692	.67854	.55177	.31539	.14453
	3	.99963	.99498	.97865	.94372	.88618	.80590	.59409	.36328
	4	.99998	.99957	.99715	.98959	.97270	.94203	.82633	.63672
	5	1.00000	.99998	.99976	.99877	.99577	.98871	.95019	.85547
	6	1.00000	1.00000	.99999	.99992	.99962	.99871	.99148	.96484
	7	1.00000	1.00000	1.00000	1.00000	.99998	.99993	.99934	.99609
9	0	.63025	.38742	.23162	.13422	.07508	.04035	.01008	.00195
	1	.92879	.77484	.59948	.43621	.30034	.19600	.07054	.01953
	2	.99164	.94703	.85915	.73820	.60068	.46283	.23179	.08984
	3	.99936	.99167	.96607	.91436	.83427	.72966	.48261	.25391
	4	.99997	.99911	.99437	.98042	.95107	.90119	.73343	.50000
	5	1.00000	.99994	.99937	.99693	.99001	.97471	.90065	.74609
	6	1.00000	1.00000	.99995	.99969	.99866	.99571	.97497	.91016
	7	1.00000	1.00000	1.00000	.99998	.99989	.99957	.99620	.98047
	8	1.00000	1.00000	1.00000	1.00000	1.00000	.99998	.99974	.99805

Table 8 (*continued*)

n	x \ p	0.05	0.10	0.15	0.20	0.25	0.30	0.40	0.50
10	0	.59874	.34868	.19687	.10737	.05631	.02825	.00605	.00098
	1	.91386	.73610	.54430	.37581	.24403	.14931	.04636	.01074
	2	.98850	.92981	.82020	.67780	.52559	.38278	.16729	.05469
	3	.99897	.98720	.95003	.87913	.77588	.64961	.38228	.17188
	4	.99994	.99837	.99013	.96721	.92187	.84973	.63310	.37695
	5	1.00000	.99985	.99862	.99363	.98027	.95265	.83376	.62305
	6	1.00000	.99999	.99987	.99914	.99649	.98941	.94524	.82813
	7	1.00000	1.00000	.99999	.99992	.99958	.99841	.98771	.94531
	8	1.00000	1.00000	1.00000	1.00000	.99997	.99986	.99832	.98926
	9	1.00000	1.00000	1.00000	1.00000	1.00000	.99999	.99990	.99902
11	0	.56880	.31381	.16734	.08590	.04224	.01977	.00363	.00049
	1	.89811	.69736	.49219	.32212	.19710	.11299	.03023	.00586
	2	.98476	.91044	.77881	.61740	.45520	.31274	.11892	.03271
	3	.99845	.98147	.93056	.83886	.71330	.56956	.29628	.11328
	4	.99989	.99725	.98411	.94959	.88537	.78970	.53277	.27441
	5	.99999	.99970	.99734	.98835	.96567	.92178	.75350	.50000
	6	1.00000	.99998	.99968	.99803	.99244	.97838	.90065	.72559
	7	1.00000	1.00000	.99997	.99976	.99881	.99571	.97072	.88672
	8	1.00000	1.00000	1.00000	.99998	.99987	.99942	.99408	.96729
	9	1.00000	1.00000	1.00000	1.00000	.99999	.99995	.99927	.99414
	10	1.00000	1.00000	1.00000	1.00000	1.00000	1.00000	.99996	.99951
12	0	.54036	.28243	.14224	.06872	.03168	.01384	.00218	.00024
	1	.88164	.65900	.44346	.27488	.15838	.08503	.01959	.00317
	2	.98043	.88913	.73582	.55835	.39068	.25282	.08344	.01929
	3	.99776	.97436	.90779	.79457	.64878	.49252	.22534	.07300
	4	.99982	.99567	.97608	.92744	.84236	.72366	.43818	.19385
	5	.99999	.99946	.99536	.98059	.94560	.88215	.66521	.38721
	6	1.00000	.99995	.99933	.99610	.98575	.96140	.84179	.61279
	7	1.00000	1.00000	.99993	.99942	.99722	.99051	.94269	.80615
	8	1.00000	1.00000	.99999	.99994	.99961	.99831	.98473	.92700
	9	1.00000	1.00000	1.00000	1.00000	.99996	.99979	.99719	.98071
	10	1.00000	1.00000	1.00000	1.00000	1.00000	.99998	.99968	.99683
	11	1.00000	1.00000	1.00000	1.00000	1.00000	1.00000	.99998	.99976
13	0	.51334	.25419	.12091	.05498	.02376	.00969	.00131	.00012
	1	.86458	.62134	.39828	.23365	.12671	.06367	.01263	.00171
	2	.97549	.86612	.69196	.50165	.33260	.20248	.05790	.01123
	3	.99690	.96584	.88200	.74732	.58425	.42061	.16858	.04614
	4	.99971	.99354	.96584	.90087	.79396	.65431	.35304	.13342
	5	.99998	.99908	.99247	.96996	.91979	.83460	.57440	.29053
	6	1.00000	.99990	.99873	.99300	.97571	.93762	.77116	.50000
	7	1.00000	.99999	.99984	.99875	.99435	.98178	.90233	.70947
	8	1.00000	1.00000	.99998	.99983	.99901	.99597	.96792	.86658
	9	1.00000	1.00000	1.00000	.99998	.99987	.99935	.99221	.95386
	10	1.00000	1.00000	1.00000	1.00000	.99999	.99993	.99868	.98877
	11	1.00000	1.00000	1.00000	1.00000	1.00000	1.00000	.99986	.99829
	12	1.00000	1.00000	1.00000	1.00000	1.00000	1.00000	.99999	.99988

Table 8 (*continued*)

n	x \ p	0.05	0.10	0.15	0.20	0.25	0.30	0.40	0.50
14	0	.48767	.22877	.10277	.04398	.01782	.00678	.00078	.00006
	1	.84701	.58463	.35667	.19791	.10097	.04748	.00810	.00092
	2	.96995	.84164	.64791	.44805	.28113	.16084	.03979	.00647
	3	.99583	.95587	.85349	.69819	.52134	.35517	.12431	.02869
	4	.99957	.99077	.95326	.87016	.74153	.58420	.27926	.08978
	5	.99997	.99853	.98847	.95615	.88833	.78052	.48585	.21198
	6	1.00000	.99982	.99779	.98839	.96173	.90672	.69245	.39526
	7	1.00000	.99998	.99967	.99760	.98969	.96853	.84986	.60474
	8	1.00000	1.00000	.99996	.99962	.99785	.99171	.94168	.78802
	9	1.00000	1.00000	1.00000	.99995	.99966	.99833	.98249	.91022
	10	1.00000	1.00000	1.00000	1.00000	.99996	.99975	.99609	.97131
	11	1.00000	1.00000	1.00000	1.00000	1.00000	.99997	.99939	.99353
	12	1.00000	1.00000	1.00000	1.00000	1.00000	1.00000	.99994	.99908
	13	1.00000	1.00000	1.00000	1.00000	1.00000	1.00000	1.00000	.99994
15	0	.46329	.20589	.08735	.03518	.01336	.00475	.00047	.00003
	1	.82905	.54904	.31859	.16713	.08018	.03527	.00517	.00049
	2	.96380	.81594	.60423	.39802	.23609	.12683	.02711	.00369
	3	.99453	.94444	.82266	.64816	.46129	.29687	.09050	.01758
	4	.99939	.98728	.93829	.83577	.68649	.51549	.21728	.05923
	5	.99995	.99775	.98319	.93895	.85163	.72162	.40322	.15088
	6	1.00000	.99969	.99639	.98194	.94338	.86886	.60981	.30362
	7	1.00000	.99997	.99939	.99576	.98270	.94999	.78690	.50000
	8	1.00000	1.00000	.99992	.99922	.99581	.98476	.90495	.69638
	9	1.00000	1.00000	.99999	.99989	.99921	.99635	.96617	.84912
	10	1.00000	1.00000	1.00000	.99999	.99988	.99933	.99065	.94077
	11	1.00000	1.00000	1.00000	1.00000	.99999	.99991	.99807	.98242
	12	1.00000	1.00000	1.00000	1.00000	1.00000	.99999	.99972	.99631
	13	1.00000	1.00000	1.00000	1.00000	1.00000	1.00000	.99997	.99951
	14	1.00000	1.00000	1.00000	1.00000	1.00000	1.00000	1.00000	.99997
16	0	.44013	.18530	.07425	.02815	.01002	.00332	.00028	.00002
	1	.81076	.51473	.28390	.14074	.06348	.02611	.00329	.00026
	2	.95706	.78925	.56138	.35184	.19711	.09936	.01834	.00209
	3	.99300	.93159	.78989	.59813	.40499	.24586	.06515	.01064
	4	.99914	.98300	.92095	.79825	.63019	.44990	.16657	.03841
	5	.99992	.99670	.97646	.91831	.81035	.65978	.32884	.10506
	6	.99999	.99950	.99441	.97334	.92044	.82469	.52717	.22725
	7	1.00000	.99994	.99894	.99300	.97287	.92565	.71606	.40181
	8	1.00000	.99999	.99984	.99852	.99253	.97433	.85773	.59819
	9	1.00000	1.00000	.99998	.99975	.99836	.99287	.94168	.77275
	10	1.00000	1.00000	1.00000	.99997	.99971	.99843	.98086	.89494
	11	1.00000	1.00000	1.00000	1.00000	.99996	.99973	.99510	.96159
	12	1.00000	1.00000	1.00000	1.00000	1.00000	.99997	.99906	.98936
	13	1.00000	1.00000	1.00000	1.00000	1.00000	1.00000	.99987	.99791
	14	1.00000	1.00000	1.00000	1.00000	1.00000	1.00000	.99999	.99974
	15	1.00000	1.00000	1.00000	1.00000	1.00000	1.00000	1.00000	.99998

Table 8 (*continued*)

n	x	p = 0.05	0.10	0.15	0.20	0.25	0.30	0.40	0.50
17	0	.41812	.16677	.06311	.02252	.00752	.00233	.00017	.00001
	1	.79223	.48179	.25245	.11822	.05011	.01928	.00209	.00014
	2	.94975	.76180	.51976	.30962	.16370	.07739	.01232	.00117
	3	.99120	.91736	.75561	.54888	.35302	.20191	.04642	.00636
	4	.99884	.97786	.90129	.75822	.57389	.38869	.12600	.02452
	5	.99988	.99533	.96813	.89430	.76531	.59682	.26393	.07173
	6	.99999	.99922	.99172	.96234	.89292	.77522	.44784	.16615
	7	1.00000	.99989	.99826	.98907	.95976	.89536	.64051	.31453
	8	1.00000	.99999	.99970	.99742	.98762	.95972	.80106	.50000
	9	1.00000	1.00000	.99996	.99951	.99690	.98731	.90810	.68547
	10	1.00000	1.00000	1.00000	.99992	.99937	.99676	.96519	.83385
	11	1.00000	1.00000	1.00000	.99999	.99990	.99934	.98941	.92827
	12	1.00000	1.00000	1.00000	1.00000	.99999	.99990	.99748	.97548
	13	1.00000	1.00000	1.00000	1.00000	1.00000	.99999	.99955	.99364
	14	1.00000	1.00000	1.00000	1.00000	1.00000	1.00000	.99994	.99883
	15	1.00000	1.00000	1.00000	1.00000	1.00000	1.00000	1.00000	.99986
	16	1.00000	1.00000	1.00000	1.00000	1.00000	1.00000	1.00000	.99999
18	0	.39721	.15009	.05365	.01801	.00564	.00163	.00010	.00000
	1	.77352	.45028	.22405	.09908	.03946	.01419	.00132	.00007
	2	.94187	.73380	.47966	.27134	.13531	.05995	.00823	.00066
	3	.98913	.90180	.72024	.50103	.30569	.16455	.03278	.00377
	4	.99845	.97181	.87944	.71635	.51867	.33265	.09417	.01544
	5	.99983	.99358	.95810	.86708	.71745	.53438	.20876	.04813
	6	.99998	.99883	.98818	.94873	.86102	.72170	.37428	.11894
	7	1.00000	.99983	.99728	.98372	.94305	.85932	.56344	.24034
	8	1.00000	.99998	.99949	.99575	.98065	.94041	.73684	.40726
	9	1.00000	1.00000	.99992	.99909	.99458	.97903	.86529	.59274
	10	1.00000	1.00000	.99999	.99984	.99876	.99393	.94235	.75966
	11	1.00000	1.00000	1.00000	.99998	.99977	.99857	.97972	.88106
	12	1.00000	1.00000	1.00000	1.00000	.99997	.99973	.99425	.95187
	13	1.00000	1.00000	1.00000	1.00000	1.00000	.99996	.99872	.98456
	14	1.00000	1.00000	1.00000	1.00000	1.00000	1.00000	.99979	.99623
	15	1.00000	1.00000	1.00000	1.00000	1.00000	1.00000	.99997	.99934
	16	1.00000	1.00000	1.00000	1.00000	1.00000	1.00000	1.00000	.99993
	17	1.00000	1.00000	1.00000	1.00000	1.00000	1.00000	1.00000	1.00000
19	0	.37735	.13509	.04560	.01441	.00423	.00114	.00006	.00000
	1	.75471	.42026	.19849	.08287	.03101	.01042	.00083	.00004
	2	.93345	.70544	.44132	.23689	.11134	.04622	.00546	.00036
	3	.98676	.88500	.68415	.45509	.26309	.13317	.02296	.00221
	4	.99799	.96481	.85556	.67329	.46542	.28222	.06961	.00961
	5	.99976	.99141	.94630	.83694	.66776	.47386	.16292	.03178
	6	.99998	.99830	.98367	.93240	.82512	.66550	.30807	.08353
	7	1.00000	.99973	.99592	.97672	.92254	.81803	.48778	.17964
	8	1.00000	.99996	.99916	.99334	.97125	.91608	.66748	.32380
	9	1.00000	1.00000	.99986	.99842	.99110	.96745	.81391	.50000
	10	1.00000	1.00000	.99998	.99969	.99771	.98946	.91153	.67620
	11	1.00000	1.00000	1.00000	.99995	.99952	.99718	.96477	.82036
	12	1.00000	1.00000	1.00000	.99999	.99992	.99938	.98844	.91647
	13	1.00000	1.00000	1.00000	1.00000	.99999	.99989	.99693	.96822
	14	1.00000	1.00000	1.00000	1.00000	1.00000	.99999	.99936	.99039
	15	1.00000	1.00000	1.00000	1.00000	1.00000	1.00000	.99990	.99779
	16	1.00000	1.00000	1.00000	1.00000	1.00000	1.00000	.99999	.99964
	17	1.00000	1.00000	1.00000	1.00000	1.00000	1.00000	1.00000	.99996
	18	1.00000	1.00000	1.00000	1.00000	1.00000	1.00000	1.00000	1.00000

Table 9. Random numbers

44	53	73	71	07	11	59	36	46	91	29	97	47	35	58	34	71	60	92	01
54	58	57	28	07	96	51	20	15	18	46	57	41	39	91	92	18	53	10	56
60	25	02	54	51	99	39	65	52	92	70	44	00	96	18	21	05	77	83	99
53	79	36	60	06	19	28	57	98	87	81	21	42	53	93	55	23	58	39	40
02	18	66	31	14	28	98	57	72	52	33	22	60	35	46	07	40	92	83	66
47	16	70	29	93	75	03	00	64	10	57	59	20	39	81	62	97	53	96	36
07	21	03	89	35	84	74	98	42	59	57	18	65	39	77	80	13	90	32	88
74	56	95	22	07	52	14	36	49	72	13	60	11	83	91	99	79	26	21	32
14	19	50	15	51	55	04	76	02	99	80	20	52	95	51	28	63	60	66	54
70	85	49	27	84	39	98	53	92	63	86	10	54	74	62	55	06	65	49	15
59	02	46	96	99	29	26	78	87	61	37	16	59	92	04	38	26	89	24	51
74	93	70	32	63	22	93	38	30	68	13	98	85	99	31	16	15	56	20	72
81	56	27	22	18	93	77	93	37	32	68	92	24	17	74	98	41	64	42	66
22	66	96	25	80	89	35	79	00	77	32	10	43	45	36	71	44	87	69	92
16	70	31	78	43	35	95	08	87	00	10	37	84	57	98	95	44	73	58	87
54	91	63	91	74	28	63	64	40	75	80	34	64	01	13	07	31	46	36	61
02	29	96	52	14	42	18	09	76	51	99	38	76	00	13	17	73	04	09	01
05	57	10	20	81	25	14	79	88	51	95	58	86	52	02	11	13	20	57	67
77	23	59	32	67	02	82	85	41	74	74	82	38	32	58	50	66	44	34	96
13	50	73	99	39	69	26	13	80	93	15	69	47	88	38	71	26	97	69	98
79	38	57	06	40	02	25	24	20	56	72	56	07	43	70	83	60	80	69	59
16	59	91	14	86	47	36	53	55	87	76	54	84	95	60	74	10	65	11	41
43	62	29	60	70	30	86	12	51	84	30	49	21	19	86	04	93	00	52	79
52	70	01	55	59	47	81	86	50	22	16	02	35	62	05	08	01	10	20	84
09	83	11	83	95	73	00	37	69	47	87	48	18	48	44	99	02	92	21	41
58	72	39	07	96	55	99	00	02	84	73	99	37	75	10	61	38	20	33	13
14	87	41	62	53	18	07	86	14	31	79	41	35	16	82	56	25	43	12	35
71	51	45	59	34	60	28	82	49	61	85	34	28	21	13	19	56	83	88	18
95	63	56	84	80	54	42	48	23	24	44	14	09	13	36	62	99	40	65	47
29	96	55	98	10	49	05	20	28	41	88	93	45	90	52	71	95	86	22	84
90	98	95	37	15	68	45	09	33	13	20	56	78	25	44	06	61	71	15	65
69	93	07	12	63	10	67	01	78	12	20	64	25	67	65	03	91	17	73	00
35	80	94	08	95	47	52	58	81	86	44	53	79	39	44	74	50	23	01	76
30	31	38	87	29	29	53	14	35	60	20	34	07	41	87	04	82	63	78	53
70	95	91	85	58	79	00	81	06	31	53	94	50	44	73	55	02	85	60	68
48	96	85	13	47	95	99	44	37	73	23	92	27	98	58	61	05	12	75	31
30	32	24	56	40	51	28	66	44	35	85	66	29	27	70	35	56	43	29	28
60	77	87	75	53	95	43	81	21	39	69	25	24	27	14	61	98	51	02	21
55	78	30	07	80	51	73	01	34	85	78	56	53	74	70	02	48	85	47	44
06	60	81	82	86	16	23	11	25	46	94	19	34	16	93	92	99	68	95	10

Table 9 (*continued*)

80	21	46	27	14	65	72	72	12	69	69	51	59	75	13	44	17	99	50	04
21	34	05	62	82	47	75	21	86	80	35	63	95	50	33	42	44	50	92	87
44	47	11	02	81	83	63	69	16	75	96	41	85	15	34	48	99	71	76	94
43	85	95	60	79	73	38	01	43	29	31	45	45	19	71	97	72	85	31	38
84	45	61	21	18	91	83	77	84	90	95	11	68	84	73	01	32	90	63	03
11	01	89	46	15	84	50	36	32	81	19	50	22	09	44	46	20	60	50	51
38	02	34	71	63	77	69	86	53	01	05	49	47	68	65	92	54	43	47	18
60	71	25	07	27	67	46	14	91	23	36	68	61	08	91	75	25	63	85	14
44	06	67	39	52	28	61	81	61	96	64	44	58	54	50	07	02	18	68	26
30	80	40	29	53	08	66	23	57	43	19	86	02	04	48	73	26	81	75	14
37	43	99	13	24	31	94	51	45	63	07	81	38	31	64	34	14	01	61	15
57	17	72	02	30	96	47	51	68	29	07	91	82	83	53	27	58	02	56	39
57	02	66	81	15	75	06	83	43	89	74	50	26	85	44	61	27	81	65	72
79	70	59	11	96	17	25	84	62	67	37	70	38	34	43	24	61	13	67	75
41	70	06	29	64	47	35	65	92	61	01	37	59	05	29	75	78	45	17	85
34	25	37	45	88	61	39	11	75	45	46	11	08	44	57	52	71	02	44	12
26	34	57	45	08	33	18	83	29	67	27	30	75	77	56	64	06	81	54	41
58	70	45	91	43	13	26	18	46	51	72	02	28	01	31	98	26	56	26	35
49	82	56	18	85	23	93	27	93	95	88	16	10	90	62	15	47	75	14	29
90	93	19	37	00	62	24	94	12	72	52	30	38	91	03	50	63	62	49	33
49	02	40	34	31	03	98	82	20	31	21	83	03	30	85	15	40	16	36	34
67	82	98	70	95	94	69	26	10	96	24	83	73	03	11	96	20	09	52	93
62	81	47	80	85	60	69	37	49	64	66	18	91	85	62	53	21	91	55	46
26	23	18	77	67	99	01	01	78	10	26	48	73	24	92	22	35	84	72	03
27	06	15	45	83	83	44	79	16	81	58	09	11	44	31	15	29	87	09	51
05	04	18	47	52	62	55	07	54	02	94	12	73	42	83	17	45	74	46	50
80	65	81	16	64	58	62	96	60	70	37	44	01	92	29	89	00	93	37	36
42	13	34	66	88	71	70	32	75	59	66	64	12	43	22	67	87	67	12	21
59	46	83	81	65	74	58	75	17	17	38	09	99	19	94	63	72	95	75	90
73	38	05	22	13	15	83	62	78	69	80	90	27	16	47	61	99	50	07	05
02	36	57	26	24	18	72	03	25	11	99	93	41	10	63	24	83	80	61	00
37	65	68	03	64	80	31	86	01	19	74	78	56	48	95	86	18	10	69	88
46	22	42	39	77	76	40	69	22	39	59	83	66	54	74	59	70	38	33	54
71	82	60	96	79	54	52	90	81	97	84	16	36	27	04	29	82	36	35	58
29	92	75	08	63	38	99	59	45	89	54	66	11	39	64	57	71	54	28	38
12	77	19	88	95	25	84	63	56	90	49	91	05	39	10	78	29	15	43	03
88	34	94	21	19	89	88	62	32	47	23	29	13	51	72	04	23	17	83	40
99	37	82	68	50	79	64	92	64	86	05	91	00	73	53	21	96	34	30	10
61	35	36	65	82	19	43	64	20	02	01	62	10	78	33	88	64	14	37	49
67	50	86	22	81	05	30	64	41	72	91	03	59	16	68	44	20	80	34	67

Table 9 (*continued*)

85	82	38	27	90	13	03	53	46	41	28	50	39	09	86	97	31	32	26	50
56	03	71	39	26	90	18	67	26	35	43	14	19	55	93	35	39	42	92	61
97	63	38	94	80	61	05	16	49	52	41	81	39	29	69	18	61	59	88	33
02	84	71	05	21	22	68	88	56	64	00	12	17	78	67	83	90	06	42	79
36	66	74	55	91	48	86	42	65	21	76	43	46	27	18	39	93	83	61	84
42	80	27	08	06	88	15	15	69	46	99	85	92	73	27	74	11	61	22	09
38	77	83	97	59	64	84	17	33	37	72	24	98	90	72	47	63	90	81	91
16	79	74	35	16	75	98	35	01	68	77	20	80	27	09	95	40	94	68	39
42	85	03	08	21	94	36	30	89	87	66	11	32	07	67	29	10	70	87	41
59	69	50	75	91	71	53	37	90	18	71	05	20	27	52	35	14	91	19	57
84	77	70	74	16	27	80	67	72	58	96	90	86	63	42	74	70	08	17	22
10	35	93	15	66	63	20	07	75	83	20	27	48	62	93	81	70	42	11	48
03	41	23	84	82	51	53	16	18	39	99	51	97	47	34	68	81	92	06	22
05	68	40	41	80	39	34	32	92	50	61	74	02	17	69	20	46	33	83	02
35	64	99	24	54	52	93	65	65	16	12	80	04	46	79	99	81	11	94	27
83	52	30	42	70	87	34	01	79	08	31	32	20	82	19	41	79	51	71	06
18	33	36	83	71	19	37	01	51	76	73	65	43	48	17	55	07	51	18	25
81	77	98	06	74	94	57	01	73	44	68	90	39	43	79	29	08	85	14	46
90	36	71	47	71	38	23	12	13	09	20	94	52	41	84	23	97	00	14	05
36	30	83	19	29	48	41	19	13	45	09	36	97	88	87	73	69	21	48	12
88	60	41	05	91	97	89	84	89	00	90	53	68	11	18	91	88	46	21	53
87	98	23	69	73	35	22	43	34	98	92	56	31	05	81	61	99	48	12	89
50	94	84	44	68	85	71	06	16	38	20	30	29	46	56	45	18	75	76	56
69	69	54	43	43	44	39	60	76	92	54	38	79	81	97	80	37	52	43	66
13	24	38	51	40	87	07	66	32	10	49	14	73	34	36	75	24	77	17	03
22	30	15	30	77	62	29	59	75	14	33	69	78	03	75	16	20	22	78	30
14	37	40	15	45	92	36	50	74	03	13	90	36	62	95	66	44	37	80	81
82	69	43	16	12	76	30	25	69	51	67	38	65	59	50	59	86	44	54	69
92	23	27	21	19	88	93	46	79	05	50	48	57	18	71	07	14	43	04	78
87	73	74	90	85	75	79	48	96	90	92	31	79	37	61	95	71	07	10	69
83	67	38	58	01	61	17	42	86	04	97	74	73	89	01	85	74	84	27	80
74	30	48	38	35	44	13	13	92	20	46	37	56	23	45	12	14	97	83	20
26	58	09	14	28	99	46	24	31	87	93	56	21	63	23	88	59	46	84	75
81	27	01	48	98	73	71	10	96	27	67	42	35	06	39	98	52	03	10	07
57	83	78	75	37	90	17	82	53	41	74	81	39	27	74	03	09	17	15	52
46	51	70	08	13	48	89	25	44	00	93	34	52	74	82	39	22	11	29	19
15	30	83	30	66	21	66	01	88	54	78	37	65	69	92	17	59	84	56	96
06	91	94	04	16	58	05	46	75	26	60	01	45	12	10	23	62	09	77	44
36	85	60	66	53	81	33	70	74	14	43	11	36	76	15	19	48	36	46	97
98	34	10	26	45	88	56	60	26	91	07	30	53	10	61	39	09	86	92	67

Table 9 (*continued*)

58	37	49	70	34	55	54	77	49	07	06	92	89	36	79	95	13	58	21	41
60	67	55	59	38	28	81	55	35	85	70	08	11	54	73	73	02	27	11	78
19	73	85	30	48	32	66	07	50	22	04	65	61	36	22	33	92	40	19	16
29	77	67	93	32	64	16	19	38	93	88	24	58	32	82	12	77	21	73	64
56	80	07	33	32	99	08	50	15	46	76	20	92	64	83	21	22	69	77	09
43	53	83	13	65	85	99	38	74	05	97	64	31	31	28	45	72	56	13	11
07	46	52	68	23	44	17	93	81	67	58	43	17	07	97	46	41	31	34	08
40	82	41	08	74	76	67	01	80	06	36	16	68	40	46	59	24	18	70	09
09	59	57	20	60	54	30	23	86	72	88	21	75	53	04	96	86	15	26	00
57	51	78	65	97	27	59	94	93	86	42	68	05	45	75	41	71	96	82	43
00	06	60	78	78	17	85	27	67	48	03	43	02	89	39	57	95	08	91	73
31	47	32	72	71	24	11	06	51	30	52	07	82	34	52	79	50	76	85	02
17	74	99	30	17	21	19	93	62	83	42	96	37	03	42	19	05	97	67	15
00	59	41	15	33	55	62	24	91	28	03	21	35	94	14	61	00	46	17	73
97	51	11	63	13	52	70	09	07	64	41	68	17	82	49	67	53	71	91	12
01	74	41	84	22	08	48	35	52	65	34	97	01	13	00	73	57	94	17	44
79	25	34	64	03	98	75	62	42	80	38	45	79	13	97	90	78	15	00	53
72	78	70	64	74	69	07	27	70	31	82	22	19	79	46	04	58	08	20	02
99	83	05	21	27	47	72	40	02	17	71	15	62	85	25	75	16	19	32	25
51	61	67	44	28	28	76	87	82	11	84	85	25	79	90	40	43	74	66	34
96	12	61	21	20	86	98	30	27	63	77	66	01	85	70	10	06	69	25	15
75	09	67	11	20	70	94	04	27	95	78	72	58	01	69	70	47	80	91	29
06	00	40	52	21	03	91	19	52	05	93	86	36	00	69	82	84	54	16	11
81	09	09	53	88	83	97	63	42	68	01	71	54	31	22	81	08	14	32	73
67	35	80	01	66	85	89	96	16	32	58	62	03	17	59	78	88	60	43	00
07	50	31	57	71	56	23	54	65	19	91	70	51	58	85	60	65	63	28	54
17	03	18	56	94	20	28	47	56	30	08	81	40	25	79	87	56	74	42	70
90	18	69	11	04	75	09	62	32	36	86	56	87	69	97	95	54	15	16	60
93	93	91	05	42	95	47	60	10	86	81	31	84	11	66	95	25	85	52	09
69	02	62	01	26	28	94	77	82	05	22	16	99	51	93	73	13	58	32	78
33	93	37	35	47	98	82	23	15	90	14	99	74	62	57	90	33	81	11	03
79	88	47	16	74	02	14	85	52	15	37	39	25	92	13	82	79	78	16	95
74	88	94	32	47	16	71	23	09	27	46	73	45	65	90	60	17	47	54	41
60	65	60	44	86	86	08	74	79	52	69	11	02	80	94	10	90	63	04	00
54	68	99	86	80	36	53	48	29	86	13	40	37	09	03	92	15	73	21	09
48	19	41	71	98	66	12	31	01	04	56	20	68	02	72	43	84	01	34	79
10	19	90	83	83	47	64	85	13	54	53	88	86	81	32	74	64	45	48	96
81	39	26	85	50	30	63	44	56	59	34	54	60	45	86	70	97	79	43	21
80	29	04	11	93	36	41	21	98	18	04	05	30	68	15	93	01	51	74	89
97	11	53	73	78	65	01	01	80	05	42	01	15	97	73	78	68	85	69	12

Answers to Exercises

201. $\Omega = \{u_0, u_1, \ldots, u_{10}\}$, where u_k denotes "k hens sit on their roosts"
202. (a) $\{2, 3, \ldots, 7\}$; (b) $\{2, 3, 4, \ldots\}$
203. $\{x \mid 0 < x < 3\}$
204. AB: both A and B occur; AB^*: A occurs but not B; A^*B^*: neither A nor B occurs
205. $E = ABC \cup ABD \cup ACD \cup BCD$; $F = ABCD^* \cup ABC^*D \cup AB^*CD \cup A^*BCD$
206. 0.5
207. 0.5
208. 1/55
209. (a) 11/850; (b) 703/1,700; (c) 1/5,525
210. 4/7
211. 24/49
212. (a) 1/120; (b) 1/15; (c) 1/12
213. 1/4
214. (a) 39/49; (b) 143/41,650
215. (a) 0.7; (b) The wanted probability is 9/21, 8/21 and 4/21 for $k = 1, 2$, and 3
216. (a) 8/11; (b) 4/5; (c) 32/55
217. 3/5
218. (a) $3p(1-p)^2$; (b) $6p^2(1-p)^2/[1-(1-p)^4]$
219. A: $2/n$; B: $2/n$
221. 0.855
222. (a) 0.432; (b) 0.008; (c) 0.992
223. A and B are independent!
224. $(1-p)/(2-p)$
225. $166/6^5$
226. 2,197/20,825
227. (a) $\left[\binom{52}{13} - \binom{48}{13} - \binom{4}{1}\binom{48}{12}\right] \Big/ \left[\binom{52}{13} - \binom{48}{13}\right] = 0.370;$

(b) $1 - \binom{48}{12} \Big/ \binom{51}{12} = 0.561$

228. 37/43

301. $F_X(x) = \begin{cases} 0 & \text{if } x < a \\ \frac{1}{2} & \text{if } a \le x < b \\ 1 & \text{if } x \ge b \end{cases}$

302. 7/9, 1/3, 2/9

303. (a) 1/12; (b) 7/12; (c) 7/12, 1/4

304. (a) 20, 60; (b) $p_X(20) = 1/3$; $p_X(60) = 2/3$

305. (a) $p_X(k) = 0.6^k 0.4$ for $k = 0, 1, \ldots$ (geometric distribution);
(b) 27/125

306. (a) $p_Y(k) = 0.6^{k-1} 0.4$ for $k = 1, 2, \ldots$ (fft-distribution);
(b) 0.6

307. (a) $p_X(k) = \binom{3}{k} 0.4^k 0.6^{3-k}$ for $k = 0, 1, 2, 3$; (b) 44/125

308. (a) $p_X(k) = \binom{4}{k}\binom{6}{3-k} \Big/ \binom{10}{3}$ for $k = 0, 1, 2, 3$; (b) 1/3

309. $(1 - \ln 2)/2 = 0.153$

310. $c = 1/72$

311. $c = 1/(2\sqrt{2})$; $P(X > 0) = 1 - 1/\sqrt{2} = 0.293$

312. $c = 4\beta^{3/2}/\sqrt{\pi}$

313. $\sqrt{2}$

314. $-\frac{1}{3} \ln 2, 0, \ln 2$

315. $-a \ln \ln 2$

316. (a) 2/5; (b) 1/5

317. (a) $1 - e^{-1.2} = 0.699$; (b) $e^{-1.6} = 0.202$; (c) $e^{-1.2} - e^{-1.6} = 0.099$;
(d) $1 - e^{-1.2} + e^{-1.6} = 0.901$; (e) 0

318. $x_\alpha = 0.83, 1.52, 2.15$ for $\alpha = 0.5, 0.1, 0.01$

319. (a) 1/2; (b) 2

320. $x_\alpha = a \ln \dfrac{1 - \alpha}{\alpha}$

322. $p_X(k) = (k - 1)/2^{k-1}$ for $k = 4, 5, \ldots$

323. $p_X(k) = 1/(n - 1)$ for $k = 1, 2, \ldots, n - 1$

401. (a) 0.58; (b) 0.11;
(c) $p_X(j) = 0.28, 0.33, 0.29, 0.08, 0.02$ for $j = 0, 1, 2, 3, 4$; $p_Y(k) = 0.20, 0.28, 0.41, 0.11$ for $k = 1, 2, 3, 4$

402. $c = 2$

403. (a) $F_{X,Y}(x, y) = \left(\frac{1}{2} + \frac{1}{\pi} \arctan x\right)\left(\frac{1}{2} + \frac{1}{\pi} \arctan y\right)$
(b) $f_X(x) = \dfrac{1}{\pi(1 + x^2)}$; $f_Y(y) = \dfrac{1}{\pi(1 + y^2)}$

404. (a) 0.06; (b) 0.18

405.

	k_1	k_2	k_3
j_1	0.03	0.15	0.12
j_2	0.04	0.20	0.16
j_3	0.03	0.15	0.12

406. 1/40
407. 5/9
408. (a) $f_X(x) = \dfrac{1}{(1+x)^2}$ if $x \geq 0$; $f_Y(y) = \dfrac{1}{(1+y)^2}$ if $y \geq 0$.

(b) $F_{X,Y}(x, y) = 1 - \dfrac{1}{1+x} - \dfrac{1}{1+y} + \dfrac{1}{1+x+y}$ if $x \geq 0$, $y \geq 0$;

$F_X(x) = 1 - \dfrac{1}{1+x}$; $F_Y(y) = 1 - \dfrac{1}{1+y}$

(c) no

409. (a) $f_X(x) = \dfrac{2+(c+1)x}{c+3} e^{-x}$ if $x \geq 0$; (b) $c = 1$

410. $p_{X,Y}(j, k) = \begin{cases} 4^{j-1}5^{k-j-1}6^{-k} & \text{if } 1 \leq j < k;\ k = 2, 3, \ldots \\ 4^{k-1}5^{j-k-1}6^{-j} & \text{if } 1 \leq k < j;\ j = 2, 3, \ldots \end{cases}$

412. 7/8

501. $p_Y(k) = 1/6$ if $k = 2, 4, \ldots, 12$

502. $p_Y(k) = \begin{cases} 1/10 & \text{if } k = 0, 4, 5, 6 \\ 2/10 & \text{if } k = 1, 2, 3 \end{cases}$

503. Y is uniformly distributed over the interval (0, 1)

504. $f_Y(y) = \dfrac{2}{\pi} \cdot \dfrac{1}{1+y^2}$ for $y > 0$, that is $Y = 1/X$ has the same distribution as X

505. $p_Y(k) = (1 - e^{-1/a})e^{-k/a}$ for $k = 0, 1, 2, \ldots$, which is a geometric distribution

506. (a) $p_V(0) = 1/4$; $p_V(1) = 1/2$; $p_V(\sqrt{2}) = 1/4$;
(b) the same distribution as in (a)

507. (a) 0, 1, 2, 3, 4

(b)

k	0	1	2	3	4
number of terms	1	2	3	2	1

(c)

k	0	1	2	3	4
$p_Z(k)$	1/12	2/9	7/18	2/9	1/12

508. $p_{X+Y}(k) = (k+1)p^2q^k$ for $k = 0, 1, 2, \ldots$

509. (a) $p_X(k) = \dbinom{5}{k} 2^{-5}$ for $k = 0, 1, \ldots, 5$

(b) $p_{X-Y}(k) = \dbinom{5}{(k+5)/2} 2^{-5}$ for $k = -5, -3, \ldots, 5$

(c) $p_{|X-Y|}(k) = 2\binom{5}{(k+5)/2}2^{-5}$ for $k = 1, 3, 5$

510. $p_Z(k) = \begin{cases} 1/n & \text{if } k = 0 \\ 2(n-k)/n^2 & \text{if } k = 1, 2, \ldots, n-1 \end{cases}$

511. $f_{X+Y}(z) = 6(e^{-2z} - e^{-3z})$ if $z \geq 0$

512. $f_{X+Y}(z) = \dfrac{1}{2\pi}(\arctan(z+1) - \arctan(z-1))$

513. $f_{X+Y}(z) = \begin{cases} 1 - e^{-z} & \text{if } 0 < z < 1 \\ e^{-z}(e-1) & \text{if } z \geq 1 \end{cases}$

514. $F_{Z_+}(z) = (z/a)^2$ if $0 \leq z \leq a$
$F_{Z_-}(z) = 2z/a - z^2/a^2$ if $0 \leq z \leq a$

515. $f_Z(z) = \sum_1^n \lambda_k \exp(-z\sum_1^n \lambda_k)$ (This is also an exponential distribution)

516. (a) $(11/36)^8 = 0.0000760$; (b) $(25/36)^8 = 0.0541$

517. $f_Z(z) = 2/(1+z)^3$ if $z > 0$

518. $F_Z(z) = \begin{cases} 0 & \text{if } z < 0 \\ (1 - e^{-z/a})(1 - qe^{-z/a})^{-1} & \text{if } z \geq 0 \end{cases}$
$f_Z(z) = (p/a)e^{-z/a}(1 - qe^{-z/a})^{-2}$ if $z \geq 0$

519. 1/4

520. $F_Z(z) = \begin{cases} 0 & \text{if } z < 0 \\ 19/20 & \text{if } z = 0 \\ 1 - \dfrac{(z-1)^2}{20} & \text{if } 0 < z \leq 1 \\ 1 & \text{if } z > 1 \end{cases}$

601. 13/3
602. $2a/3$
604. 1.5
605. 2
606. $a/(a+1)$
607. 1/2
608. $\ln(1+a)$
609. 9, 1/9
610. 3/2, 3/4
611. (a) $2/3, 1/\sqrt{18}$; (b) $6m\sigma - 3\sigma^2 = 2\sqrt{2}/3 - 1/6$
(c) $1 - (m-\sigma)^2 = 1/2 + 2\sqrt{2}/9$ (note that $m + 2\sigma > 1$)
612. 4/45
613. 6.2, 0.05
614. $p_X(0) = 0.5$; $p_X(1) = 0.2$; $p_X(2) = 0.3$;
$p_Y(1) = 0.6$; $p_Y(2) = 0.4$; $E(X) = 0.8$; $E(Y) = 1.4$
The product XY is 0, 1, 2 with probabilities 0.5, 0.2, 0.3;
$E(XY) = 0.8$; $C(X, Y) = -0.32$
615. $E(X) = E(Y) = 15$; $V(X) = V(Y) = 50$; $E(XY) = 200$; $C(X, Y) = -25$; $\rho(X, Y) = -1/2$
617. $V(X) = 1$; $V(Y) = 2/3$; $\rho(X, Y) = 0$; X and Y are uncorrelated but not independent

618. (a) $2\sqrt{2RT/(M\pi)}$; (b) $3RT/(2N)$
619. (a) $1/k + 1 - (1-p)^k$; (b) 5
620. $E(Y) = \int_{-\infty}^{100} x f_X(x)\,dx + \int_{100}^{\infty} (x - 100) f_X(x)\,dx = E(X) - 100(1 - F_X(100))$ has minimum for $m = 109$

701. (a) 0, 1, 4; (b) b_1 is false, b_2 and b_3 are true
702. 0 and $3\sigma/2$
703. -6 and 180
705. 9
706. 36 and $\sqrt{3}$
707. 25
708. 286 and 8.7
709. $E(Y) \approx -1.0, -0.52, 0.48, 1.5$; $D(Y) \approx 0.043$ in all cases
710. 1.0
712. 1/2 and 1/12
713. (a) $E(X) = a + 1/2$; $V(X) = 1/12$; (b) $V_{\text{appr}}(Y) = \dfrac{1}{12}(a + 1/2)^{-4}$;
(c) $V(Y) = \dfrac{1}{a(a+1)} - (\ln(1 + 1/a))^2$; 0.19, 0.84, 0.9967

801. 0.0309, 0.197, 0.0309
802. (a) 3.090; (b) -3.090; (c) 1.960; (d) 1.282
803. 2 and 3
804. 1.000, 0.841, 0.159, 0.00135
805. 13.02
806. 0.97725, 0.97722
807. 0.9522; (a) 0.002; (b) 0.006; (c) 0.008
808. 4.98, 0.10
809. 0.20
810. $N(0, 5)$, $N(2, 5)$
811. (a) $N(250, 25)$, $N(50, 25)$, $N(125, 6.25)$;
(b) 0.0694, 0.0228, 0.0455
812. (a) $N(1{,}360, 1{,}525)$, $N(80, 1{,}525)$; (b) 0.153; (c) 0.0202
813. (a) $N(8{,}640, 30^2 \cdot 12)$; (b) 0.9938; (c) 0.0000
814. $N(1, 2)$
815. (a) $N(0, 0.2^2/n)$; (b) 0.317; (c) 0.0455; (d) $n \geq 4{,}331$
816. (a) 0.0124; (b) 0.00616; (c) 27 pills
817. 0.968, 0.724, 0.841
818. (a) 65, 0.2; (b) 0.933
819. 0.159
820. e^b
821. 0.006
822. 0.9995

901. Bin(15, 1/2)
902. (a) 3/4; (b) Bin(5, 3/4); (c) 135/512

903. (a) Bin(8, 1/2); (b) 35/128
904. 0.0706
905. 0.549, 0.198
906. (a) Bin(12, 0.80); (b) Bin(12, 0.20); (c) 0.653; (d) 0.653
907. 0.771
908. (a) Bin(288, 1/3); (b) 96, 64; (c) yes; (d) 0.713, 0.974
909. (a) Bin(48, 1/16); (b) 3, 45/16;
(c) normal approximation should not be used; use Poisson approximation;
(d) 0.647, 0.185
910. (a) 0.652; (b) 0.450; (c) 0.0167
911. (a) Bin(n_A, 0.95), Bin(n_B, 0.90); (b) 0.0104
912. 0.0279
913. $E(X) = 60$ in both cases. $V(X) = npq = 24$ both for $N = 200$ and $N = 1{,}000$; $V(X) = npq\dfrac{N-n}{N-1} = 24\dfrac{N-100}{N-1}$, which for $N = 200$ and $N = 1{,}000$ becomes 12.1 and 21.6 (The first variance refers to drawing with replacement, the second to drawing without replacement.)
914. (a) yes; (b) yes; (c) yes; (d) yes
915. (a) 2/3; (b) 0.590 (use binomial approximation)
916. 0.977 (use Poisson approximation)
917. 0.0183
918. 0.132, 0.679, 0.224, 0.137
919. (a) 0.407; (b) 0.331; (c) 0.147
920. 2
921. (a) 0.000; (b) 0.285
922. 0.0769
923. $p_{Y_1,Y_2,Y_3}(k_1, k_2, k_3) = \dfrac{n!}{k_1!k_2!k_3!}(1/2)^{k_1}(1/6)^{k_2}(1/3)^{k_3}$ where $k_1 + k_2 + k_3 = n$
924. Bin(n, 1/2)
925. $p_X(k) = \binom{2n-k}{n}\left(\dfrac{1}{2}\right)^{2n-k}$, $k = 0, 1, \ldots, n$
926. Po(λp)

1101. The arithmetic mean, median, variance, standard deviation, and coefficient of variation are, respectively, 54.0, 52.5, 119, 11 and 20
1102. The arithmetic mean and standard deviation are, respectively, 14.9 and 4.03
1103. The arithmetic mean and standard deviation are, respectively, 174.6 and 32.7
1104. The arithmetic mean and variance are, respectively, 1.78 and 0.2043
1105. The arithmetic mean and standard deviation are, respectively, 9.506 and 0.200

1201. $X \sim N(g, 15^2/2)$
1202. (a) p_1^* can be 0 and 1; p_2^* can be 0, 1/2 and 1; (b) yes;
(c) $V(p_1^*) = pq$; $V(p_2^*) = pq/2$; the efficiency of p_1^* relative to p_2^* is 1/2
1203. (b) $2/n$
1204. (a) 0.82; (b) 0.85
1205. (a) $L(\theta) = \theta(1-\theta)^{4-1} \cdot \ldots \cdot \theta(1-\theta)^{1-1} = \theta^6(1-\theta)^{18}$;
(b) $L(\theta)$ is largest for $\theta = 1/4$; this is the ML estimate

1206. (a) $L(\theta) = \theta^2 \cdot 2.16^{-\theta-1}$; $L(\theta) = 0.40, 0.41$ and 0.34 for $\theta = 2, 3$ and 4;
(b) $L(\theta)$ is largest when $\theta = 3$; this is the ML estimate
1207. The ML estimate is $-n/\sum_1^n \ln x_i$
1208. The ML estimate is $\sum_1^n x_i^2/2n$
1209. (a) $1/(\theta - 1)$; (b) The LS estimate is 3
1210. The LS estimate is 80.77
1211. 1.73
1212. (a) 0.184, 0.232, 0.158; (b) 0.017
1213. 0.067
1214. (a) 0.40; (b) $\sqrt{\lambda}/10$
1215. (a) 0.64; (b) $\sqrt{pq/25}$; (c) 0.096
1216. (a) 0.64; (b) 0.048; (c) 0.066
1217. 80.9, 0.10
1218. $m^* \approx 7.6$; $\sigma^* \approx 2.4$
1219. (a) $\bar{x}^2$; (b) $\bar{x}^2 - \dfrac{\sigma^2}{n}$
1220. (b) 0.69
1221. $N^* = 5$
1222. $\dfrac{n}{n-1} x\left(1 - \dfrac{x}{n}\right)$
1223. (b) $1/n$

1301. (a) 0.21
(b) The number of intervals that miss the target is Bin(15, 0.10). Table 8 shows that $P(X = 0) < P(X = 1) > P(X = 2) > \ldots$; hence it is most probable that one single interval misses the target
1302. (271, 39,500)
1303. $(334, \infty)$
1304. (43.2, 47.2)
1305. (8.04, 8.16)
1306. (7.98, 8.22)
1307. $n \geq 11$
1308. (11.8, 112)
1309. About 200 observations
1310. (0.16, 0.20)
1311. (9.9, 13.7)
1312. (0.12, 0.32)
1313. (5.5, 16.5) (Hence the medicine seems to increase the blood pressure!)
1314. (45.7, 52.9)
1315. (0.024, 0.056)
1316. (a) $n \approx 40{,}000$; (b) $n \approx 6{,}000$
1317. (a) $N \leq$ about 2,500; (b) $n \geq$ about 1,690
1318. (a) (0.024, 0.048); (b) (0, 0.046); (c) (0, 4,600)
1319. (a) 225; (b) 250
1320. (a) (361, 439); (b) (36.1, 43.9)
1321. (a) $2^{-n}, 2^{-n}, 1 - 2^{-n+1}$; (b) $1 - 2^{-n+1}$
1322. (a) 4.425; (b) $\sigma\sqrt{5/12}$; (c) 0.054; (d) (4.30, 4.55)

1323. (1.42, 2.62)
1324. (0.23, 0.30)
1325. $1 - c^{-n}$

1401. (a) Bin(15, 0.5); 0.0592
1402. 20
1403. 0.0115
1404. (a) $a = -1000 \ln(1 - \alpha)$
(b) no, for $x_1 = 75$ is larger than $a = -1000 \ln 0.95 = 51.3$
(c) yes
1405. $P = P(X \leq 45 \text{ if } \theta = 1{,}000) = e^{-45/1{,}000} = 0.044$ which is smaller than 0.05. Hence the result is significant at level of significance 0.05
1406. Since H_1 contains small values of p, person A ought to reject H_0 if it takes a long time before he wins. We obtain $P = \sum_{k=11}^{\infty} 0.2 \cdot 0.8^{k-1} = 0.8^{10} = 0.107$. Since $P > 0.10$, the hypothesis H_0 cannot be rejected.
1407. Only assertion 2
1408. A suitable test quantity is the arithmetic mean $\bar{x}$. The critical region consists of all $\bar{x}$ such that $u = (\bar{x} - 4.0)/(0.2/\sqrt{10}) > \lambda_{0.05} = 1.64$. In this case $\bar{x} = 4.10$ and $u = 1.58$; hence the result is not significant at level of significance 0.05
1409. $h(m) = 1 - \Phi[1.64 - (m - 4.0)/(0.2/\sqrt{10})]$; $h(3.8) \approx 0$; $h(4.3) = 0.9990$
1410. $n \geq 64$
1411. (a) $u = 0.67 < t_{0.05}(9) = 1.83$; H_0 cannot be rejected
(b) $u = -1.24 > -1.83$; H_0 cannot be rejected
1412. $u = -6.36$; we get $|u| > t_{0.025}(17) = 2.11$; H_0 is rejected
1413. $u = 2.76 > t_{0.025}(7) = 2.36$; H_0 is rejected
1414. $u = 11.8$; we get $|11.8| > \lambda_{0.025} = 1.96$; H_0 is rejected
1415. (a) $d(\bar{x}) = 0.0030$
(b) $u = 13.15$; we get $|13.15| > \lambda_{0.025} = 1.96$; H_0 is rejected
1416. $P = (5/6)^{12} = 0.11$; not significant
1417. $P = \sum_{0}^{8} \binom{120}{k} (1/6)^k (5/6)^{120-k} = 0.0024$ (normal approximation); the hypothesis that the die is symmetrical can be rejected at level of significance 0.01
1418. (a) 0.34; (b) 0.036
1419. $26/252 = 0.103$
1420. $P = \sum_{19}^{\infty} e^{-10.0} 10.0^k/k! = 0.0072$ (use Table 7 at the end of the book); H_0 is rejected at level of significance 0.01
1421. $Q = 11.5 > \chi^2_{0.01}(3) = 11.3$; H_0 is rejected
1422. $Q = 3.77 < \chi^2_{0.05}(2) = 5.99$; H_0 cannot be rejected
1423. (a) 0.89; (b) 7 in each sample
1424. $n = 130$. Reject H_0 if the sum of the 130 observations is greater than $520 + 3.09\sqrt{520} = 590.5$
1425. Not significant
1426. 14

1501. (a) $\alpha' + \beta(x + 1) - (\alpha' + \beta x) = \beta$; $\beta^* = 0.7410/1.860 = 0.398$
(b) If $x = \bar{x}$ we get $\alpha + \beta(x - \bar{x}) = \alpha$; $\alpha^* = \bar{y} = 0.537$; $D(\alpha^*) = \sigma/\sqrt{7}$

(c) If $x = 0$ we get $\alpha + \beta(x - \bar{x}) = \alpha - \beta\bar{x}$; $(\alpha - \beta\bar{x})^* = \alpha^* - \beta^*\bar{x} = 0.059$; $D(\alpha^* - \beta^*\bar{x}) = 0.958\sigma$

(d) 0.025

1502. (a) $(-0.21, 1.41)$; (b) $(-0.32, 1.88)$

1503. The hypothesis cannot be rejected

1504. $(1.65, 2.57)$

1505. The hypothesis is rejected at level of significance 0.05

1506. $v_0^* = y_0/t_0 - t_0 \sum y_i t_i^2 / \sum t_i^4 = 0.313$; $(0.290, 0.336)$

Answers to Selected Exercises

P1. $p/(7 - 6p)$
P2. Mean f and variance $2f$
P3. 0.206
P4. In both cases the sum assumes the values 2, 3, ..., 12 with probability 1/36, 2/36, 3/36, 4/36, 5/36, 6/36, 5/36, 4/36, 3/36, 2/36, 1/36, respectively
P5. 3
P6. (a) 10; (b) $5x^4$ and $(10/3)(y - y^4)$
P7. 0.85
P8. 15/34
P9. X has mean $1/(\alpha + 3)$ and Y has mean $1 - 1/(\alpha + 3)$
P10. $p_X(2) = 1/10$, $p_X(3) = 3/10$, $p_X(4) = 6/10$
P11. 1/6
P12. 1/4
P13. $(5/6)^{k-2}(1/6)$, $k \geq 2$
P14. (a) $Z = m - X - 30$; (b) $N(m - 30, 7.5^2)$; (c) 492.3
P15. (a) $N(m - 30, 8.08^2)$; (b) 493.3
P16. $\sqrt{m}$ and 1/2
P17. (a) yes; (b) no
P18. $\sqrt{\pi/\beta}/2$ and $(1 - \pi/4)/\beta$

P19.

$X \backslash Y$	0	1	2	3	$p_X(j)$
0	$\frac{1}{8}$	$\frac{1}{8}$	0	0	$\frac{1}{4}$
1	0	$\frac{1}{4}$	$\frac{1}{4}$	0	$\frac{1}{2}$
2	0	0	$\frac{1}{8}$	$\frac{1}{8}$	$\frac{1}{4}$
$p_Y(k)$	$\frac{1}{8}$	$\frac{3}{8}$	$\frac{3}{8}$	$\frac{1}{8}$	1

Note that $X \sim \text{Bin}(2, 1/2)$, $Y \sim \text{Bin}(3, 1/2)$

P20. 0.00037
P21. 0.0178
P22. $f_Z(z) = \begin{cases} 4z, & 0 \le z < 1/2 \\ 4(1-z), & 1/2 \le z < 1 \end{cases}$
P23. 1/4
P24. $-1/63$ $\left(\text{Note that} \iint\limits_{0<x+y<1} x^a y^b (1-x-y)^c \, dx\, dy = a!b!c!/(a+b+c+2)!.\right)$
P25. $E(Z) \approx 0$ and $V(Z) \approx 1/(4n_1) + 1/(4n_2)$
P26. 5/9
P27. (a) 5/6; (b) 29/30; (c) 4/5
P28. 1/2
P29. $F_X(x) = 4x(\sqrt{3} - x)/3, 0 \le x \le \sqrt{3}/2$
P30. (a) 0.0864; (b) 0.173
P31. 0.27
P32. (a) 5/324; (b) 35/648
P33. 0.083
P34. (a) 1/3; (b) 1/5
P35. 0.785
P36. (a) $p_X(k) = \binom{k-1}{r-1} p^r q^{k-r}, k \ge r$; (b) r/p and rq/p^2
P38. $f_Y(y) = y \exp(-y^2/2), y \ge 0$
P39. $n/2^{n-1}$
P40. 11/2

S1. $n \ge 96$
S2. $a^* = -(1/n)\sum \ln x_i$
S3. (a) (1.15, 2.69); (b) (0.68, 1.93)
S4. 8%
S5. $\bar{x}/3$
S6. (37.6, 67.8)
S7. For $a = 0, 1, \ldots, 7$ the power is, respectively, 0, 0, 1/21, 3/21, 6/21, 10/21, 15/21, 1
S8. The hypothesis should be rejected
S9. The ML estimate is $\sum x_i y_i / \sum x_i^2$; its variance is $1/\sum x_i^2$
S10. The ML and LS estimates are both 0.79
S11. $(1/4 + \sum x_i^2/30)^{1/2} - 1/2 = 1.29$
S12. The hypothesis is rejected at the level of significance 1%
S13. (b) (122, 156)
S14. H_0 cannot be rejected at the level of significance 5%
S15. (24.8, 28.9)
S16. H_0 is rejected
S17. (0, 1.78)
S19. (a) Let y be the number of x's equal to 1 or 2. The ML estimate of θ is $y/(2n)$
S20. The hypothesis $d = 0$ is rejected at the level of significance 5%
S21. $c = 3/n$
S22. 30.38
S23. The estimate θ_1^*
S24. 1.83

S25. The assertion cannot be rejected at the level of significance 5%
S26. $h(m) = 1 - \Phi(2 - 4m) + \Phi(-2 - 4m)$
S27. H_0 is rejected at the level of significance 5%
S28. 0.344 and 0.177
S29. $\theta^* = 2.01$; $V(\theta^*) = 0.05\theta^2$
S30. 5.8
S31. (a) 15; (b) 3
S32. 0.891
S33. 590
S34. (a) $h(m) = 1 - \Phi(7.96 - 3m) + \Phi(4.04 - 3m)$; (b) 19
S35. The difference between A and B is significant at the level of significance 5%. (Exclude pairs (H, H) and (H^*, H^*) before performing a sign test.)
S36. (780, 1,220)
S37. (1.589, 1.605)
S38. $n = 15$, $a = 9.73$
S40. $(2n + 1)/(2n) \cdot \max(x_i)$. (Note that the x's are *not* uniformly distributed.)

Index

Abstract model 3
Accelerated motion 201, 202
Accidents 248
Accuracy 107
Addition formula 12
Addition of random variables 83
Addition theorem
  for three events 14
  for two events 13
Adjusted *ML* estimate 199
Allocation
  optimal 300
  proportional 300
Alternative hypothesis 257
Anaesthesia, anaesthetics 61, 244, 265
Analogue model 3
Androsterone 287
Animal experiment 2
Arithmetic mean
  of data set 185
  of random variables 119
Axioms of probability 12

Banach's match box problem 166
Bar diagram 181
  cumulative 181
Bayes' theorem 24
Bayesian statistics 219
Bernoulli's theorem 154
Bertrand's paradox 74
Biased estimate 194
Bilirubin and protein 285, 287
Binomial approximation (of hypergeometric distribution) 157
Binomial distribution 48, 147
Birthday problem 34
Bivariate normal distribution 71
Block 303
Boole's inequality 14
Bose–Einstein statistics 74
Buffon's needle problem 76

Calibration 280
Capacitance of condensers 181
Caries and fluoride 174
Cauchy distribution 98
Central limit theorem 141
Chebyshev's inequality 121
Check list (for planning) 294
Children in family 67
χ^2 distribution (chi-square distribution) 227
χ^2 method (chi-square method) 270
Class limit 182
Classical definition of probability 15
Classical problems of probability 32, 74, 126
Coefficient of correlation 110
Coefficient of regression 281
Coefficient of variation
  of data set 186
  of random variable 103

Combinations 31
Combinatorics 29
Comparative experiment 301
Comparative experimental investigation 301
Comparative investigation 170, 301
Comparative nonexperimental investigation 301
Comparative survey 301
Complement theorem 12
Complete investigation 170
Completely randomized experiment 302
Composite hypothesis 255
Conditional probability 21
Confidence interval 223
  one-sided 224
  two-sided 224
Confidence level 223
Confidence limit 223
Confidence method 262
Consistent estimate 194
Continuity correction 152
Continuous random variable 49
Continuous sample space 7
Contour curves 69, 109
Control of medicine 168
Convolution 84, 85
Convolution formula
  for independent continuous random variables 86
  for independent discrete random variables 84
Correction
  continuity correction 152
  finite population correction 156
Correlation 110
  coefficient of 110
Covariance 109
Critical region 255
Cumulative bar diagram 181
Cumulative histogram 184
Cumulative relative frequency 181

Defective units 1, 16, 171
Degrees of freedom 227, 229
Density function 50
  joint 68
  marginal 69
  skew 50
  symmetric 50
Dependence (of random variables) 108
Descriptive statistics 179
Difference (between random variables) 89
Discrete random variable 44
Discrete sample space 7
Dispersion (measure of) 102, 185
Distribution 46
  binomial 48, 147
  bivariate normal 71
  Cauchy 98
  χ^2 (chi-square) 227
  exponential 54, 86, 87, 92, 98, 105
  extreme value 63
  fft 48, 97
  gamma 58, 228
  general normal 135
  geometric 48
  hypergeometric 49, 155
  logistic 64
  lognormal 144
  marginal 67
  multinomial 68, 161
  normal 56, 131
  one-point 47
  Poisson 49, 158
  posterior 218
  prior 218
  Rayleigh 57
  standard normal 132
  t distribution 229
  two-point 47
  uniform 15, 47, 53, 70, 87, 98, 104
  Weibull 57
Distribution function 42, 66
  empirical 217
  joint 66
  marginal 67, 69
Disturbing factor 297
Drawing
  with replacement 19
  without replacement 18

Efficiency (of estimate) 195
Empirical distribution function 217
Error
  of the first kind 260
  of the second kind 260
  random 107
  systematic 107
ESP 253, 257
Estimate
  adjusted *ML* 199
  biased 194

consistent 194
efficient 195
interval 223
least squares 201
LS 201
maximum likelihood 199
ML 199
point 192
unbiased 194
Estimated regression line 283
Events 5
independent 25, 26
mutually disjoint 9
mutually exclusive 9
Expectation
of function of random variable 99
of function of several random variables 100
of random variable 97
Experiment
completely randomized 302
randomized block 303
reproducible 296
Exponential distribution 54, 86, 87, 92, 98, 105
Extrapolation 289
Extreme value distribution 63

Fair game 107, 122
Favourable cases 15
Fermi–Dirac statistics 74
fft-distribution 48, 97
Finite population 169
Finite population correction 156
Finite sample space 7
Fluoride and caries 174
Fractile (of random variable) 52
Frequency function 50
Frequency interpretation (of probability) 11
Frequency table 180
Frequency tabulated data 180
Function of random variable 80

Gamma distribution 58, 228
Gamma function 58
Gauss's approximation formulae 123, 125
General normal distribution 135
Geometric distribution 48
Graphical method (probability paper) 210
Graphical presentation (of data) 180
Grouped data 181, 187

Handshakes 31
Harvest damage 295, 298
Histogram 183
cumulative 184
Homogeneity test 271
Household survey 168
Housing 267, 272
Hypergeometric distribution 49, 155
Hypothesis
alternative 257
composite 255
simple 255
Hypothesis testing 253

Improbable event 28
Independent events 25, 26
Independent random variables 71
Independent trials 27
Inference 177
Infinite population 169
Interval estimate 223
Interval estimation 223
Interval of variation (of data set) 186
Interviews 245, 247
Inventory 28
Investigation
comparative 170, 301
noncomparative 170, 294

Joint density function 68
Joint distribution function 66
Joint probability function 66

Kolmogorov's system of axioms 12

L function (likelihood function) 199
Larger of two random variables 89
Law of large numbers 121
Least squares estimate 201
Least squares method 201
Level of significance 255
L function (likelihood function) 199
Lifetimes 72, 73, 90, 112, 200
Likelihood function 199
Limit theorems
Bernoulli's theorem 154

Limit theorems (*cont.*)
  Central limit theorem 141
  Law of large numbers 121
Linear regression 280
Linear transformation (of random variable) 81
Location (measure of) 101, 185
Logarithmic transformation (of random variable) 82, 124
Logistic distribution 64
Lognormal distribution 144
Lottery 100
Lower quartile 52
LS (least squares) estimate 201
LS (least squares) method 201

Marginal density function 69
Marginal distribution function 67, 69
Marginal probability function 67
Market research 157, 247
Maximum likelihood 198
Maximum likelihood estimate 199
  adjusted 199
Maximum likelihood method 198
Mean
  of data set 185
  of random variable 97
  of sample 196
Measure of dependence 108
Measure of dispersion
  of data set 185
  of random variable 102
Measure of location
  of data set 185
  of random variable 101
Measurements 1, 41, 43, 120, 140, 172, 173, 193, 206, 224, 225, 240, 258, 261, 264, 269
Median
  of data set 185
  of random variable 52, 101
Menu 32
Method of least squares 201
Method of maximum likelihood 198
Missing values 297
Mixed population 24
Mixture (of random variables) 60
ML (maximum likelihood) estimate 199
  adjusted 199
ML (maximum likelihood) method 198
Model 3
Multidimensional random variable 65
Multinomial distribution 68, 161
Multiple regression 290
Multiplication principle 32
Mutually disjoint events 9
Mutually exclusive events 9

Noncomparative investigation 170, 294
Nonlinear regression 289
Normal approximation
  in the theory of hypothesis testing 264
  in the theory of interval estimation 241
  of binomial distribution 152
  of hypergeometric distribution 157
  of Poisson distribution 161
Normal distribution 56, 131
  bivariate 71
  general 135
  standard 132
Normal probability paper 210
Null hypothesis 254

Odds 108
One-point distribution 47
One-sided confidence interval 224
One-sided test of significance 256
Optimal allocation 300
Outcome 5

P method 256
Paired samples 238
Parallel system 126
Parameter 175
Parameter space 175
Percentile (of random variable) 52
Permutations 31
Petersburg paradox 126
Planning (check list) 294
P method 256
Point estimate 192
Poisson approximation
  of binomial distribution 153
  of hypergeometric distribution 158
Poisson distribution 49, 158
Poker 33
Population 168
  finite 169
  infinite 169

Possible cases 15
Posterior distribution 218
Power 257
Power function 257
Precision 107
Prediction (of point on theoretical regression line) 283
Prior distribution 218
Prize in food package 36, 127
Probability 10
  classical definition of 15
  conditional 21
  subjective 14, 108
  total 23
Probability distribution 46
Probability function 45
  joint 66
  marginal 67
  skew 45
  symmetric 45
Probability paper 210
Probability space 12
Production by machines 24
Proportional allocation 300
Protein and bilirubin 285, 287

Quadratic transformation (of random variable) 83
Quantile 52
Quartile 52
  lower 52
  upper 52
Quotient (of random variables) 91

Radioactive decay 6, 55, 159, 193, 195, 209
Random error 107
Random model 3
Random numbers 295, 305
Random sample 176
Random sampling 295
Random trial 3
Random variables 41
  continuous 49
  discrete 44
  independent 71
  multidimensional 65
  standardized 105
  two-dimensional 66
  uncorrelated 110
Random walk 148
Randomization 302
Randomized block experiment 303
Range (of data set) 186
Ratio (of random variables) 91
Raw data 180
Rayleigh distribution 57
Reciprocal (of random variable) 124
Regression
  linear 280
  multiple 290
  nonlinear 289
  simple linear 281
Regression coefficient 281
Regression line
  estimated 283
  theoretical 281
Regression variable 281
Relative frequency 154, 158, 180
Rencontre 35, 128
Repeated tests of significance 275
Repeated trials 27
Representative selection 296
Reproducible experiment 296
Residual sum of squares 284
Risk 27
Round-off errors 54, 143
Ruin problem 34
Russian roulette 32
rv (random variable) 41

Sample mean 196
Sample (random) 176
Sample space 5
  continuous 7
  discrete 7
  finite 7
Sample standard deviation 198
Sample survey (stratified) 299
Sample variable 192
Sample variance 197
Sampling investigation 170
  scheme for 175
Sampling survey 170
Sign test 268
Significance level 255
Significance test 255
  one-sided 256
  strengthened 260
  two-sided 256
Significant*, significant**, significant*** 255
Simple hypothesis 255
Simple linear regression 281
Simple random sampling 295

Simple sampling 295
Skew density function 50
Skew probability function 45
Smaller of two random variables 89
Smiling-face scale 169
Square root transformation (of random variable) 82
Standard deviation
  of data set 185
  of random variable 103
  of sample 198
Standard error 209
Standard normal distribution 132
Standardized random variable 105
Statistic 192
Statistical inference 177
Statistical investigation 167
Stratified sample survey 299
Stratum 299
Strengthened test of significance 260
Subjective probability 14, 108
Sugar-content 234, 236, 238, 263
Sum of random variables 83
Sum of squares
  about the arithmetic mean 186
  residual 284
Symmetric density function 50
Symmetric probability function 45
Systematic error 107
Systematic selection 306

t distribution 229
Tabulation (of data) 180
Target-shooting 27, 70
Taste testing 266
t distribution 229
Tensile strength 90
Test of homogeneity 271
Test of significance 255
  one-sided 256
  strengthened 260
  two-sided 256
Test quantity 255
Testing hypotheses 253
Theoretical regression line 281
Time series 290
Total probability theorem 23
Treatment 301
2×2 table 272
Two-dimensional random variable 66
Two-point distribution 47
Two-sided confidence interval 224
Two-sided test of significance 256

Unbiased estimate 194
Uncorrelated random variables 110
Uniform distribution 15, 47, 53, 70, 87, 98, 104
Upper quartile 52
Urn model 17

Variability 2
Variance
  of data set 185
  of product of random variables 125
  of random variable 103
  of sample 197
Variation 2
  coefficient of 103, 186
  continuous 169
  discrete 169
  interval of 186
Venn diagram 8

Waiting time 54, 60
Weibull distribution 57
Woollen fabric 161
Work study 242